Physik
67,- €
'19

Springer Theses

Recognizing Outstanding Ph.D. Research

Aims and Scope

The series "Springer Theses" brings together a selection of the very best Ph.D. theses from around the world and across the physical sciences. Nominated and endorsed by two recognized specialists, each published volume has been selected for its scientific excellence and the high impact of its contents for the pertinent field of research. For greater accessibility to non-specialists, the published versions include an extended introduction, as well as a foreword by the student's supervisor explaining the special relevance of the work for the field. As a whole, the series will provide a valuable resource both for newcomers to the research fields described, and for other scientists seeking detailed background information on special questions. Finally, it provides an accredited documentation of the valuable contributions made by today's younger generation of scientists.

Theses are accepted into the series by invited nomination only and must fulfill all of the following criteria

- They must be written in good English.
- The topic should fall within the confines of Chemistry, Physics, Earth Sciences, Engineering and related interdisciplinary fields such as Materials, Nanoscience, Chemical Engineering, Complex Systems and Biophysics.
- The work reported in the thesis must represent a significant scientific advance.
- If the thesis includes previously published material, permission to reproduce this must be gained from the respective copyright holder.
- They must have been examined and passed during the 12 months prior to nomination.
- Each thesis should include a foreword by the supervisor outlining the significance of its content.
- The theses should have a clearly defined structure including an introduction accessible to scientists not expert in that particular field.

More information about this series at http://www.springer.com/series/8790

Pablo Fernández Menéndez

Neutrino Physics in Present and Future Kamioka Water-Čerenkov Detectors with Neutron Tagging

Doctoral Thesis accepted by
the Autonomous University of Madrid, Spain
and University of Tokyo, Japan

Springer

Author
Dr. Pablo Fernández Menéndez
Neutrino Group & Neutrino Platform (CERN)
IFIC, CSIC and Universitat de València
Paterna, Spain

Supervisor
Prof. Luis A. Labarga
Department of Theoretical Physics
Autonomous University of Madrid
Madrid, Spain

ISSN 2190-5053 ISSN 2190-5061 (electronic)
Springer Theses
ISBN 978-3-319-95085-3 ISBN 978-3-319-95086-0 (eBook)
https://doi.org/10.1007/978-3-319-95086-0

Library of Congress Control Number: 2018948651

Printed on acid-free paper

This Springer imprint is published by the registered company Springer International Publishing AG part of Springer Nature
The registered company address is: Gewerbestrasse 11, 6330 Cham, Switzerland

Pa Mario, pa Irene y pa'quellos que-yos puea prestar.

Pensa uomo, pensa.

Supervisor's Foreword

I am delighted to introduce the Doctoral Thesis of Pablo Fernández in this "Springer Thesis Series" book.

Neutron tagging is a major improvement in neutrino experimental physics with water-Čerenkov detectors. In June 2015, the Super-Kamiokande (SK) Collaboration approved SuperK-Gd (former GADZOOKS!), an upgrade consisting of dissolving a gadolinium salt to 0.2% in its water. With it, SK will be capable of tagging neutrons at a very high exigency ($\sim$80%). The feasibility of the project was proved with the EGADS demonstrator, also at the Kamioka Observatory. This Doctoral work has contributed extensively to the success of EGADS: mainly at construction, at calibrations, and at continuous monitoring.

This thesis has developed a highly efficient and realistic Gd-neutron tagging detection method for SuperK-Gd based on the current operation of the Super-Kamiokande detector. It thoroughly studies potential radioactive contamination in the system, mainly in the Gd salt, and its impact on relevant measurements by SuperK-Gd, namely, the diffuse supernova neutrino background, the neutrinos from the Si-burning phase of a close-enough star, reactor neutrinos oscillations, and the measurement of the solar neutrino spectrum at very low energy.

The thesis has developed a series of algorithms based on Gd-neutron tagging that discriminate neutrino from antineutrino interactions in SuperK-Gd rather efficiently and algorithms to characterize CC and NC neutrino interactions. The thesis has developed a second step in the reconstruction of the interacting neutrino's energy based on the neutron multiplicity measured in the final sate by the (Gd-) neutron tagging. It reduces very significantly the amount of non-visible energy, thus providing a better match between the measured and the true energy of the neutrino.

The thesis has studied for the first time the impact of Gd-neutron tagging in a global oscillation analysis of atmospheric neutrinos by SuperK-Gd. The global sensitivity study, at $\sin^2\theta_{23} = 0.575$, $\delta_{CP} = 4.189$, and $m^2_{32} = 0.0025$ eV^2, for 2520 days of SK-IV, showed that the inverted mass hierarchy sensitivity rejection is improved as compared with the current official analysis (2.7 units of χ^2 with Gd, 1.6 current officials). In addition, the sensitivity to the CP phase was similarly improved. Corresponding studies have been performed for the T2K-long baseline experiment, and for the next-generation Hyper-Kamiokande experiment with Gd loading, showing also significant improvements.

The current phase of Super-Kamiokande, SK-IV, has already some, low efficiency ($\sim 19\%$), neutron tagging capability by deuterium production from neutron captures on hydrogen. This thesis incorporates for the first time (the Gd-developed) neutron tagging information for atmospheric oscillation analysis using the currently available H-neutron tagging in Super-Kamiokande. It proved the significant improvements that neutron tagging in water-Cerenkov detectors brings to the atmospheric neutrino oscillation analysis, providing, for instance, better constraints on the neutrino mass hierarchy. The global oscillation analysis of SK data shows a preference for normal hierarchy of 4.6 units of χ^2 (Note: results are preliminary), whereas in the current official analysis the preference in 4.3 units of χ^2.

In the following second part, I provide some more personal but scientific information on Pablo Fernandez:

Pablo Fernandez was in the Kamioka Observatory for the first time during July 2011 as undergraduate student (4th year of Physics). His main activity was at the pre-calibration of the 240 photomultipliers for the EGADS demonstrator. He showed an excellent performance. In addition, Pablo worked with the Calibration Group of Super-Kamiokande on a little work titled *Spontaneous Fission Neutron Sources* that he presented at the Group's weekly meeting. During the Academic Year 2011/2012, while his curricular 5th year of Physics, he continued collaborating with SuperK-Gd. He carried out two seminal studies analyzing and quantifying some relevant backgrounds involved in this new technique. The works, *Study of neutron production in pure water with 0.2% $Gd_2(SO_4)_3$ due to radioactive contamination* and *Background study due to ^{238}U contamination in $Gd_2(SO_4)_2$ compound,* were presented at the world-wide Lowe meetings held on March 13 and May 11, 2012. He came back to Kamioka in July 2012, just graduated, to participate in the biyearly LINAC calibration of Super-Kamiokande and the accompanying "Nickel" calibrations (9 MeV γs from capture on Ni of neutrons from a ^{252}Cf source). During the academic year 2012/2013, he did his Master's studies while continuing his contributions to SK. At that time, he got a UAM's 4-years "Teaching Assistantship" that allowed him to carry on his doctoral work until completion.

He did collaborate very strongly at all key steps of the EGADS demonstrator, remarkably the already mentioned PMT pre-calibration program, the full instrumenting of the EGADS tank, and the cleaning and refurbishment after the rusted cable incident.

He did carry out several research works that are most relevant to the SuperK-Gd project: (1) he studied in depth the radioactive contamination of several test samples of $Gd_2(SO_4)_3$ measured with high-purity Ge detectors in the Canfranc Underground Laboratory (Spain). (2) He studied the impact of the encountered typical radioactive contaminations on the most important physics measurements expected with SuperK-Gd mainly Supernova Relic Neutrino, solar analyses, and pre-supernova early warning neutrinos, and impact on the current SK physics program, mainly very low energy solar neutrinos. (3) He established the maximum radioactive contaminations in the salt of Gd such to not jeopardize those measurements. (4) He did help in the development of a Ra removal system with an accompanying Rn monitoring system. (5) He did develop a maximum likelihood method for optimizing the neutrino/antineutrino separation by SuperK-Gd in the atmospheric and T2K neutrino samples. (6) He did develop methods for significant discrimination between NC and CC reactions in multi-GeV neutrino interactions in SuperK-Gd. (7) He did propose the use of the Gd-tagged neutrons in SuperK-Gd—and also H-tagging, already available in SK—to improve the reconstruction of the energy of the interacting neutrino; he did develop a method for it and achieved remarkable results. (8) He did study the improvements in the atmospheric oscillation analysis induced by all those features (and also with H-neutron tagging).

He presented on behalf of the SK collaboration with the results of the SK-Gd group, with emphasis of course on his own works, on several Workshops and major Scientific Conferences. He gave the talk *Identifying electron antineutrino with Super-Kamiokande: GADZOOKS!, status and some of its current challenges* at the IMFP2013 Workshop (Santander, Spain, May 2013) and a poster with the same title at the E.U.'s ITN Project Invisibles School (Durham, U.K., July 2013). He presented the talk *Status of GADZOOKS!: Neutron Tagging in Super-Kamiokande* at ICHEP 2014 (Valencia, Spain, July 2014), the poster tilted *GADZOOKS! (SuperK-Gd): status and physics potential* at ICRC 2015 (The Hague, The Netherlands, August 2015), the poster titled *Benefits of Gd for High Energy Neutrinos* at NEUTRINO 2016 (London, U.K., July 2016), and the invited plenary talk "Gd-*doping and the impact on SuperK and T2K*" at the NNN2016 workshop (Beijing, China, November 2016).

Madrid, Spain
May 2018

Luis A. Labarga

Abstract

This thesis is focused on the upgrade of the Super-Kamiokande detector, consisting in the addition of a salt of gadolinium into the water of the detector to enable a very high efficient capability to detect the neutrons produced in the detector: the SuperK-Gd project (former GADZOOKS!). This feature will improve largely the scientific power of SK because the neutron production is related to the matter–antimatter character of the interacting neutrino; charged current interacting neutrinos tend to produce less neutrons than their analogous antineutrinos. It also provides a new and powerful selection criterion for proton decay searches due to the fact that in a proton decay, should it exist, basically no neutrons are produced in the final state. In addition to these anticipated benefits, in this work, other features are explored finding out new improvements due to neutron tagging in various physics analyses.

For the realization of SuperK-Gd, an exhaustive R&D program has been carried out. It pivoted on the EGADS prototype and included many studies of different nature to ensure the feasibility and performance of Gd-doped water-Čerenkov detectors.

This thesis covers works on EGADS, construction, calibration and monitoring, Gd-neutron tagging implementation, capability, and impact on the neutrino physics on a wide variety of physics phenomena within a broad energy range and from different origins, including detectability of Diffuse Supernova Neutrino Background (DSNB), supernova burst neutrinos, pre-supernova neutrinos (from Si-burning phase), reactor antineutrinos and solar neutrinos, and also neutrino oscillation parameters using atmospheric and long baseline neutrinos. Since in the not-too-distant future experimental neutrino physics will need to build even larger detectors to address the remaining or newly arisen unknown properties of neutrinos, this thesis also studies their sensitivities for atmospheric and long baseline neutrinos for the Hyper-Kamiokande project with neutron tagging.

In order to prove the relevance of neutron tagging in large water-Čerenkov detectors and confirm the studies with actual data, a complete and global analysis of SK atmospheric neutrino data is performed including the previously developed neutron tagging tools applied to the fourth phase of Super-Kamiokande, which is already capable of detecting a low, but already useable, fraction of the neutrons produced through hydrogen neutron captures. It provides the most updated oscillation analysis on the atmospheric oscillation parameters of Super-Kamiokande (Note: results are preliminary).

Acknowledgements

In the realization of this thesis, several people have been certainly important and crucial. First, I would like to thank my advisor Luis Labarga and the Kamioka Observatory director and spokesperson of the Super-Kamiokande experiment, Masayuki Nakahata, who introduced me to the Super-Kamiokande collaboration and accepted my participation in the SuperK-Gd project since I was an undergraduate back in 2011. Also, from the very early days of my research, Profs. Mark Vagins, Yusuke Koshio, and Michael Smy have helped me a lot in improving my skills and knowledge in physics and its communication.

Once properly adapted to the SK collaboration, and since the beginning of my Ph.D. members of all the Super-Kamiokande and Hyper-Kamiokande collaborations have, in greater or lower level, allowed me to learn a lot of fundamental physics. I would like to emphasize the knowledge acquired from Lluis Martí, Sekiya-san, Ikeda-san, Yokozawa-san, Mori-san, Nakano-san, Yano-san, Yang Zhang, Takeuchi-san, Shiozawa-san, Ed Kerns, Chris Walter, Roger Wendell, Zepeng Li, Chris Kachulis, José Palomino, Hirota-san, Mine-san, Moriyama-san, Bill Kropp, and Jeff Griskevich. With a very special mention to Roger Wendell, a great person from whom I have acquired a huge amount of knowledge, and without his dedication and patience an important part of this work would not have been possible.

I am very thankful for the continuous support of my family, Manuel, Celestina, and Encarna, which have been crucial not only to complete this job but also to my personal development, helping me in every aspect of my life and coping with me and my head. Apart from these living people, I specially remember Argentina and Mario. The first one was very important in my childhood and, most probably, without her endless efforts I could not be where I am. The latter is, who I think, the major contributor of my tireless attitude for finding things out and for my love for mathematics and physics.

My friends have also contributed to this work with continuous support and interest. I cannot avoid mentioning Adrián, Antonio, Carlos, César, Eddy, Fer, Jose, and Juan. And of course, my mind-mate Irene, who is the person most understands me and cares for me. She has been crucial producing uncomfortable, but incredibly enriching discussions about almost everything and from who I have learnt countless things. I thank all of them for accepting this disturbed mind of mine, that, I know, is not always easy to deal with.

Contents

Chapter 1
Introduction

Neutrinos are the only particles showing deviation from the Standard Model of Fundamental Interactions (SM) [Weinberg67, Glashow70]. It states that they are massless but they are not. This was discovered by the Super-Kamiokande (SK) experiment in 1998 from the observation of a deficit in the atmospheric neutrino data, correlated to the distance of their production point [Fukuda98a]. Those oscillations come from changes from one neutrino flavour to another during their flight to SK, implying that their mass eigenstates are different to their flavour eigenstates.

There are many natural and human-made neutrino sources. Some are known as our Sun (and most probably the rest of the stars), supernovae, cosmic rays interacting with the atmosphere, nuclear reactors, neutrino accelerators and radioactive decays of unstable isotopes, and some others are only conjectured like black holes [Halzen95], dark matter [Choi15] and gamma ray bursts [Vietri98]. In addition, there is the widely accepted cosmic neutrino background generated after the very first second from the Big Bang [Faessler16]. This diversity in origin also determines their characteristics, such as flavour or energy, providing information about a broad range of physics phenomena and properties.

Although neutrinos are the most abundant of the known massive particles in our universe, adding up to 0.3% of its total mass, they are extremely hard to detect due to their tiny interaction cross sections, only through nuclear weak force. For that reason, in order to detect them and study their properties, huge experiments with very large amount of active matter and very low background are needed. One of the most successful technology employed in neutrino experiments is the so called water-Čerenkov. They measure the radiation emitted by charged particles traveling in the water (for instance those originated in a neutrino interaction) with momentum larger than their Čerenkov threshold. This is the case of all the ongoing and projected experiments treated in this thesis.

Large water-Čerenkov detectors have provided numerous and crucial discoveries and results for neutrino physics. These kind of experiments, and specially Super-Kamiokande, are responsible of the major discoveries in neutrinos physics up to date. The precursor of SK, the KamiokaNDE experiment, first detected

P. Fernández Menéndez, *Neutrino Physics in Present and Future Kamioka Water-Čerenkov Detectors with Neutron Tagging*, Springer Theses,
https://doi.org/10.1007/978-3-319-95086-0_1

supernova neutrinos from SN1987a [Hirata87], and Super-Kamiokande is responsible for the discovery of neutrino oscillations using atmospheric neutrinos, it solved the solar neutrino problem by measuring with high precision solar neutrinos [Fukuda98b, Fukuda99a], providing also the first indication of the day-night asymmetry of solar neutrinos due to terrestrial effects [Renshaw14]. Furthermore, Super-Kamiokande has been the far detector of the K2K neutrino beam [Ahn03, Ahn06], confirming the atmospheric oscillation parameters and, currently, it is the T2K far detector providing the world's most stringent bounds for the CP violating phase [Abe15a]. Super-Kamiokande is also a leader in the measurement of the supernova relic neutrinos (DSNB), having the best limits to its flux [Bays12, Zhang14]. In addition and besides neutrino physics, SK has also the best limits for a numerous variety of exotic nucleon decays, such as proton decay [Miura16] or neutron-antineutron oscillations [Abe15b], both predicted by Grand Unified Theories (GUT) beyond the Standard Model (SM) of particles.

The core of this work is focused on the near future upgrade of the Super-Kamiokande detector consisting in the addition of a salt of gadolinium into the water of the detector to enable a very high efficient capability to detect the neutrons produced in the detector. This upgrade is called SuperK-Gd, (previously GADZOOKS!) [Vagins04]. This feature will largely improve the detection power of SK because the neutron production is related to the matter-antimatter character of the interacting neutrino: charged current interacting neutrinos tend to produce less neutrons than their analogous antineutrinos. It also provides a new and powerful selection criterion for proton decay searches due to the fact that proton decays, if exist, are accompanied by almost no neutrons in the final states. In addition to these anticipated benefits, this work explores other features and finds out new improvements from neutron-tagging to various physics analyses.

This work covers Gd-neutron tagging implementation, capability and impact on the neutrino physics, comprising a wide variety of physics phenomena within a broad energy range and from different origins. It presents detectability studies for Diffuse Supernova Neutrino Background (DSNB), supernova burst neutrinos, pre-supernova neutrinos (from Si-burning phase), reactor antineutrinos and solar neutrinos, and also sensitivity studies to neutrino oscillation parameters using atmospheric and long baseline neutrinos.

For the realisation of SuperK-Gd, an exhaustive R&D program and feasibility studies have been done, being the most important the EGADS program. The EGADS prototype for SuperK-Gd is a scaled down SK detector where most of the technology needed for the addition of Gd to a water-Čerenkov detector was developed. In parallel to the R&D at EGADS, another studies have been performed to assure the feasibility of Gd-doped water-Čerenkov detectors and their performance.

For the not-too-distant future of neutrino physics in water-Čerenkov detectors even larger detectors will be needed to be built to address the remaining unknown properties of neutrinos. This thesis also presents a sensitivity study for atmospheric and long baseline neutrinos for the Hyper-Kamiokande project with neutron-tagging.

In order to proof the relevance of neutron-tagging in large water-Čerenkov detectors with actual data and confirm the studies carried out, a complete and global analysis of SK atmospheric neutrino data is performed including the previously developed neutron-tagging tools applied to SK's fourth phase which is capable of detecting a low fraction of the neutrons produced through hydrogen-neutron captures. This provides the most updated and with the most stringent limits on the atmospheric oscillation parameters of SK (Note: results are preliminary).

In Chap. 2, an overview of neutrino physics is given, from its theoretical prediction to the current state of neutrino physics experiments, containing an explanation of how each neutrino source can be used to solved different neutrino physics and how they are connected with some of the most fundamental problems in physics. This Chapter also includes a description of the neutrino-nucleus interactions, which are of great importance for water-Čerenkov detectors.

In Chap. 3, a brief description of the Super-Kamiokande experiment is presented, which will be used as detector framework for the rest of the work involving SuperK-Gd and Hyper-K.

Chapter 4 deals with the projected and approved upgrade of SK by dissolving a gadolinium salt into its water to enhanced the detection of the final state neutrons of the interactions occurring in the detector. Along this Chapter, the Gd-neutron tagging technique will be fully described, as well as the main R&D program towards the realisation of SuperK-Gd.

In Chap. 5 the neutron production mechanisms from neutrino physics is reviewed and, then, the Gd-tagged neutron reconstruction is developed, as well as various tools which use neutrons as base, such the neutrino-antineutrino separation, the neutral and charged current interactions discrimination and the neutrino energy corrections with neutrons. Finally, an important issue, as the study of the radioactive contamination added with the Gd salt is explained.

Chapter 6 uses all the items explained in the previous Chapter to perform studies of the physics potential of SuperK-Gd and also the extension to the next generation detector, Hyper-Kamiokande.

In Chap. 7, the Gd-neutron tagging tools are applied to SK-IV data with H-neutron tagging and a global atmospheric oscillation analysis is performed with all SK data. Finally, the current status and future prospects of neutrino physics, the most important results and the conclusions extracted from them are shown in Chap. 8.

Based mainly on the results of this doctoral work, the author has presented the following conferences, posters and documents:

- Oral presentation, *Identifying $\overline{\nu}$ with Super-Kamiokande: GADZOOKS!, status and some of its current challenges* at International Meeting of Fundamental Physics, Santander, Spain.
- Poster, *Identifying $\overline{\nu}$ with Super-Kamiokande: GADZOOKS!, status and remaining issues* at INVISIBLES13 workshop, Durham, United Kingdom.
- Oral presentation and proceedings, *Status of GADZOOKS!: Neutron-tagging in Super-Kamiokande* at International Conference of High Energy Physics (ICHEP14), Valencia, Spain.
- Poster and proceedings, *GADZOOKS!: status and physics potential* at International Cosmic Ray Conference (ICRC15), The Hague, The Netherlands.

- Poster and proceedings, *Gd-doping and the impact on SuperK and T2K* at Neutrino16, London, United Kingdom.
- Oral presentation, *Gd-doping and the impact on SuperK and T2K* at Next Generation Nucleon Decay and Neutrino Experiments (NNN16), Beijing, China.
- *GADZOOKS! proposal*: document presented to SK and T2K collaborations reviewing the feasibility and impact of SuperK-Gd.

References

[Weinberg67] S. Weinberg, A model of leptons. Phys. Rev. Lett. **19**, 1264–1266 (1967). https://doi.org/10.1103/PhysRevLett.19.1264

[Glashow70] S.L. Glashow et al., Weak Interactions with Lepton-Hadron sSymmetry. Phys. Rev. D **2**, 1285–1292 (1970). https://doi.org/10.1103/PhysRevD.2.1285

[Fukuda98a] Y. Fukuda et al., Evidence for oscillation of atmospheric neutrinos. Phys. Rev. Lett. **81**, 1562–1567 (1998). https://doi.org/10.1103/PhysRevLett.81.1562

[Halzen95] F. Halzen et al., Neutrinos from primordial black holes. Phys. Rev. D **52**, 3239–3247 (1995). https://doi.org/10.1103/PhysRevD.52.3239

[Choi15] K. Choi et al., Search for neutrinos from annihilation of captured low-mass dark matter particles in the Sun by Super-Kamiokande. Phys. Rev. Lett. **114**(14), 141301 (2015). https://doi.org/10.1103/PhysRevLett.114.141301

[Vietri98] M. Vietri, On the energy of neutrinos from gamma-ray bursts. Astrophys. J. **507**(1), 40 (1998)

[Faessler16] A. Faessler et al., *Can one Measure the Cosmic Neutrino Background?*, 2016

[Hirata87] K. Hirata et al., Observation of a neutrino burst from the supernova SN 1987a. Phys. Rev. Lett. **58**, 1490–1493 (1987). https://doi.org/10.1103/PhysRevLett.58.1490

[Fukuda98b] Y. Fukuda et al., Measurements of the solar neutrino flux from super-kamiokande's first 300 days. Phys. Rev. Lett. **81**, 1158–1162 (1998). https://doi.org/10.1103/PhysRevLett.81.1158

[Fukuda99a] Y. Fukuda et al., Constraints on neutrino oscillation parameters from the measurement of day-night solar neutrino fluxes at super-kamiokande. Phys. Rev. Lett. **82**, 1810–1814 (1999). https://doi.org/10.1103/PhysRevLett.82.1810

[Renshaw14] A. Renshaw et al., First indication of terrestrial matter effects on solar neutrino oscillation. Phys. Rev. Lett. **112**(9), 091805 (2014)

[Ahn03] M.H. Ahn et al., Indications of neutrino oscillation in a 250 km long-baseline experiment. Phys. Rev. Lett. **90**, 041801 (2003). https://doi.org/10.1103/PhysRevLett.90.041801

[Ahn06] M.H. Ahn et al., Measurement of neutrino oscillation by the k2k experiment. Phys. Rev. D **74**, 072003 (2006). https://doi.org/10.1103/PhysRevD.74.072003

[Abe15a] K. Abe et al., Measurements of neutrino oscillation in appearance and disappearance channels by the T2K experiment with 6.6 10^{20} protons on target. Phys. Rev., 2015. **D91**(7), 072010. https://doi.org/10.1103/PhysRevD.91.072010

[Bays12] K. Bays et al., Supernova relic neutrino search at super-kamiokande. Phys. Rev. D **85**, 052007 (2012). https://doi.org/10.1103/PhysRevD.85.052007

[Zhang14] H. Zhang et al., Supernova relic neutrino search with neutron tagging at Super-Kamiokande-IV. Astropart. Phys. **60**, 41–46 (2014). https://doi.org/10.1016/j.astropartphys.2014.05.004

[Miura16] M. Miura, *Search for Proton Decay via* $p \rightarrow e^+\pi^0$ *and* $p \rightarrow \mu^+\pi^0$ in 0.31 Megaton·Years Exposure of the Super-Kamiokande Water Cherenkov Detector, 2016

[Abe15b] K. Abe et al., The Search for $n - \bar{n}$ oscillation in Super-Kamiokande I. Phys. Rev. D **91**, 072006 (2015). https://doi.org/10.1103/PhysRevD.91.072006

[Vagins04] M.R. Vagins et al., Antineutrino spectroscopy with large water Čerenkov detectors. Phys. Rev. Lett. **93**, 171101 (2004). https://doi.org/10.1103/PhysRevLett.93.171101

Chapter 2
Introduction to Neutrino Physiscs

Neutrinos are, currently, the most unknown particles in the SM, therefore, the physics related to them, involve many fields and there are various independent experiments dedicated to measure their properties, such as their masses or their oscillation parameters. Neutrinos are crucial in theoretical models beyond the Standard Model [Bilenky15, Ringwald16] and may be key to solve the flavour problem [King04b] or used as motivation to a hidden symmetry in quark sector to explain the strong CP problem [Cao15]. Another peculiarity of neutrinos is that they lack of right-handed partners and are neutrally charged (electric and colour) being a reasonable candidate to be a Majorana particle [Majorana28].

In addition, they are important also in cosmology because their number of species determine the element abundances and could be they key to understand the matter-antimatter asymmetry of our universe [Chen16].

This thesis and this chapter will be focused on the study of astrophysical neutrinos and atmospheric and long baseline neutrino oscillation parameters. Nevertheless, a brief introduction and description of the whole neutrino physics picture is presented.

2.1 Main Neutrino Sources

There are various neutrino sources, booth natural and artificial. The formers have the benefit that usually produce neutrinos in large quantities and help us understanding the subjacent physics process which may, in some cases, be very complex. On the other hand, artificial sources produce fewer neutrinos but with a much better known properties, such as energy, direction, flavour and travelled distance (Fig. 2.1).

The upcoming paragraphs will be describing the main neutrino sources that will be treated along this thesis, looking at the main characteristics of the neutrinos produced and arguing their physics potential.

P. Fernández Menéndez, *Neutrino Physics in Present and Future Kamioka Water-Čerenkov Detectors with Neutron Tagging*, Springer Theses,
https://doi.org/10.1007/978-3-319-95086-0_2

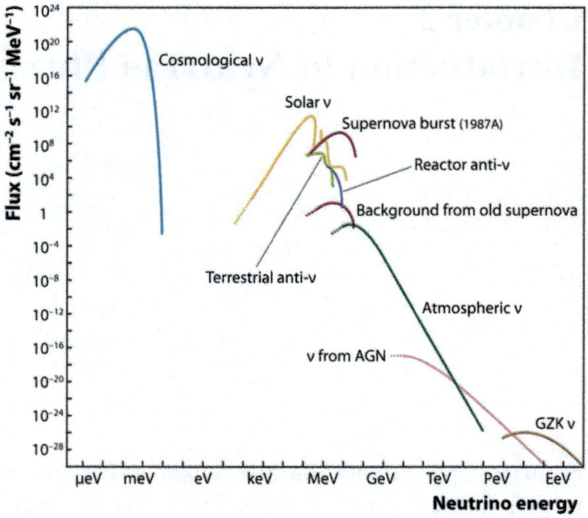

Fig. 2.1 Spectrum of all neutrino sources except accelerator neutrinos. CνB (Cosmological Neutrino Background with T = 1.9 K after the Big Bang), solar neutrinos, neutrinos from SN1987A burst (few seconds), reactor neutrinos, terrestrial neutrinos, atmospheric neutrinos, the so-called GZK (Greisen-Zatsepin-Kuzmin) neutrinos and AGN (Active Galactic Nuclei) neutrinos [Spiering12]

2.1.1 Atmospheric Neutrinos

Atmospheric neutrinos are produced in the interaction of cosmic rays with the Earth's atmosphere [Honda15]. Cosmic rays are made up of high energy particles arriving at the Earth from many places in the Universe, these particles are, mainly, protons, $\sim$5% are Helium nuclei and the remaining fraction are heavier nuclei (Fig. 2.2).

These particles interact with the Earth's atmosphere and produce π and K mesons, which are unstable and decay into other particles, being neutrinos a significant fraction of them. In Eq. (2.1), the decays with the largest branching ratios (BR) are shown.

$$\begin{array}{cc}
K^+ \rightarrow \mu^+ + \nu_\mu \rightarrow e^+ + \nu_e + \overline{\nu}_\mu + \nu_\mu & (BR = 63.6\%) \\
K^- \rightarrow \mu^- + \overline{\nu}_\mu \rightarrow e^- + \overline{\nu}_e + \nu_\mu + \overline{\nu}_\mu & \\
\hline
K^+ \rightarrow \pi^+ + \pi^0 & (BR = 20.7\%) \\
K^+ \rightarrow \pi^- + \pi^0 & \\
\hline
K^+ \rightarrow \pi^+ + \pi^+ + \pi^- & (BR = 5.6\%) \\
K^+ \rightarrow \pi^- + \pi^- + \pi^+ & \\
\hline
K^+ \rightarrow \pi^0 + e^+ + \nu_e & (BR = 5.1\%) \\
K^+ \rightarrow \pi^0 + e^- + \overline{\nu}_e & \\
\hline
\pi^+ \rightarrow \mu^+ + \nu_\mu \rightarrow e^+ + \nu_e + \overline{\nu}_\mu + \nu_\mu & (BR \sim 100\%) \\
\pi^- \rightarrow \mu^- + \overline{\nu}_\mu \rightarrow e^- + \overline{\nu}_e + \nu_\mu + \overline{\nu}_\mu &
\end{array} \tag{2.1}$$

The neutrino energy spectrum strongly depends on that of the primary cosmic-ray flux, which decreases with energy, at, approximately, a rate of $E^{-2.7}$, for energies larger than 1 GeV.

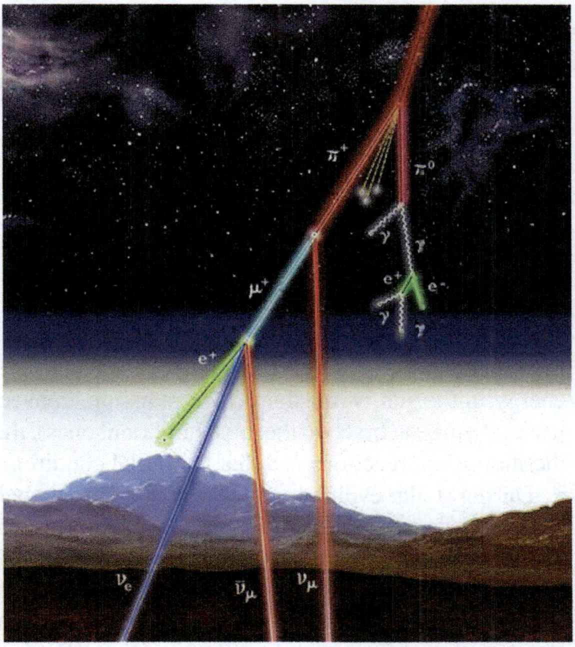

Fig. 2.2 Schematic diagram of neutrino production from cosmic rays interacting with the atmosphere

The neutrino production can be, naively, obtained from the previous interactions, noting that there are about twice as many $\overset{(-)}{\nu}_{\mu}$ than $\overset{(-)}{\nu}_{e}$. Additionally, neutrinos are produced homogeneously in all the Earth's atmosphere, causing the up-going and down-going neutrino fluxes to be almost the same if oscillations are not taken into account. The actual and detailed computation of atmospheric neutrinos is a very difficult task because of the different compositions of the atmosphere depending on the location, the variety of production mechanisms, the wide energy range and also due to the effect on the cosmic rays of the solar activity [Richard15] and the Earth's magnetic field.

2.1.2 *Accelerator Neutrinos*

These are human-made neutrino sources, usually obtained by accelerating protons in a synchrotron and, then, made collide with a target to produced large quantities of mesons, mainly π and κ [Kopp07], rather analogous to the production of atmospheric neutrinos. Of the mesons produced those with the same charge are selected through a series of magnets and will mostly decay into muon neutrinos or muon antineutrinos, depending on the operation mode of the accelerator, ν or $\overline{\nu}$ mode. The resulting neutrinos are sent to the near and far detectors in which, some of these neutrinos will be detected almost unmistakably due to their well known energy and direction.

This kind of facilities provides neutrinos with much better determined characteristics than those coming from other sources. This makes possible neutrino oscillation analyses more efficient, achieving high sensitivities with much lower statistics.

2.1.3 Solar Neutrinos

The Standard Solar Model is a very successful theory developed, mainly, by Bahcall [Bahcall64, Bahcall92] which models the evolution, activity and processes inside the Sun. The basic assumptions for the model are that it is in hydrostatic equilibrium implying that the sun evolves with a local balance between pressure and gravity, the energy in the sun is transferred via radiation, convection, conduction and neutrino losses (being the first two the most important ones), the energy is only produced from thermonuclear reactions and that the initial condition was a homogeneous sun.

During stellar evolution, hydrogen is burnt into helium in the hot and dense core, there are two mechanisms for this conversion, the *pp* chain and CNO cycle, being the former the dominant process in the sun and the one that produces Hep and ^{8}B neutrinos, those detected by Super-Kamiokande [Fukuda99b, Fukuda99a].

Using this theory, the solar neutrino flux was calculated to be proved by the Homesteak experiment in the 1960s by Davis [Davis Jr68], where it was found to be approximately $1/3$ of the predicted flux (Fig. 2.3).

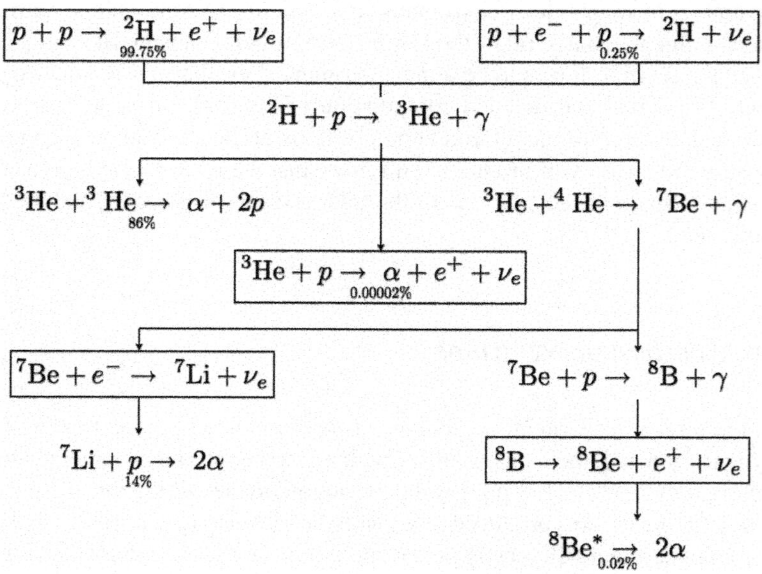

Fig. 2.3 Scheme of the solar neutrino production due to the nuclear interactions in the Sun

Despite the robustness of the solar model and the quality on the experimental side, it was thought that something should be wrong in one or both of them. This was known as the solar neutrino problem, which was understood as soon as neutrino oscillations entered the game. Neutrino oscillations in the sun must be done taking into account the matter effects of the sun, in which, the electron density is huge in the core ($N_{e,core} = 150$ g/cm^3) and decreases towards the surface. Neutrinos are thought to travel through the sun adiabatically, originated ν_e in the core of the sun will emerge from the surface having oscillated to $\tilde{\nu}_2$, the second neutrino mass eigenstate with sun matter effects. Neutrinos will travel in this state until the hit the Earth where there could be an enhancement of the ν_e as they travel through the Earth, this leads to a slightly larger ν_e flux during the night as compared with daytime, know as the solar neutrino day-night effect.

The MSW effect, [Mikheyev85], for certain values of the mass squared splitting, the energy dependence of the ν_e survival probability can lead to a distortion of the observed energy spectrum. For the current mass squared splitting and mixing angle preferred values by solar neutrino data, the MSW resonance in the Sun will occur at ~3 MeV. This, highly distorting the observed solar neutrino spectrum by inducing an upturn when going from high to low energies. The observation of this upturn in the solar neutrino energy spectrum, would be a direct signature of MSW type solar neutrino oscillations (Fig. 2.4).

SK's latest flux measurement for ^{8}B neutrinos is $\Phi_{^8B}(SK) = (2.345 \pm 0.014 \pm 0.036) \times 10^6\,\text{cm}^{-2}\text{s}^{-1}$ [Abe16b].

2.1.4 *Reactor Neutrinos*

Nuclear reactors generate energy by the fission of heavy nuclei, usually ^{235}U, these fragments have way too many neutrons, and are unstable, decaying through a series of α and β-decays into stable nuclei with a lower ratio of neutrons to protons [Declais94].

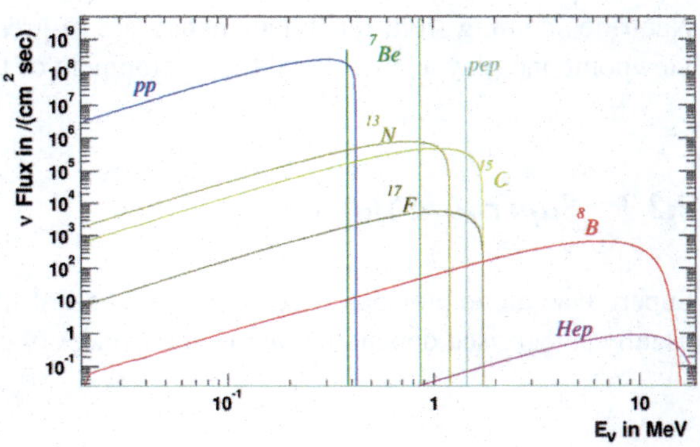

Fig. 2.4 Solar neutrino spectrum

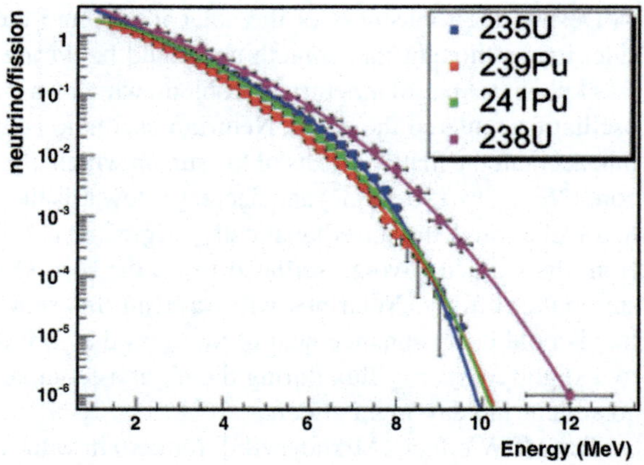

Fig. 2.5 Neutrino energy spectrum from the different isotopes used as fuel in nuclear reactors

In β-decays, many $\overline{\nu}_e$ are produced. Their energy spectrum can be calculated from the estimate of the number of neutrinos produced per fission, the power output, and taking into account all the additional fissionable isotopes produced from the primary fissions, such as ^{239}Pu, ^{241}Pu or ^{238}U (Fig. 2.5).

This last term is the most difficult term to compute and which carries most of the sizeable error of the computation. This is currently compensated with the installation of a near detector as in the case of accelerator neutrinos, where, without neutrino oscillations, the reactor antineutrino energy spectrum can be better measured, canceling out most of the systematic errors.

2.2 Supernova Neutrinos

Supernovae are one of the most energetic events occurring in our universe and which produce an enormous amount of neutrinos. Since just one or two of these explosions happen in a hundred years, and close enough to detect their neutrinos, they are treated separately from the previous section concerning the most common neutrino sources. Neutrinos coming from this phenomenon are of great importance from a physical viewpoint and play a key role in the development of this work.

2.2.1 Supernova Burst

Supernovae are explosions occurring at the end of life of a massive star. They are usually categorised depending on the abundances of certain spectral lines (Fig. 2.6).

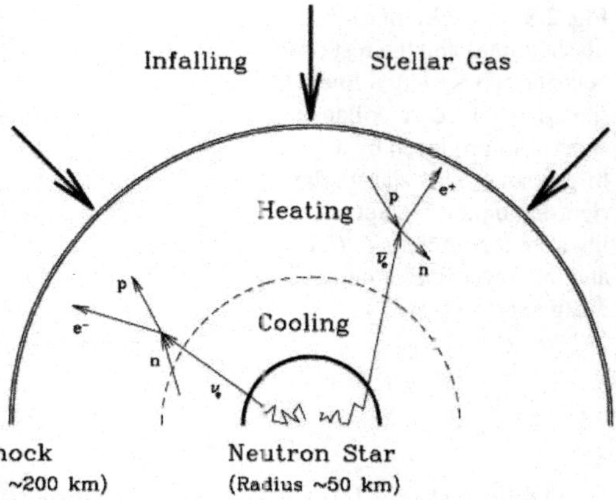

Fig. 2.6 Diagram of the dynamics of matter and neutrinos during Supernova core-collapse

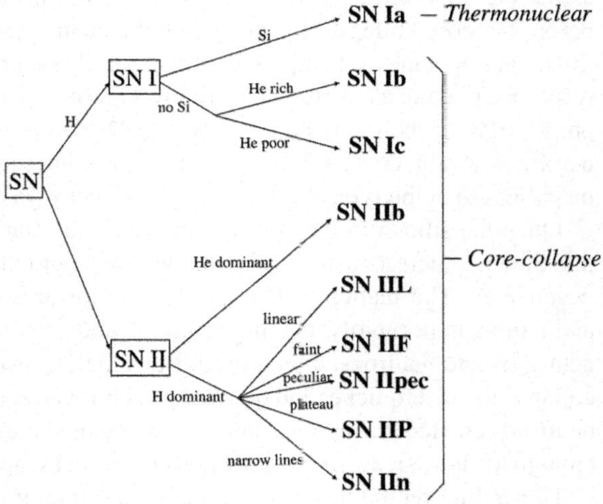

Fig. 2.7 Scheme of the supernova types depending on the former star

In terms of neutrinos, the most interesting supernovae are those whose mass is more than eight solar masses and their core collapses produce large quantities of neutrinos of all kinds, i.e. TypeIb, TypeIc and TypeII. During a core-collapse explosion, around 99% of the gravitational binding energy liberated is carried away by neutrinos, leaving afterwards a neutron star or a black hole [Burrows13].

The evolution of such stars begins with the burning of hydrogen, turning it into helium as seen in Fig. 2.3, for the solar case. Once it runs out of most hydrogen in the core, the star collapses increasing the core's density, and allowing the helium fusion into carbon. This process continues with heavier elements making the star being separated in layers of different elements (Fig. 2.7).

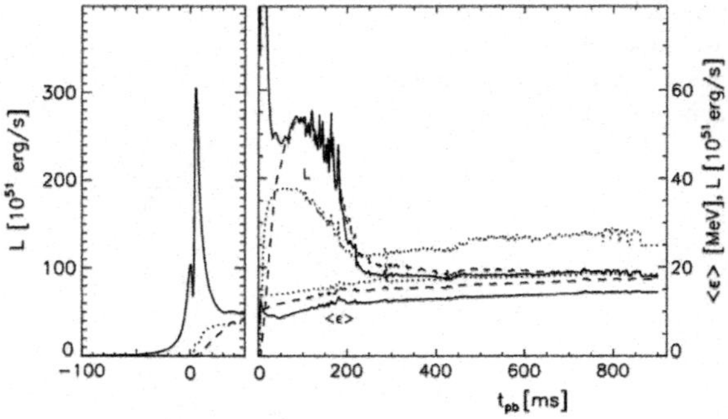

Fig. 2.8 ν_e (solid line), $\overline{\nu}_e$ (dashed line) and the rest of neutrino types (dotted line) spectra from a core-collapse supernova produced by a fifteen solar mass star. In the right-hand plot, the upper lines are the luminosity (L) and the lower lines are the mean energy ($\langle\epsilon\rangle$)

The last stage, before the core-collapse, is the Si-burning phase, also known as the pre-supernova phase. This phase occurs in the final days of a massive star before the core-collapse. It starts when the central temperature exceeds 3×10^9 K. This stage includes a complex combination of photo-disintegration and α-capture reactions, playing a key role in the future supernova explosion. During the Si burning phase $\sim$1% of the total energy of core-collapse are emitted through pre-supernova neutrinos. Antineutrinos from Si burning stage have an average energy of 1.85 MeV, much less than the typical 10–20 MeV of supernova neutrinos.

Once the silicon in the core is exhausted, the star begins to collapse again, and this time no nuclear fusion will compensate the gravitational force only the electron degeneracy. The matter is then accelerated towards the centre, increasing density and temperature rapidly, making electrons and protons to be quickly converted into neutrinos and neutrons, which diminishes the electron degeneracy. This causes the collapse to occur quicker and quicker. This process is called neutronisation burst and neutrinos created during this stage are inside the core, which reaches densities large enough to slow, or, even trap, the neutrinos from escaping during a few milliseconds.

Depending on the theoretical model and mass of the star, the neutrino emission and the last details of the star's life may vary, but, in general, most of the gravitational energy of the star is released in the form of neutrinos with same abundance in all flavours. These neutrinos and antineutrinos evolve very differently due to the charged leptons associated, such that the muon and tau neutrinos interact only via NC interactions, as seen in Fig. 2.12. This makes that $\langle E_\nu \rangle < \langle E_{\overline{\nu}} \rangle < \langle E_{\overset{(-)}{\nu}_\mu} \rangle \approx \langle E_{\overset{(-)}{\nu}_\tau} \rangle$ (Fig. 2.8).

After exploding, the majority of the light elements and the outer layers are dispersed, leaving a hugely dense neutron core. Of these stars, those beginning with less than twenty solar masses, are left as neutron stars, whereas for those in between twenty and fifty solar masses, a black hole is produced after the supernova explosion. For stars above fifty solar masses, the collapse causes directly a black hole with no supernova explosion [Totani98a].

2.2.2 Diffuse Supernova Neutrino Background

The basics of the core-collapse model were checked with the unique, up to date, nearby supernova explosion seen by neutrino detectors. This is the famous SN1987a, occurred in the Large Magellanic Cloud, close to our Milky Way, around 50 kpc away. About three hours before the photons from the supernova explosion arrived to the optical telescopes, twenty four neutrinos were detected in KamiokaNDE (8) [Hirata87], IMB (11) [Velde88] and Baksan (5) [Alexeyev88], a huge signal compared with the solar neutrino one.

The SN1987a measurement meant a much better knowledge of the neutrino flux, spectrum and time profile, which was crucial for a more accurate prediction of the Diffuse Supernova Neutrino Background (DSNB) [Beacom10]. The DSNB is the overall neutrino flux of all supernovae that happened in the history of our universe.

The DSNB flux is calculated as the convolution of core-collapse supernova rate and the neutrino emission per supernova, accounting for the different redshifts of supernovae. Equation (2.2) shows the expression for its calculation.

$$\frac{\mathrm{d}\Phi_{DSNB}}{\mathrm{d}E_\nu} = c\int_0^\infty R_{SN}(z)(1+z)\phi\Big(E_\nu(1+z)\Big)\left|\frac{\mathrm{d}t}{\mathrm{d}z}\right|\mathrm{d}z \tag{2.2}$$

where R_{SN} is the the core-collapse supernova rate at redshift z, ϕ is the neutrino emission per supernova, $|\frac{\mathrm{d}t}{\mathrm{d}z}|^{-1} = H_0(1+z)\sqrt{\Omega_\Lambda + \Omega_m(1+z)^3}$, which depends on the Hubble constant H_0, is the cosmic matter density Ω_m and the cosmological constant Ω_Λ. The neutrino spectrum and the neutrino rate per supernova are the most uncertain inputs in this calculation, making the first measurement of these neutrinos essential to understand key processes during supernovae.

On the other hand, the supernova rate is well known and proportional to the star formation rate (R_{SF}), which is a function of the star mass. The supernova calculation rate follows Eq. (2.3).

$$R_{SN}(z) = R_{SF}\frac{\int_8^{50}\psi(M)\mathrm{d}M}{\int_{0.1}^{100}M\psi(M)\mathrm{d}M} \approx \frac{R_{SF}(z)}{143M_{sun}} \tag{2.3}$$

where ψ is the initial mass distribution of the stars and M is the mass of the star.

Figure 2.9 shows the expected fluxes and spectra for a series of theoretical models for different effective temperatures (Fig. 2.9).

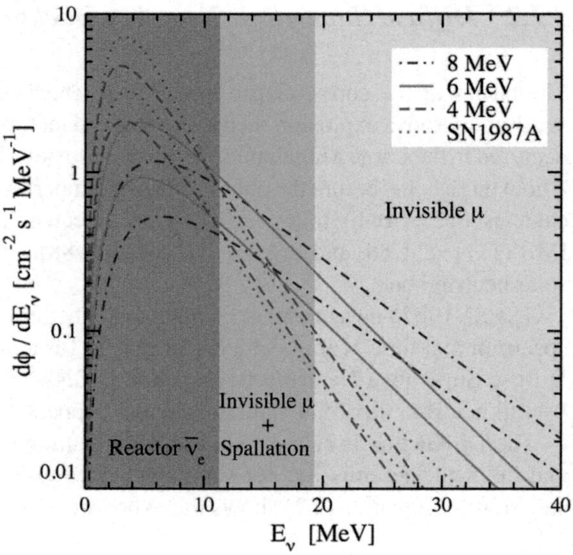

Fig. 2.9 Predicted DSNB $\overline{\nu}_e$ spectrum for different models and showing the three main backgrounds for this measurement and the energy region where they apply

2.3 Neutrino Mass and Neutrino Oscillations

In this section, the current understanding of the neutrinos as particles of the Standard Model is reviewed briefly and divided into two parts: the neutrino mass and particle characteristics, and the neutrino oscillations, in Sects. (2.3.1 and 2.3.2) respectively. At the end, in Sect. 2.3.3, the latest results from different experiments of the different parameters and properties are reviewed.

2.3.1 Neutrino Masses

The existence of neutrinos was first theoretically proposed by W. Pauli as explanation for the broad spectrum of e in β-decays [Pauli30, Fermi08], turning the one body decay problem, which violates energy conservation, into to a two body decay which does not. This was confirmed experimentally by Cowan and Reines in [Cowan56] with the discovery of the electron neutrino [Cowan56] (Fig. 2.10).

Years latter the corresponding muon and tau neutrinos were proposed and discovered [Danby62, Kodama01], leaving the picture of three generations of neutrinos, each of them associated to a charged lepton.

Neutrinos are until today the lightest massive particle known Standard Model (SM) contains. They are fermions with nor electric neither colour charge, being able for them to interact only throw the gravitational and the nuclear weak forces. Due to the smallness of their masses, just the latter is relevant and reachable by current experiments.

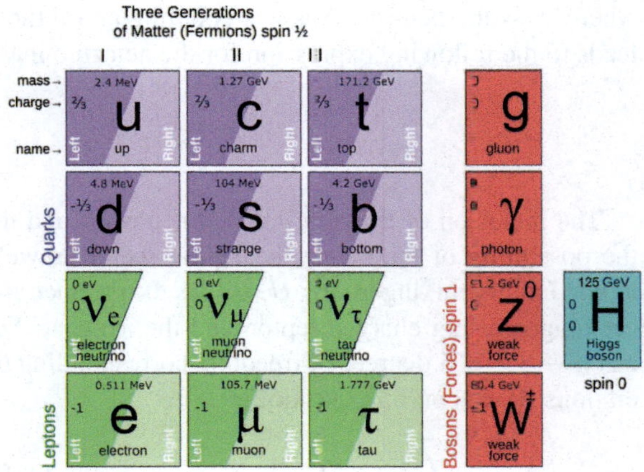

Fig. 2.10 Particle and force content of the standard model

Neutrinos have spin one half ($S_\nu = 1/2$) and weak hypercharge -1 ($Y_W = -1$). In the Standard Model gauge group, $SU(3)_C \times SU(2)_L \times U(1)_Y$,they are embedded in the left-handed lepton doublet of $SU(2)_L$, along with their charged partner, and from a singlet of $SU(3)_C$. The lagrangian corresponding to this part of the SM is as shown in Eq. (2.4).

$$\mathcal{L}_{SM} = i\overline{\nu}_L \not{D} \nu_L + (h.c.) = \overline{\nu}_L \Big(i\not{\partial} + \frac{e}{2\cos\theta_W}\not{B} - \frac{e}{2\sin\theta_W}\not{W} \Big) \nu_L + (h.c.) \quad (2.4)$$

where ν_L is the chiral left-handed, flavour (e, μ, τ) neutrino field.

Neutrinos have been proven to be non-massless throw their oscillations from one flavour to another, impossible for massless neutrinos.

The fact that neutrinos have mass, makes necessary to extend the SM with a mechanism to provide neutrinos with mass. There are basically, two different ways to realise the scenario of three massive neutrinos: the existence of a right-handed neutrino or assuming that neutrinos are Majorana particles. Here, both of them are explained separately in order to be as clear as possible since more elaborated models can be found in the literature, for instance in [King04a].

The most straightforward attempt is the addition of a right-handed neutrino (ν_R) for each generation, such that neutrinos acquire mass in a similar way to the rest of the fermions of the SM, through the Higgs mechanism. This right-handed neutrino would be a singlet of $\mathbf{SU(2)}_L$ and of $\mathbf{SU(3)}_C$ with $Y_W = 0$. Therefore, it would only interact through gravity or through some other new force. The Dirac mass term obtained then for neutrinos after the $\mathbf{SU(2)}_L \times U(1)_Y$ is spontaneously broken symmetry and the Higgs field acquires a vacuum expectation value.

$$\mathcal{L}^{Dirac}_{\nu,mass} = -\frac{y_\nu v}{\sqrt{2}} \overline{\nu_L} \nu_R + (h.c.) \quad (2.5)$$

where y_ν is the neutrino Yukawa coupling and v the v.e.v. of the Higgs field. This leads to the following expression for the neutrino masses of each generation.

$$m_\nu^D = \frac{y_\nu v}{\sqrt{2}} \tag{2.6}$$

The inclusion of the right-handed neutrinos and the Yukawa couplings Y_ν open the possibility of mixing in the lepton sector as well as in the quark sector. The lepton unitary mixing matrix U_{PMNS} is the product of the left-handed unitary matrices diagonalising charged lepton and the neutrino Yukawa couplings. This matrix has $(3-1)^2 = 4$ degrees of freedom, corresponding the 3 to the three neutrino generations [Pontecorvo57, Maki62].

$$U_{PMNS} = \begin{pmatrix} c_{13}c_{12} & c_{13}s_{12} & s_{13}e^{-i\delta} \\ -c_{23}s_{12} - s_{23}s_{13}c_{12}e^{i\delta} & c_{23}c_{12} - s_{23}s_{13}s_{12}e^{i\delta} & s_{23}c_{13} \\ s_{23}s_{12} - c_{23}s_{13}c_{12}e^{i\delta} & -s_{23}c_{12} - c_{23}s_{13}s_{12}e^{i\delta} & c_{23}c_{13} \end{pmatrix} \tag{2.7}$$

where $c_{ij} = \cos\theta_{ij}$ and $s_{ij} = \sin\theta_{ij}$.

Despite this approach looks very reasonable, there exists one problem concerning the neutrino Yukawa couplings. They are orders of magnitude smaller than those for the other charged fermions of the same generation. This suggest that there may be something else that makes the neutrino masses so small without having to deal with such small Yukawa couplings.

An alternative to the addition of right-handed neutrinos to the SM, is to assume neutrinos as Majorana particles, that is they are its own antiparticle. Neutrinos are the only particles in the SM allowed to be Majorana because of their neutral charges. This allows to have another possible mass term, in addition to the Dirac mass term, the Majorana mass term.

$$\mathcal{L}_{\nu,mass}^{Majorana} = \frac{1}{2}\nu_L^T M C \nu_L + (h.c.) \tag{2.8}$$

where C is the charge conjugation operator and M is the Majorana mass matrix. The matrix M can be diagonalised by a unitary transformation U_M, obtaining $M_{diag} = U_M^T M U_M$. Therefore, for Majorana neutrinos, one also has a lepton mixing matrix $U_{PMNS+Maj.}$, shown in Eq. (2.9). This results in a mixing matrix with $3-1$ more degrees of freedom, the Majorana extra phases (α_1, α_2).

$$U_{PMNS+Maj.} = U_{PMNS} \cdot \begin{pmatrix} 1 & 0 & 0 \\ 0 & e^{i\alpha_1} & 0 \\ 0 & 0 & e^{i\alpha_2} \end{pmatrix} \tag{2.9}$$

The Majorana phases cancel out for any observable in neutrino oscillations and, therefore, neutrino oscillation experiment only see the parameters of the U_{PMNS} matrix.

The fact that no new particles are required makes this mechanism appealing, but the mass term breaks explicitly the $\mathbf{SU(2)}_L$ and, as the Dirac mass term, fails to explain the smallness of the neutrino masses. Because both of these methods are not fully satisfactory, a third mechanism involving the previous one is proposed in various versions, that is the seesaw mechanism. The simplest case is the Type I seesaw mechanism, where three heavy right-handed Majorana neutrinos (one for each generation) are added to the particle content.

$$\mathcal{L}_{nu,mass}^{I\ seesaw} = -m_D \overline{\nu_L} \nu_R + \frac{M_N}{2} \nu_R^T C \nu_R + (h.c.) \tag{2.10}$$

Rearranging the neutrino fields, one gets the following mass matrix,

$$M = \begin{pmatrix} 0 & m_D \\ m_D & M_N \end{pmatrix} \tag{2.11}$$

This means that neutrinos are a mixture of Dirac and Majorana particles, but the most interesting scenario is that where $m_D \ll M_N$ and giving naturally the left-handed small masses assuming right-handed neutrinos have large masses, near GUT scale. In this case, the right-handed neutrinos have mass $\sim M_N$ and the left-handed neutrinos m_ν.

$$m_\nu = -\frac{m_D^2}{M_N} \tag{2.12}$$

All versions of this mechanism work in a similar way, assuming the existence of a very large particle which is responsible for the smallness of left-handed neutrinos. This method is very interesting also because it relates the neutrino physics to the GUT scale.

2.3.2 Neutrino Oscillations

In this section, the formalism of neutrino oscillations is explained in vacuum and in matter.

As we have seen in the previous section, neutrinos are transformed by a non-diagonal lepton mixing matrix U. This idea was first proposed by Gribov and Pontecorvo in [Gribov69]. Following this reasoning, flavour neutrino eigenstates $|\nu_l\rangle$ $(l = e, \mu, \tau)$ transform to the neutrino mass eigenstates $|\nu_i\rangle$ $(i = 1, 2, 3)$.

$$|\nu_l\rangle = \sum_l (U_{PMNS}^{li})^* |\nu_i\rangle \tag{2.13}$$

This mixing matrix has only 4o of freedom which correspond to the three neutrino mixing angles and the CP phase. This can be parametrised as the product of three-dimensional rotations in the following manner.

$$U_{PMNS} = U_{23}(\theta_{23}) I_\delta U_{13}(\theta_{13}) I_\delta^\dagger U_{12}(\theta_{12}) \quad (2.14)$$

The neutrino propagation through vacuum is governed by the Schrödinger equation with no potentials acting on the neutrino mass eigenstates.

$$i\mathrm{d}|\nu_l\rangle t = H_0^l|\nu_l\rangle \Rightarrow |\nu_i(t)\rangle = e^{-iE_it}|\nu_i(0)\rangle \quad (2.15)$$

where H_0^l is the neutrino dynamics Hamiltonian and E_i its eigenstate of $|\nu_i\rangle$.

As seen before, the mass of the neutrino will be much smaller than its total energy, being able to approximate its energy by doing a first order Taylor expansion.

$$E_i = \sqrt{p^2 + m_i^2} \approx p\left(1 + \frac{m_i^2}{2p^2} + \mathcal{O}\left(\frac{m_i^4}{p^4}\right)\right) = p + \frac{m_i^2}{2p} + \mathcal{O}\left(\frac{m_i^4}{p^3}\right) \quad (2.16)$$

One also needs to consider that neutrinos will be generated as a neutrino flavour eigenstate, since it will come from a weak interaction. With all this one can compute the amplitude of having a $|\nu_l\rangle$ in another flavour state, $|\nu_{l'}\rangle$, after some time t.

$$A_{\nu_l\rightarrow\nu_{l'}}(t) = \sum_{i,j}(U^{li}_{PMNS})^* U^{l'j}_{PMNS} e^{-iE_it}\langle\nu_j(p)|\nu_i(p)\rangle = \sum_i (U^{li}_{PMNS})^* U^{l'i}_{PMNS} e^{-iE_it} \quad (2.17)$$

It will come clear now, in the computation of the oscillation probabilities, that neutrino masses, cannot be observed in neutrino oscillation experiments, but only the differences of their squares. The notation, $\Delta m^2_{ij} = m_i^2 - m_j^2$, is assumed.

The neutrino oscillation probability in vacuum from one flavour eigenstate ν_l to another flavour eigenstate $\nu_{l'}$, is given by Eq. (2.18).

$$P_{\nu_l\rightarrow\nu_{l'}}(t) = |A_{\nu_l\rightarrow\nu_{l'}}(t)|^2 = \left|\sum_i U^{l'i}_{PMNS}(U^{li}_{PMNS})^* e^{-iE_it}\right|^2 = \sum_{i,j} U^{l'i}_{PMNS}(U^{li}_{PMNS})^*(U^{l'j}_{PMNS})^* U^{lj}_{PMNS} e^{-i(E_i-E_j)t} = \sum_{i,j} \mathfrak{U}^{ij}_{ll'} e^{-i\frac{\Delta m^2_{ij}}{2p}t} \quad (2.18)$$

where the product of matrices, $\mathfrak{U}^{ij}_{ll'} = U^{l'i}_{PMNS}(U^{li}_{PMNS})^*(U^{l'j}_{PMNS})^* U^{lj}_{PMNS}$, is defined.

Due to the smallness of the neutrino masses and in the ultra-relativistic case (like is the case in neutrino oscillations with $E_\nu > 1MeV$), the momentum of the neutrino can be very well approximated by its energy and the speed of the neutrino is very close to the speed of light.

$$\frac{t}{2p} \approx \frac{L}{c}\frac{c}{2E} = \frac{L}{2E} \quad (2.19)$$

where L is the distance travelled by the neutrino and E is its energy. With this, the probability can be rewritten in terms of the diagonal, real and imaginary parts in the following way.

$$P_{\nu_l \to \nu_{l'}} = \sum_i \mathfrak{U}^{ii}_{ll'} + 2\sum_{i<j} \mathrm{Re}(\mathfrak{U}^{ij}_{ll'}) \cos\left(\frac{\Delta m^2_{ij}}{2E_i} L\right) + 2\sum_{i<j} \mathrm{Im}(\mathfrak{U}^{ij}_{ll'}) \sin\left(\frac{\Delta m^2_{ij}}{2E_i} L\right) =$$
$$\delta_{ll'} - 4\sum_{i<j} \mathrm{Re}(\mathfrak{U}^{ij}_{ll'}) \sin^2\left(\frac{\Delta m^2_{ij} L}{4E}\right) + 2\sum_{i<j} \mathrm{Im}(\mathfrak{U}^{ij}_{ll'}) \sin\left(\frac{\Delta m^2_{ij} L}{2E}\right) \tag{2.20}$$

In the case of antineutrino oscillations, there is only a difference in the definition of their states.

$$|\overline{\nu}_l\rangle = \sum_l U^{li}_{PMNS} |\overline{\nu}_i\rangle \tag{2.21}$$

Therefore, the oscillation probabilities are equivalent to the neutrino case but exchanging $(U^{li}_{PMNS})^* \rightarrow U^{li}_{PMNS}$. Obviously, this change only affects the imaginary terms of the oscillation probability, responsible for the neutrino CP violation (Fig. 2.11).

All the previous is valid for vacuum neutrino oscillations, but the presence of matter can greatly affect the neutrino oscillation probabilities [Mikheyev85]. This happens due to the coherent forward scattering as neutrinos interact with the constituents of the matter both through charged-current (exchange of $W^{\pm}$ bosons) or neutral-current (exchange of Z^0 bosons). This interactions enter as a new term in the Hamiltonian which will affect the propagation of neutrinos by Eq. (2.15) (Fig. 2.12).

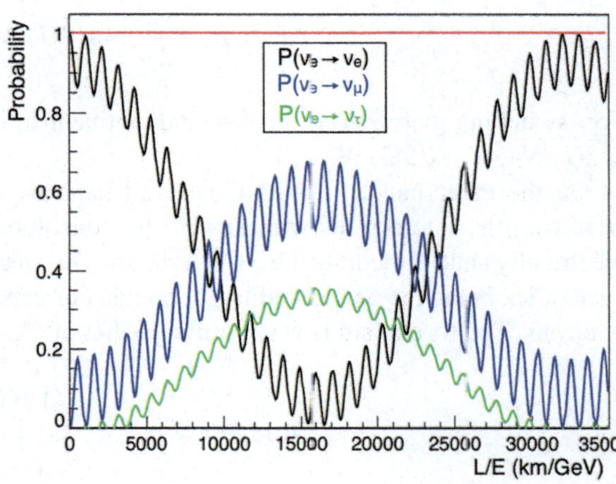

Fig. 2.11 Neutrino oscillation probabilities assuming initial state is an electron neutrino

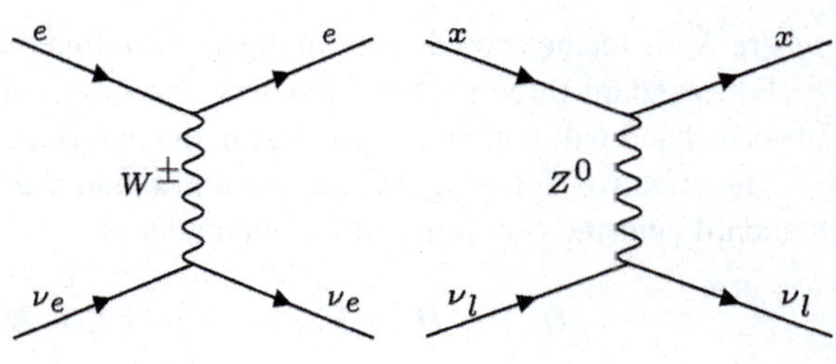

Fig. 2.12 Tree-level Feynman diagrams for coherent forward scattering of neutrinos on any fermion, *x*, for NC interactions (left) and on electrons for the CC scattering (right)

For the CC case only electron neutrinos and antineutrinos are affected because of their weak interaction with electrons.

The Lagrangian for the neutrino CC coherent forward scattering with matter has the following form.

$$\mathcal{L}_{CC} = -\frac{G_F}{\sqrt{2}}|\overline{\nu}_e\rangle\gamma^{\mu}(1-\gamma^5)\langle\nu_e||e\rangle\gamma_{\mu}(1-\gamma^5)\langle e| \tag{2.22}$$

As electrons form part of the medium, they can be seen as a distribution of electrons and not as single particles. Then, the electron degrees of freedom can be integrated out and the electron number density N_e introduced.

$$\mathcal{L}_{CC} = -\frac{G_F N_e}{\sqrt{2}}|\overline{\nu}_e\rangle\gamma^{\mu}(1-\gamma^5)\langle\nu_e| \tag{2.23}$$

Being the resulting CC neutrino Hamiltonian,

$$H_{CC} = \sqrt{2}G_F N_e \begin{pmatrix} 1 & 0 & 0 \\ 0 & 0 & 0 \\ 0 & 0 & 0 \end{pmatrix} \tag{2.24}$$

and the total neutrino energy

$$E \approx p + \frac{m^2}{2p} + \sqrt{2}G_F N_e \tag{2.25}$$

For switching between neutrinos and antineutrinos it is enough to exchange $\sqrt{2}G_F N_e \rightarrow -\sqrt{2}G_F N_e$.

On the other hand, in the NC case all neutrino flavours are affected equally, occurring the interactions with any of the constituents of matter. However, in an electrically neutral medium the NC interactions between electrons and protons cancel each other, being the only remaining term that corresponding to the interactions with neutrons. The formalism is very similar to that of CC scattering.

$$H_{NC} = -\sqrt{2}G_F N_n \begin{pmatrix} 1 & 0 & 0 \\ 0 & 1 & 0 \\ 0 & 0 & 1 \end{pmatrix} \tag{2.26}$$

where N_n is the neutron density in the surrounding matter.

For standard three neutrino flavours, the phase introduced by this term can be absorbed and will not have any effect in the neutrino oscillations.

The effective (ignoring NC term) hamiltonian needed for the computation of the standard neutrino oscillations through matter is

$$H_{eff} = H_0 + H_{CC} = H_0 + \sqrt{2}G_F N_e \text{diag}(1,0,0) \tag{2.27}$$

One can now proceed as in the vacuum case by defining $\tilde{\nu}_i$ the matter eigenstates which diagonalise the effective Hamiltonian H_{eff}, satisfying then the analogous transformation as in Eq. (2.13).

$$|\nu_l\rangle = \sum_l (\tilde{U}^{li})^* |\tilde{\nu}_i\rangle \tag{2.28}$$

Being $\tilde{U}$ the effective lepton mixing matrix with matter effects.

One caveat before the computation is that the electron density is not constant throughout the Earth, meaning that the transformation matrix, $\tilde{U}$, is time-dependent. Then, the analogous to Eq. (2.15) has the following expression in matrices terms.

$$id|\nu\rangle t = H_{eff}|\nu\rangle \Rightarrow id(\tilde{U}^\dagger|\tilde{\nu}\rangle)t = \tilde{U}^\dagger \tilde{H}_{eff}\tilde{U}\tilde{U}^\dagger|\nu\rangle \Rightarrow id|\tilde{\nu}\rangle t = \left(\tilde{H}_{eff} - i\tilde{U}d\tilde{U}^\dagger t\right)|\tilde{\nu}\rangle \tag{2.29}$$

where ν and $\tilde{\nu}$ are all the flavour and matter neutrino eigenstates respectively and $\tilde{H}_{eff}$ is the diagonalised effective Hamiltonian with the matter potential. The final solution to Eq. (2.29) is rather difficult, being the analytic solution obtained after several approximations.

The different Earth's densities are usually treated as an overlay of constant density layer, namely the mantle ($\rho_{mantle} = 5.5$ g/cm^3) and the core ($\rho_{core} = 11.5$ g/cm^3).

Matter effects affect the neutrino mixing, being crucial for the measurement of the different oscillation parameters. In fact, one of the main properties of the oscillations in matter, is the opposite behaviour of the term depending on the mass hierarchy (sign(Δm^2_{31})). The oscillograms in Table 2.1, show the Earth's matter oscillation probability difference between neutrinos and antineutrinos as defined next.

$$\Gamma^{\nu-\overline{\nu}}_{l,MH} = \sum_{x=e,\mu} \left(P^{(\nu)}_{x\to l,MH} - P^{(\overline{\nu})}_{x\to l,MH}\right) \tag{2.30}$$

Table 2.1 only shows the values for those neutrinos passing through the Earth, that is with $\cos\theta_{zenith} < 0$, in order to better focus on the importance of the matter effects. The Earth's matter resonance is seen at energies from 1 GeV to 10 GeV and it is also seen clearly how the zenith angle affects that resonance, being dimmer towards the horizontal direction and stronger when neutrinos travel through the mantel and the core of the Earth. It is seen that both possible neutrino mass orderings have opposite behaviour.

The mass ordering of the neutrinos is one of the main remaining uncertainties in neutrino physics. The measurement of this property is mainly based on the analysis of the oscillation matter effects of atmospheric neutrinos passing through the Earth (Fig. 2.13).

Table 2.1 Sum in all flavours of the difference of neutrino-antineutrino oscillation probabilities as a function of energy and cosine of zenith angle, with final electronic (top) and muonic (bottom) flavours and for both neutrino mass orderings, normal (left) and inverted (right)

	Normal hierarchy	Inverted hierarchy
ν_e	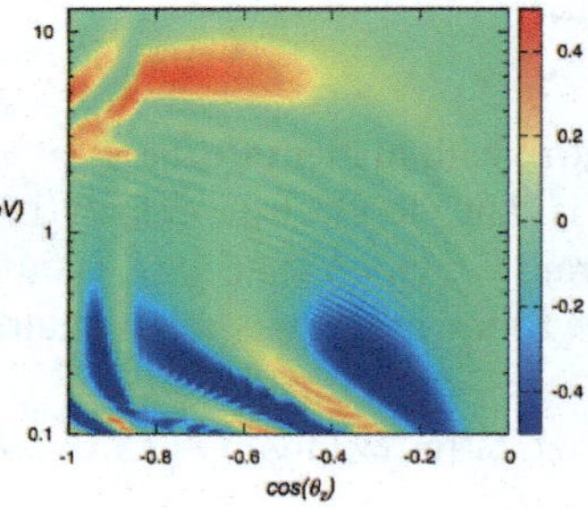	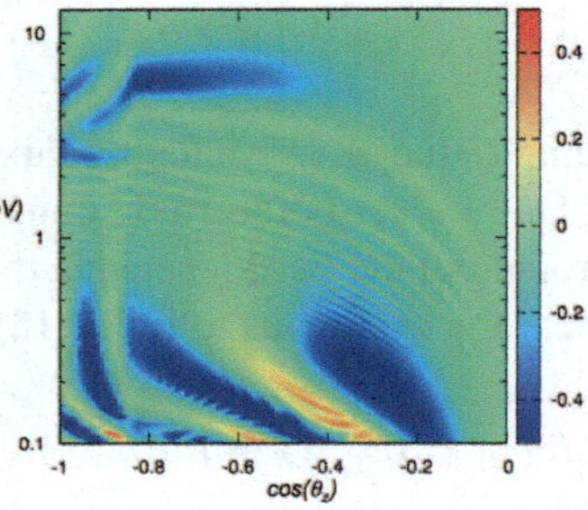
ν_μ	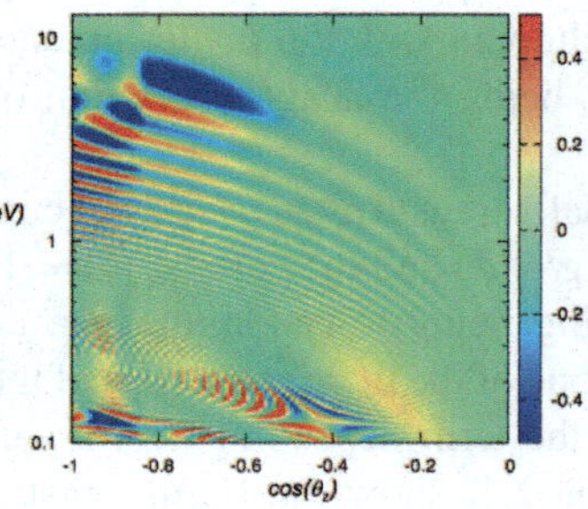	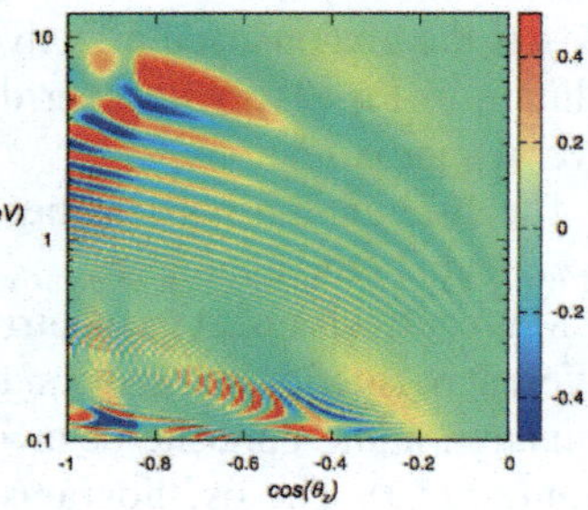

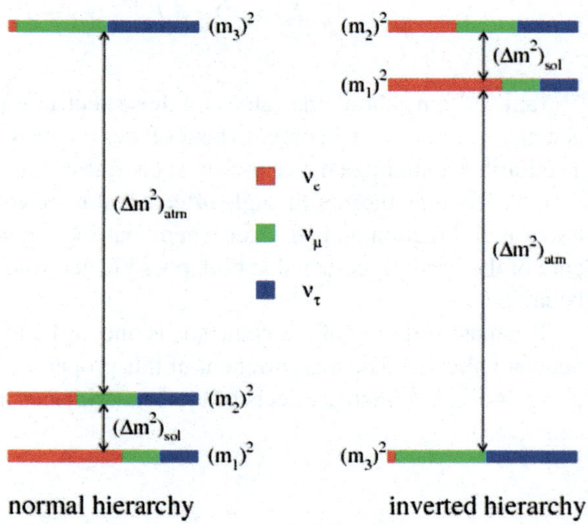

Fig. 2.13 Scheme of the two possible neutrino mass hierarchies depending on the sign of Δ^2_{31}. In the left for normal (third mass eigenstate is the heaviest) and right for inverted (third mass eigenstate is the lightest)

Table 2.2 Neutrino oscillation best-fit values derived from a global fit of the current neutrino oscillation data, where Δm^2 is defined as $\Delta m^2 = m_3^2 - (m_2^2 + m_1^2)/2$

Parameter	Mass ordering	Best-fit	3σ (2σ for δ_{CP}) range
Δm_{21}^2 (10^{-5} eV^2)	NH/IH	7.37	6.93–7.97
$\|\Delta m^2\|$ (10^{-3} eV^2)	NH	2.50	2.37–2.63
	IH	2.46	2.33–2.60
$\sin^2\theta_{12}$	NH/IH	0.297	0.250–0.354
$\sin^2\theta_{23}$	NH	0.437	0.379–0.616
	IH	0.569	0.383–0.637
$\sin^2\theta_{13}$	NH	0.0214	0.0185–0.0246
	IH	0.0218	0.0186–0.0248
$\delta_{CP}/2$	NH	1.35	0.92–1.99
	IH	1.32	0.83–1.99

2.3.3 Experimental Status

In this section, the latest results for the neutrino properties are summarised, from the mass to the oscillation parameters.

The measurement of the neutrino mass is addressed by measuring the spectrum of electrons near the endpoint in ^{3}H β-decay experiments and from cosmological and astrophysical data, being the most stringent upper limits for the mass of $\overline{\nu}_e$ the result from the Troitzk experiment, $m_{\overline{\nu}_e} < 2.05$ eV at 95% confidence level. Gathering data from Planck, supernovae and Baryon Acoustic Oscillations the limit on the sum of the neutrino masses is lowered, $\sum_j m_j < 0.23$ eV at 95% confidence level.

Concerning the oscillation parameters, solar and reactor neutrinos are most sensitive to θ_{12} and Δm_{21}^2, coming the best results from KamLAND. The parameter θ_{13} is very precisely obtained from short baseline reactor experiments, such as Daya Bay, RENO or Double Chooz, and also, from the long baseline experiments T2K and Noνa. The latter, together with the SK atmospheric neutrino results, provide the constraints on θ_{23} and Δm_{32}^2, but not enough to determine the octant or the mass ordering. Finally, the CP violating phase has the largest uncertainty of all oscillation parameters, and long baseline neutrino experiments are the responsible of the best limits on its value. The relatively large value of θ_{13}, enables the possibility for SK, T2K and Noνa to measure the three least known parameters, i.e., the octant of θ_{23}, the neutrino mass ordering and the CP phase [Patrignani16] (Table 2.2).

2.4 Neutrino-Nucleus Interactions

In this section, the basic neutrino interactions occurring inside a water-Čerenkov detector are reviewed. Each type of interaction becomes relevant for certain physical phenomena depending on its energy range, [Formaggio12]. Therefore, the complete

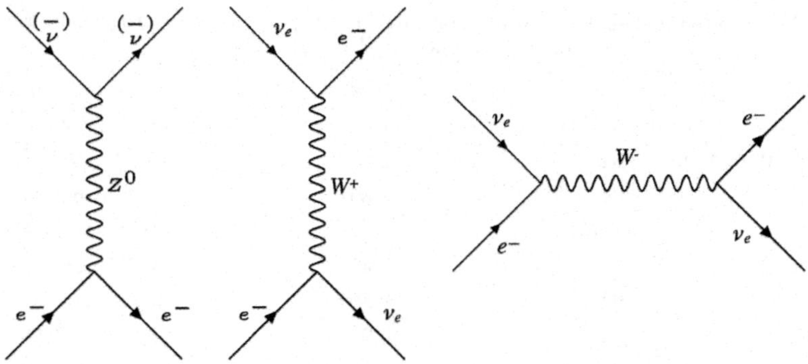

Fig. 2.14 Tree level Feynman diagrams for the neutral current (left) $\nu - e$ scattering and charged current (centre and right) $\nu_e - e$ scattering

understanding of the neutrino-nucleus interactions will be of great importance across the development of this work.

Elastic Scattering

The first of the interactions is also the simplest of them and crucial for the detection of low energy neutrinos, mainly solar neutrinos. In this process, an electron is scattered off by an electron via the exchange of a virtual vector boson. This type of interaction was the first used for the measurement of neutrinos and the confirmation of weak neutral currents [Hasert73] (Fig. 2.14).

$$\nu + e^- \rightarrow \nu + e^- \tag{2.31}$$

The expression of the differential cross-section for this process is the following.

$$\frac{d\sigma(\nu + e^- \rightarrow \nu + e^-)}{dT_e} = \frac{G_F^2 m_e}{2\pi}\left((g_V + g_A)^2 + (g_V - g_A)^2\left(1 - \frac{T_e}{E_\nu}\right) + (g_V {-}^2 g_A^2)\frac{m_e T_e}{E_\nu^2}\right) \tag{2.32}$$

where G_F is the Fermi coupling constant, m_e the electron mass, E_ν the neutrino energy and T_e the kinetic energy of the scattered electron. The axial-vector and vector couplings g_A and g_V can be expressed in terms of the weak angle θ_W ($\sin^2\theta_W = 0.2317$) as follows.

$$\begin{aligned} g_A &= \tfrac{1}{2},\ g_V = 2\sin\theta_W + \tfrac{1}{2} \text{ for } \nu_e \\ g_A &= \tfrac{-1}{2},\ g_V = 2\sin\theta_W - \tfrac{1}{2} \text{ for } \nu_\mu \text{ and } \nu_\tau \end{aligned} \tag{2.33}$$

The recoil electrons are scattered in the forward direction as the neutrino energy increases. The angle between the incident neutrino and the outgoing recoil electron is given by the next formula.

$$\cos\theta = \frac{1 + m_e/E_\nu}{\sqrt{1 + 2m_e/T_e}} \tag{2.34}$$

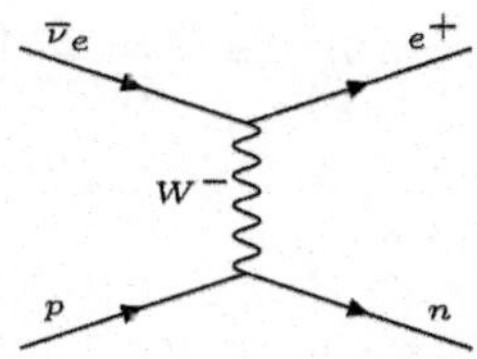

Fig. 2.15 Tree level Feynman diagram of inverse-β decay

Inverse-β Scattering

This is the simplest nuclear interaction that we can study, the antineutrino-proton scattering, also known as inverse-β decay. This process is very important at low energies, from its energy threshold, $E_{\overline{\nu}} = 1.806$ MeV, up to tenths of MeV. Low energy antineutrinos from reactors, DSNB and close enough supernova bursts are detected through this interaction (Fig. 2.15).

The reaction takes part between an incoming antineutrino and a hydrogen nucleus.

$$\overline{\nu}_e + p \rightarrow e^+ + n \tag{2.35}$$

Its differential cross-section has the next expression.

$$\frac{d\sigma(\overline{\nu}_e + p \rightarrow e^+ + n)}{d\cos\theta} = \frac{G_F^2 |V_{ud}|^2 E_e p_e}{2\pi}\left(f_V^2(1 + \beta_e \cos\theta) + 3f_A^2(1 - \frac{\beta_e}{3}\cos\theta)\right) \tag{2.36}$$

where f_V and f_A are the vector and axial-vector form factors, E_e, p_e, β_e and θ are the energy, momentum, velocity and scattering angle of the outgoing positron.

The fact that the signature of this interaction produces a neutron is the central motivating piece for the development of Gd-doped water-Čerenkov detectors. By tagging the final state neutron, it is possible from distinguish neutrinos interacting elastic scattering, from antineutrinos interaction inverse-β decay. This will be further explained in Sect. 5.1.

Quasi-elastic Scattering

Climbing in the energy of the incoming neutrino, the available phase space for the outcomes makes the interactions more complex and diverse.

From 100 MeV to ~2 GeV, the quasi-elastic scattering dominates the neutrino-nucleus cross-sections. This makes this process crucial for the study of long baseline and low energy atmospheric neutrinos and, therefore, for the study of the neutrino mass ordering and the leptonic CP violating phase.

The differential neutrino cross section with respect to the momentum transferred q has the following expression for the charged current version.

$$d\sigma_\nu q^2 = \frac{M^2 G_F^2 \cos\theta_C}{8\pi E_\nu^2}\left(A(q^2) + B(q^2)\frac{s-u}{M^2} + C(q^2)\frac{(s-u)^2}{M^4}\right) \tag{2.37}$$

where M the nucleon mass, neutron or proton respectively. G_F is the Fermi constant, E_ν the neutrino energy, θ_C the Cabibbo angle and s and u the corresponding Mandelstam variables. The factors A, B and C are defined as follows.

$$A(q^2) = \frac{m^2 - q^2}{4M^2}\left(\left(4 - \frac{q^2}{M^2}\right)\left(|F_A|^2 - |F_{V,1}|^2\right) - \frac{q^2}{M^2}|\eta F_{V,2}|^2\left(1 + \frac{q^2}{4M^2}\right)\right.$$
$$\left. - \frac{4q^2 \eta F_{V,1} F_{V,2}}{M^2} - \frac{m}{M}\left(\left(F_{V,1} + \eta F_{V,2}\right)^2 + |F_A|^2\right)\right) \tag{2.38}$$

$$B(q^2) = \frac{q^2}{M^2}\left(F_A\left(F_{V,1} + \eta F_{V,2}\right)\right) \tag{2.39}$$

$$C(q^2) = \frac{1}{4}\left(|F_A|^2 + |F_{V,1}|^2 - \frac{q^2}{4M^2}|\eta F_{V,2}|^2\right) \tag{2.40}$$

m is the outgoing lepton mass, η is the anomalous magnetic dipole moment, $\eta = \mu_p - \mu_n$, and the axial vector and vector form factors F_A, $F_{V,1}$ and $F_{V,2}$ have the next expressions.

$$F_{V,1}(q^2) = \frac{1}{\left(1 - \frac{q^2}{4M_V^2}\right)^2} \frac{\left(1 - \frac{q^2}{4M^2}(1 + \eta)\right)}{1 - \frac{q^2}{4M^2}} \tag{2.41}$$

$$\eta F_{V,2}(q^2) = \frac{1}{\left(1 - \frac{q^2}{4M_V^2}\right)^2} \frac{\eta}{1 - \frac{q^2}{4M^2}} \tag{2.42}$$

$$F_A(q^2) = -1.232\left(1 - \frac{q^2}{M_A^2}\right) \tag{2.43}$$

where M_A and M_V are the axial vector and vector masses with values of 1.12 GeV and 0.84 GeV respectively.

For the neutral current analogous, the expression for the differential cross section with respect to the transferred momentum is very similar to Eq. (2.37).

$$d\sigma_\nu q^2 = \frac{M^2 G_F^2}{8\pi E_\nu^2}\left(A(q^2) + B(q^2)\frac{s - u}{M^2} + C(q^2)\frac{(s - u)^2}{M^4}\right) \tag{2.44}$$

where the factors A, B and C are modified accordingly to the different expressions of the axial vector and vector form factors.

Resonant Single Meson and Coherent Pion Productions

Single mesons, from neutrino interactions, are produced mainly through baryon resonances with the invariant mass of the hadronic system being less than 2.0 GeV/c^2. The main species produced are pions, although other mesons are also produced, like κ and η. In this interactions, neutrinos with enough energy excite a nucleon producing a baryon resonance. This quickly decays, most often to a nucleon and single pion.

$$\nu_l + N \rightarrow l + N^* \qquad N^* \rightarrow \pi + N' \tag{2.45}$$

where N and N' correspond to neutrons or protons. The charged current interactions for single pion production are the following.

$$\begin{aligned} \nu_l + p \rightarrow l^- + p + \pi^+, \ \overline{\nu}_l + p \rightarrow l^+ + p + \pi^- \\ \nu_l + n \rightarrow l^- + p + \pi^0, \ \ \overline{\nu}_l + p \rightarrow l^+ + n + \pi^0 \\ \nu_l + n \rightarrow l^- + n + \pi^+, \ \ \overline{\nu}_l + n \rightarrow l^+ + n + \pi^- \end{aligned} \tag{2.46}$$

These processes involve various resonances, but the most important for current neutrino experiments are those occurring at the lowest energies, dominated by the $\Delta(1232)$ resonance.

Baryonic resonances can also decay into multiple pions through other inelastic scatterings.

Additionally, neutrinos can also coherently produce single pion final states by scattering the entire nucleus (A). This produces almost no nuclear recoil and a pion almost parallel to the direction of the interacting neutrino. The charged current mechanisms for the coherent pion production are the next.

$$\nu_l + A \rightarrow l^- + A + \pi^+, \ \overline{\nu}_l + A \rightarrow l^+ + A + \pi^- \tag{2.47}$$

And analogously for the neutral current interactions, substituting the charged lepton by its neutrino partner and charged pions by the neutral ones.

Neutrino interactions above GeV energies can also produce strange quarks, thus kaons, in the final state. Next, the charged current processes of κ production are listed.

$$\begin{aligned} \nu_l + n \rightarrow l^- + \kappa^+ + \Lambda^0 \\ \nu_l + p \rightarrow l^- + \kappa^+ + p \\ \nu_l + n \rightarrow l^- + \kappa^0 + p \\ \nu_l + n \rightarrow l^- + \kappa^+ + n \\ \nu_l + p \rightarrow l^- + \kappa^+ + \Sigma^+ \\ \nu_l + n \rightarrow l^- + \kappa^+ + \Sigma^0 \\ \nu_l + n \rightarrow l^- + \kappa^0 + \Sigma^+ \end{aligned} \tag{2.48}$$

The antineutrino and neutral current analogous have similar expressions with the appropriate electric charge balancing (Fig. 2.16).

Deep Inelastic Scattering

In deep inelastic scattering, the neutrino scatters off a quark in the nucleon via the exchange of a virtual $W^{\pm}$ or Z boson, producing a lepton and a hadronic shower in the final state.

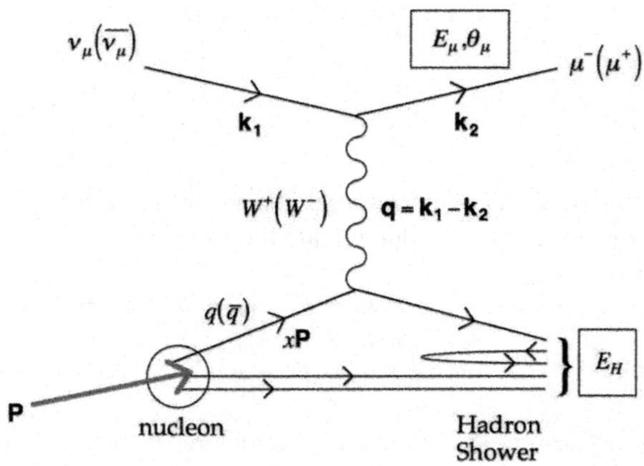

Fig. 2.16 Tree level Feynman diagram for CC DIS muon neutrino interaction

$$\begin{aligned} \nu_l + N \to l^- + X,\ \overline{\nu}_l + N \to l^+ + X \text{ (CC)} \\ \nu_l + N \to \nu_l + X,\ \overline{\nu}_l + N \to \overline{\nu}_l + X \text{ (NC)} \end{aligned} \tag{2.49}$$

where N is the target nucleon and X represents the resulting hadronic shower. The Bjorken scaling variable is defined as follows.

$$y = \frac{E_{had}}{E_\nu},\ x = \frac{-q^2}{2ME_\nu y} \tag{2.50}$$

The differential cross section of neutrino DIS with respect to the Bjorken variables has the next expression.

$$\frac{d^2\sigma^{\nu,\overline{\nu}}}{dx dy} = \frac{G_F^2 M_N E_\nu}{\pi\left(1 - q^2/M_{W,Z}^2\right)^2}\left(y^2 x F_1(x,q^2) + \left(1 - y - \frac{M_N x y}{2E_\nu}\right) F_2(x,q^2) \pm \left(y - \frac{y^2}{2}\right) x F_3(x,q^2)\right) \tag{2.51}$$

where M_N is the nucleon mass and $F_i(x, q^2)$ are the nucleon structure functions. The $\pm$ sign corresponds to the charged or neutral current processes respectively.

As these interactions mainly take place when the neutrino energy is larger than $\sim$10 GeV, secondary interactions of the hadronic shower within the nuclear media are important. These are very hard to model and they contribute to large multiplicities of hadrons in the final state.

References

[Bilenky15] S.M. Bilenky, Neutrino in standard model and beyond. Phys. Part. Nucl. **46**(4), 475–496 (2015). https://doi.org/10.1134/S1063779615040024

[Ringwald16] A. Ringwald, SM*A*S*H, in *12th Patras Workshop on Axions, WIMPs and WISPs (AXION-WIMP 2016)* Jeju Island, South Korea, 20–24 June 2016, https://inspirehep.net/record/1492283/files/arXiv:1610.05040.pdf

[King04b] S.F. King et al., Neutrino physics and the flavor problem. J. Korean Phys. Soc. **45**, S443–S446 (2004)

[Cao15] Q.-H. Cao et al.,*Strong CP Problem, Neutrino Masses and the 750 GeV Diphoton Resonance*. 2015

[Majorana28] E. Majorana, Teoria simmetrica dell'elettrone e del positrone. Il Nuovo Cimento (1924–1942) **14**(4), 171. ISSN 1827–6121, (1928). https://doi.org/10.1007/BF02961314

[Chen16] P. Chen et al., Leptogenesis and residual CP symmetry. JHEP **03**, 206 (2016). https://doi.org/10.1007/JHEP03(2016)206

[Spiering12] C. Spiering, Towards high-energy neutrino astronomy. A historical review. Eur. Phys. J. **H37**, 515–565 (2012). https://doi.org/10.1140/epjh/e2012-30014-2

[Honda15] M. Honda et al., Atmospheric neutrino flux calculation using the NRLMSISE-00 atmospheric model. Phys. Rev. **D92**(2), 023004 (2015). https://doi.org/10.1103/PhysRevD.92.023004

[Richard15] E. Richard, et al., *Measurements of the Atmospheric neutrino Flux by Super-Kamiokande: Energy Spectra, Geomagnetic Effects, and Solar Modulation*, 2015. arXiv:1510.08127

[Kopp07] S.E. Kopp, Accelerator-based neutrino beams. Phys. Rept. **439**, 101–159 (2007). https://doi.org/10.1016/j.physrep.2006.11.004

[Bahcall64] J.N. Bahcall, Solar neutrinos. I: theoretical. Phys. Rev. Lett. **12**, 300–302 (1964). https://doi.org/10.1103/PhysRevLett.12.300

[Bahcall92] J.N. Bahcall et al., Standard solar models, with and without helium diffusion, and the solar neutrino problem. Rev. Mod. Phys. **64**, 885–926 (1992). https://doi.org/10.1103/RevModPhys.64.885.64.885

[Fukuda99b] Y. Fukuda et al., Measurement of the solar neutrino energy spectrum using neutrino-electron scattering. Phys. Rev. Lett. **82**, 2430–2434 (1999). https://doi.org/10.1103/PhysRevLett.82.2430.82.2430

[Fukuda99a] Y. Fukuda et al., Constraints on neutrino oscillation parameters from the measurement of day-night solar neutrino fluxes at super-kamiokande. Phys. Rev. Lett. **82**, 1810–1814 (1999). https://doi.org/10.1103/PhysRevLett.82.1810.82.1810

[Davis Jr68] R. Davis Jr. et al. Search for neutrinos from the sun. Phys. Rev. Lett. **20**(21), 1205 (1968)

[Mikheyev85] S. Mikheyev et al., Resonance enhancement of oscillations in matter and solar neutrino spectroscopy. Yadernaya Fizika **42**, 1441–1448 (1985)

[Abe16b] K. Abe et al. Solar Neutrino Measurements in Super-Kamiokande-IV. Phys. Rev. **D94**(5), 052010 (2016). https://doi.org/10.1103/PhysRevD.94.052010

[Declais94] Y. Declais et al., Study of reactor anti-neutrino interaction with proton at Bugey nuclear power plant. Phys. Lett. B **338**, 383–389 (1994). https://doi.org/10.1016/0370-2693(94)91394-3

[Burrows13] A. Burrows, Colloquium: perspectives on core-collapse supernova theory. Rev. Mod. Phys. **85**, 245 (2013). https://doi.org/10.1103/RevModPhys.85.245

[Totani98a] T. Totani et al., Future detection of supernova neutrino burst and explosion mechanism. Astrophys. J. **496**, 216–225 (1998). https://doi.org/10.1086/305364

[Hirata87] K. Hirata et al., Observation of a neutrino burst from the supernova SN 1987a. Phys. Rev. Lett. **58**, 1490–1493 (1987). https://doi.org/10.1103/PhysRevLett.58.1490

[Velde88] J.C.V.D. Velde et al., Neutrinos from sn1987a in the imb detector. Nucl. Instrum. Methods **A264**, 28–31 (1988)

[Alexeyev88] E. Alexeyev et al., Detection of the neutrino signal from sn 1987a in the lmc using the inr baksan underground scintillation telescope. Phys. Lett. B **205**(2–3), 209–214 (1988)

[Beacom10] J.F. Beacom, The diffuse supernova neutrino background. Ann. Rev. Nucl. Part. Sci. **60**, 439–462 (2010). https://doi.org/10.1146/annurev.nucl.010909.083331

[Pauli30] W. Pauli. *Letter to the radioactive ladies and gentlemen* (Physical Institute of the Federal Institute of Technology, 1930)

[Fermi08] E. Fermi, Tentativo di una teoria dei raggi β. Il Nuovo Cimento (1924–1942) **11**(1), 1 (2008). ISSN 1827–6121. https://doi.org/10.1007/BF02959820

[Cowan56] C.L. Cowan, et al., Detection of the free neutrino: a confirmation. Science **124**(3212), 103–104. ISSN 0036-8075. https://doi.org/10.1126/science.124.3212.103, http://science.sciencemag.org/content/124/3212/103

[Danby62] G. Danby et al., Observation of high-energy neutrino reactions and the existence of two kinds of neutrinos. Phys. Rev. Lett. **9**, 36–44 (1962). https://doi.org/10.1103/PhysRevLett.9.36.9.36

[Kodama01] K. Kodama, et al. Observation of tau neutrino interactions. Phys. Lett. B **504**(3), 218–224 (2001). ISSN 0370-2693. https://doi.org/10.1016/S0370-2693(01)00307-0, https://www.sciencedirect.com/science/article/pii/S0370269301003070

[King04a] S.F. King, Neutrino mass models. Rept. Prog. Phys. **67**, 107–158 (2004). https://doi.org/10.1088/0034-4885/67/2/R01

[Pontecorvo57] B. Pontecorvo, Mesonium and antimesonium. J. Exptl. Theoret. Phys. (U.S.S.R.) **33**, 549–551 (1957)

[Maki62] Z. Maki et al., Remarks on the unified model of elementary particles. Prog. Theor. Phys. **28**(5), 870–880 (1962). https://doi.org/10.1143/PTP.28.870, http://ptp.oxfordjournals.org/content/28/5/870.abstract

[Gribov69] V. Gribov et al., Neutrino astronomy and lepton charge. Phys. Lett. B **28**(7), 493 – 496 (1969). ISSN 0370-2693. https://doi.org/10.1016/0370-2693(69)90525-5, http://www.sciencedirect.com/science/article/pii/0370269369905255

[Patrignani16] C. Patrignani et al., Review of particle physics. Chin. Phys. **C40**(10), 100001 (2016). https://doi.org/10.1088/1674-1137/40/10/100001

[Formaggio12] J.A. Formaggio et al., From eV to EeV: neutrino cross sections across energy scales. Rev. Mod. Phys. **84**, 1307–1341 (2012). https://doi.org/10.1103/RevModPhys.84.1307

[Hasert73] F. Hasert et al., Observation of neutrino-like interactions without muon or electron in the gargamelle neutrino experiment. Phys. Lett. B **46**(1), 138–140 (1973)

Chapter 3
The Super-Kamiokande Detector

In this Chapter, the SK detector is described and explained in detail since it is the precursor and base experiment for SuperK-Gd. SK is described as a neutrino detector itself for astrophysical, solar and atmospheric neutrinos and also as far detector of the T2K neutrino beam.

3.1 Detector Overview

The SK detector is a cylindrical tank with 39.3 m in diameter and 41.4 m in height and filled with 50 kton of ultrapure water. It is located under Mountain Ikenoyama, in the Kamioka mine (Gifu Prefecture, Japan), with 1000 m of rock overburden (or 2700 m water equivalent) to reduce as much as possible the cosmic ray muon background. Although the inner detector volume is 32 kton, the fiducial volume, used for most of the physics measurements, is 22.5 kton, 2 m away from the inner detector wall.

The SK experiment began its data taking in April 1996 and shut down for maintenance in July 2001.This first phase of the experiment is called SK-I. During refilling after maintenance, an accident occurred in November 2001, in which more than half of the PMTs were destroyed due to the implosion of a PMT. After the accident, the SK detector was rebuilt with half of the original PMT density in the inner detector. An FRP case was also attached to each PMT this time, to prevent chain reaction implosions. The operation was resumed in October, 2002, phase called SK-II. In October 2005, the experiment was completely rebuilt and resumed data taking with the full number of PMTs in July 2006, this phase is SK-III. In September 2008 starts the current phase, SK-IV, with upgraded front-end electronics (Fig. 3.1 and Table 3.1).

P. Fernández Menéndez, *Neutrino Physics in Present and Future Kamioka Water-Čerenkov Detectors with Neutron Tagging*, Springer Theses,
https://doi.org/10.1007/978-3-319-95086-0_3

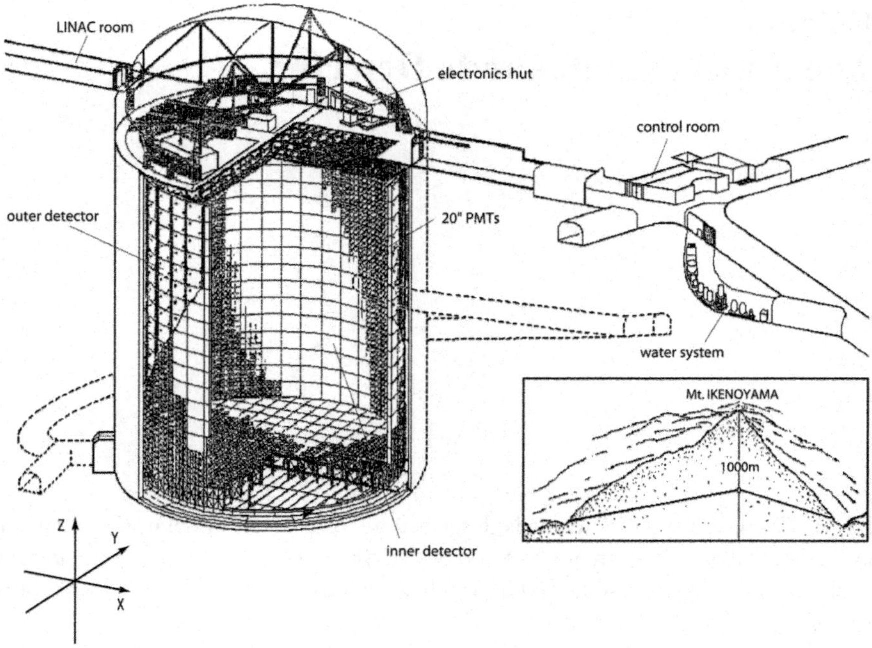

Fig. 3.1 Schematic view of SK emplacement

3.2 Čerenkov Radiation

According to special and general relativity postulates, the speed of light in vacuum c is an absolute limit on velocity [Einstein05], however, it is also true that light travels slower through other media.

In the 19th century, Heaviside first realised that a charged particle passing through a medium at higher speed than the speed of light in that medium, should emit some electromagnetic radiation, [Heaviside08]. He formulated part of the theory of the Čerenkov radiation, predicting the angle that would be formed between the charged particle and the emitted light, shown in Eq. (3.1).

Table 3.1 Super-Kamiokande phases and main features

Phase	Period	Number of PMTs		FRP case	Electronics
		ID (Coverage)	OD		
SK-I	Apr. 1996–Jul. 2001	11146 (40%)	1884	No	ATM
SK-II	Oct. 2002–Oct. 2005	5182 (19%)	1884	Yes	ATM
SK-III	Jul. 2006–Sep. 2008	11129 (40%)	1884	Yes	ATM
SK-IV	Sep. 2008–Today	11129 (40%)	1884	Yes	QBEE

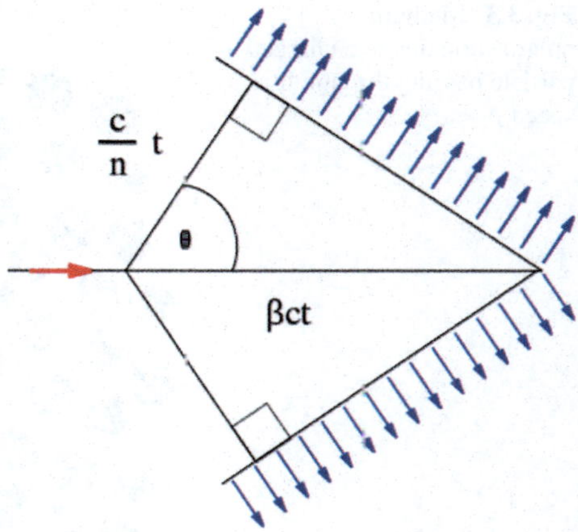

Fig. 3.2 Schematic diagram of a charged particle passing through a medium of refractive index larger than 1 at $\beta > n$ and emitting Čerenkov radiation

$$\cos\theta_{\check{C}} = \frac{1}{n\beta} = \frac{1}{n}\sqrt{1 - \frac{m}{|\vec{p}|}} \tag{3.1}$$

where m and $\vec{p}$ are the mass and momentum of the charged particle, β is the quotient between the speed of the charged particle and the speed of light in vacuum, and n is the refraction index of the medium. The light emission has cylindrical symmetry with axis the track of the particle. This makes the emitted light form a cone.

Čerenkov radiation occurs when the following is satisfied (Fig. 3.2).

$$\beta \geq n \tag{3.2}$$

This radiation was then discovered by Vavilov and Čerenkov in 1937 while observing water being bombarded by radiation, [Čerenkov37]. Čerenkov together with Frank and Tamm, who developed the theory of the phenomenon, received the Nobel prize in 1958.

The theoretical formalism of this phenomenon is obtained by solving Maxwell's equations of a charged current $\vec{j}$, s in a polarisable medium. The charged current is that associated with the charged particle with charge e and $\beta \geq n$.

$$\vec{j} = e\delta(\vec{r} - \vec{v}t)\vec{v} \tag{3.3}$$

And the medium will have a polarisation vector **P**.

$$\mathbf{P} = (n^2 - 1)\mathbf{E} \Rightarrow \mathbf{D} = n^2\mathbf{E} \tag{3.4}$$

Once solved the equations, the number of photons per unit length and unit wavelength is given by the next expression.

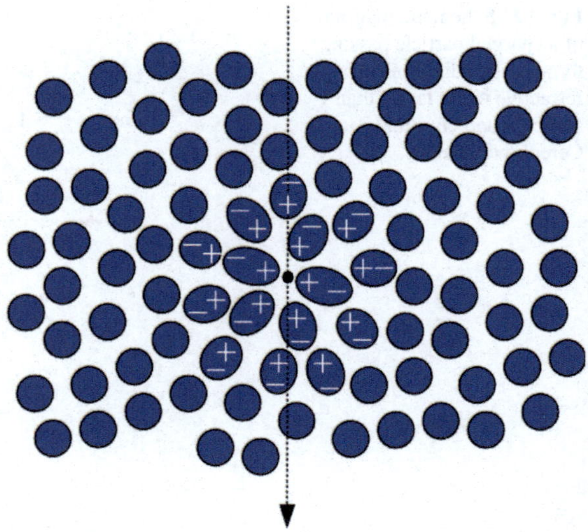

Fig. 3.3 Medium polarisation due to a charged particle passing through it at speed $\beta < n$

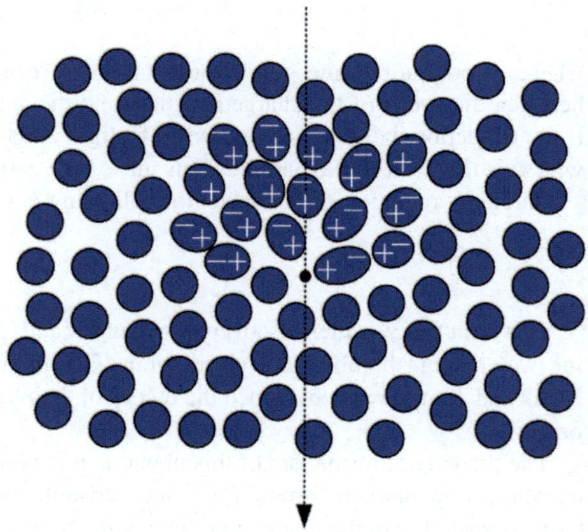

Fig. 3.4 Medium polarisation due to a charged particle passing through it at speed $\beta > n$

$$\frac{\mathrm{d}^2N}{\mathrm{d}x\mathrm{d}\lambda} = \frac{2\pi\alpha}{\lambda^2}\left(1 - \frac{1}{\beta^2 n(\lambda)^2}\right) = \frac{2\pi\alpha}{\lambda^2}\sin\theta_{\check{C}} \tag{3.5}$$

where α is the fine structure constant and $n(\lambda)$ is the refraction index of the medium.

In water-Čerenkov detectors the refraction index is ∼1.34 and the charged particles, ideally produced after the interaction of a neutrino, are in the ultra-relativistic regime, $\beta \sim 1$. Hence, the typical Čerenkov angle is ∼41.7° (Figs. 3.3, 3.4 and 3.5).

For water, the typical wavelength is of the around of 375 nm, the full spectrum is shown in Fig. 3.7.

3.3 Detector Description

The Super-Kamiokande detector can be divided into two major parts, those are the tank containing the ultrapure water and the photomultiplier tubes acquiring the light produced inside the detector. In the upcoming sections, these are described and their most important characteristics reviewed [Fukuda03].

3.3.1 Water Tank

The SK detector is optically divided into inner (ID) and outer (OD) detectors, being the latter used, mainly, as veto to reduce the cosmic muon background. The inner detector is currently (SK-IV) instrumented with 11129 photomultiplier tubes (PMT) of 20 in. in diameter facing inwards, accounting for a photo-coverage of 40%. The outer detector, on the other hand, is instrumented with 1884 8-in. PMTs facing outwards the detector. For the optical separation, black polyethylene terephthalate sheets (black sheets) are used behind the inner PMTs and tyvek sheets behind the outer PMTs (Figs. 3.5 and 3.6).

Around the inner surface of the tank, 26 sets of horizontal and vertical Helmholtz coils are arranged to compensate for the Earth's magnetic field, which otherwise would affect the photoelectron trajectories between dynodes inside the PMTs.

3.3.2 Photomultipliers

The ID PMTs were developed by Hamamatsu Photonics in collaboration with KamiokaNDE collaborators (model R3600). These PMTs are most sensitive to pre-

Fig. 3.5 Čerenkov radiation spectrum in water

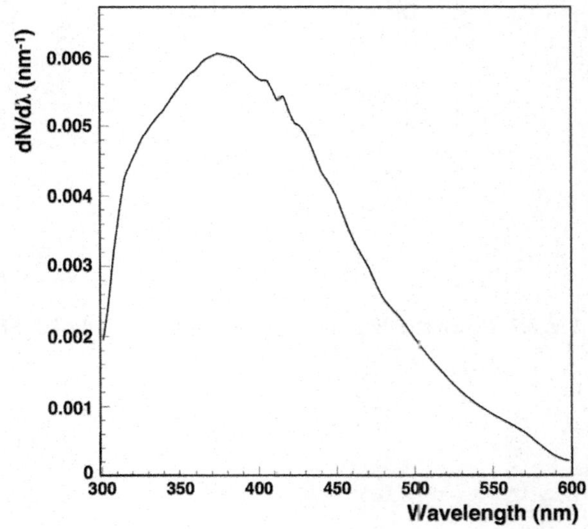

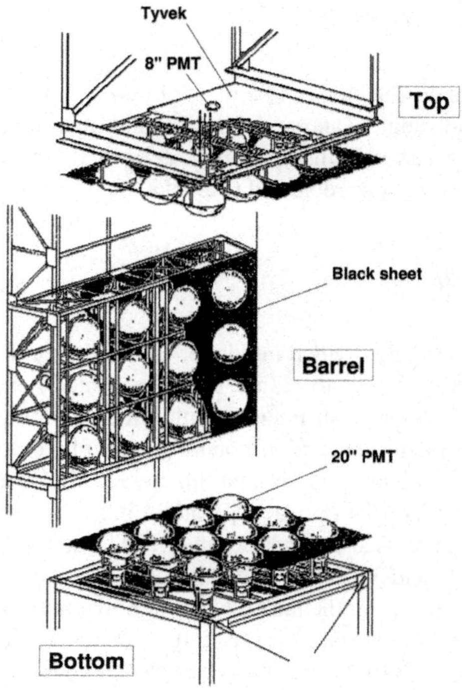

Fig. 3.6 Supporting structure of SK PMTs

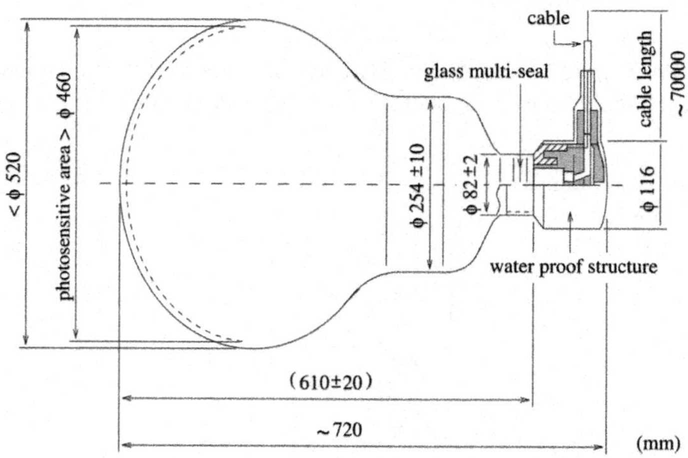

Fig. 3.7 Schematic diagram of the inner structure of the 20" SK's ID PMTs

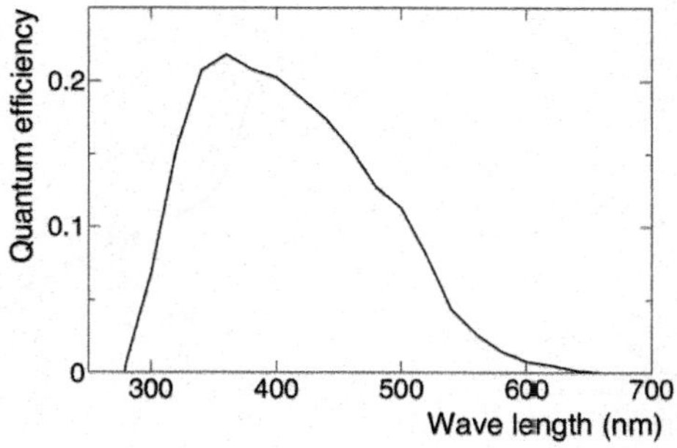

Fig. 3.8 Wavelength dependence of PMT photocathode quantum efficiency

Table 3.2 Specification of SK ID PMTs

Shape	Hemispherical
Photocathode area	50 cm diameter
Window material	Bialkali (Sb-K-Cs)
Quantum efficiency	22% at $\lambda = 390$ nm
Dynode	11 stage Venetian blind type
Gain	10^7 at $\sim$2000 V %
Dark current	200 nA at 10^7 gain
Dark pulse rate	3 kHz at 10^7 gain
Cathode non-uniformity	<10%
Anode non-uniformity	<40%
Transit time	90 ns at 10^7 gain
Transit time spread	2.2 ns (1σ) for 1 p.e. equivalent signals
Weight	13 kg
Pressure tolerance	6 kg/cm^2 water proof

dominant wavelengths in the Čerenkov spectrum, that is in the range of 300–600 nm, as shown in Fig. 3.7.

The photocathode of the PMTs is coated with bi-alkali (Sb-K-Cs) being its maximum quantum efficiency is 22% and optimised so it is reached at $\sim$390 nm, as seen in Fig. 3.8, very close to the water Čerenkov spectrum maximum.
The main characteristics of these PMTs are listed in Table 3.2.

The ID PMTs have an uniform gain of 10^7 at a high voltage ranging from 1500 to 2000 V (Fig. 3.9).

On the other hand, in the OD, the PMTs used are Hamamatsu R1408. Their quantum efficiency spectrum does not match so well that of the Čerenkov radiation. Therefore, in order to acquire more photons, a wavelength shifter plate is attached to their photocathodes. Although this worsens the time resolution, this is not a serious problem in the OD.

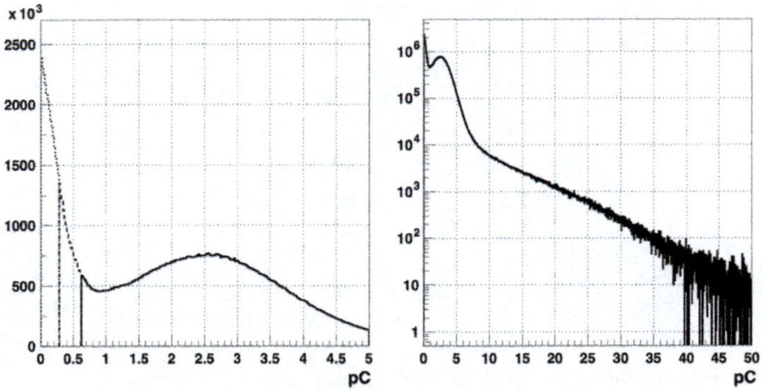

Fig. 3.9 The 1 p.e. distribution in pC. In logarithmic scale in the right and in linear scale in the left, where the solid line shows the data with normal PMT gain, the dashed line shows the data with double gain and half threshold, and the dotted line corresponds to the linear extrapolation between them

3.4 Water and Air Purification Systems

The understanding of the characteristics and properties of the SK water, is crucial for all the physics measurements, as the photon propagation depends hugely on the water quality. Therefore, the water coming from the Kamioka mine is purified in a dedicated system, passing through several resins and filters, and continuously recirculated at a rate of 60 ton/h. The main purpose is to keep the light attenuation length as high as possible and to remove radioactive impurities in the water (Fig. 3.10).

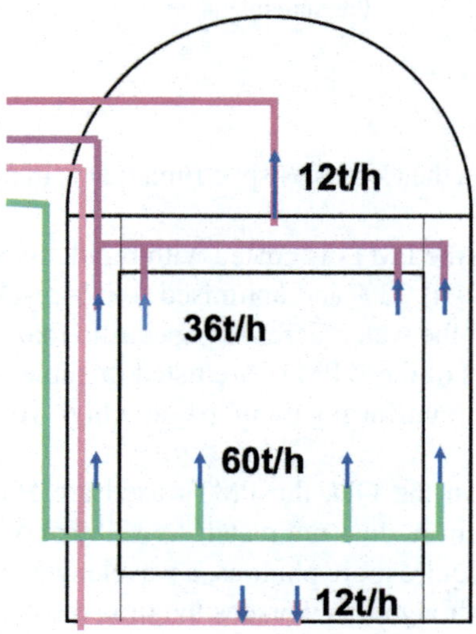

Fig. 3.10 Circulation scheme of pure water in the SK tank

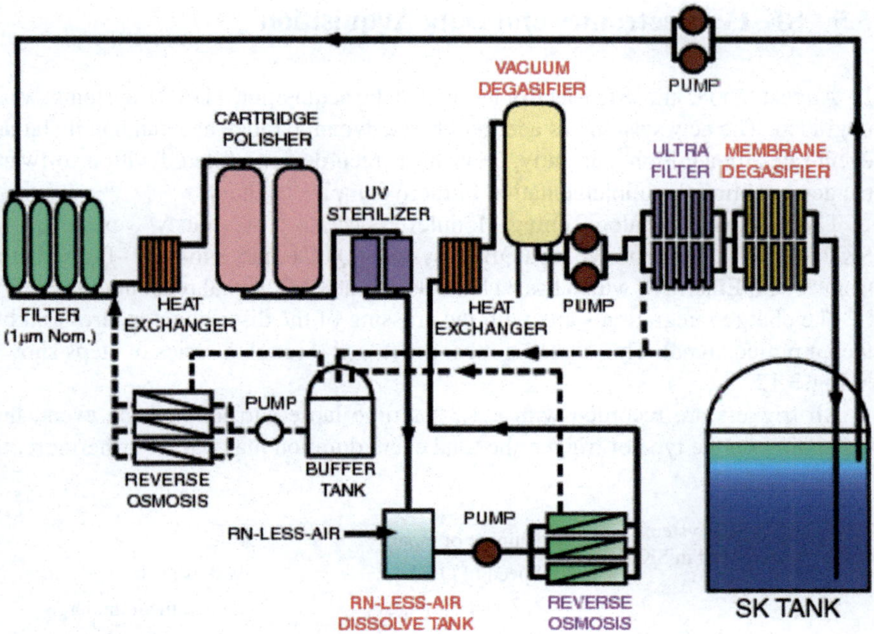

Fig. 3.11 Water purification system at SK

The SK water system consists in: a 1 micron filter, which removes the biggest particles, such as dust; a heat exchanger that cools the water up to about 13 °C, in order to reduce the PMT noise and the growth of bacteria; a cartridge polisher for removing heavy ions; UV lamps to avoid any possible bacteria present in the water; an ultra-filter that removes impurities of molecular weight heavier than 100.

In addition, and for avoiding high concentrations of radon, radon-free air is dissolved into the water. Then, the water passes through vacuum and membrane degasifiers for removing as much gas as possible from the water (Fig. 3.11).

To monitor the water temperature, thermometers with 0.0001 °C precision are placed at eight different positions in both ID and OD. Below 10 m from the bottom of the tank, water is always convecting resulting in a uniform temperature throughout the lower part of the ID. Above that region, the water temperature gradually rises, existing a difference of 0.2 °C difference between the top of the ID and the convection zone.

As it happens in most underground facilities, the air becomes highly contaminated with radon (Ra), produced as the uranium (U), of the surrounding rock, decays. In order to keep the Ra levels small for the low energy neutrino analyses, radon-free air from outside is continuously pumped, after passing through a filtration system, into the SK mine.

3.5 SK-IV Electronics and Data Acquisition

In August 2008, the SK electronics and data acquisition (DAQ) systems were upgraded. The new system has a larger charge dynamic range and can handle larger event rates. In addition, currently, every hit is recorded and filtered with a software trigger, enabling the implementation of more complex triggers.

The ATM (Analog-Timing-Module) system, previously existing in SK-I to SK-III, was replaced by a QBEE system (QTC (charge-to-time)-Based Electronics with Ethernet), which has 3 channels for the PMT signal readout (Table 3.3).

The charge integration starts with the crossing of the discriminator threshold by the amplified signal. This signal is then, processed through a series of steps shown in Fig. 3.12.

All triggers are recorded with a 1.3 μs time lapse surrounding the event, but depending on the type of trigger, the total event duration may vary. Furthermore, in

Table 3.3 QBEE system characteristics used in SK-IV

Number of inputs	24
Speed of QTC	~900 ns/cycle
Number of ranges	3 (small, medium, large)
Discriminator	−0.3 to −14 mV
Charge dynamic range	0.2–25,000 pC
Charge resolution	~0.2 pC
Charge linearity	<1%
Timing resolution	0.3 ns (2 pC) and <0.2 ns (>10 pC)
Power consumption	<1 W/channel

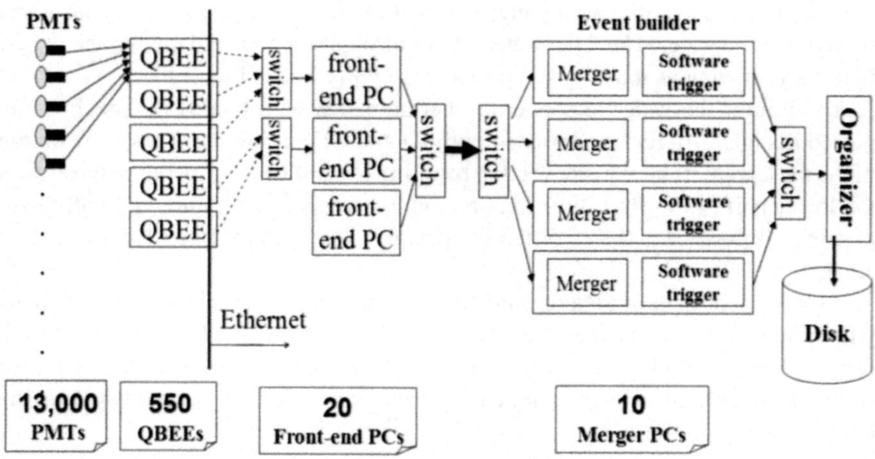

Fig. 3.12 Data acquisition diagram for SK-IV

Table 3.4 List of the SK-IV triggers

SK-IV triggers	Threshold (hits/200 ns)	Event duration (μs)
OD	22	–
SLE	$34 \xrightarrow[05/2015]{} 31$	$-0.5 \rightarrow 1.0$
LE	47	$-5 \rightarrow 35$
HE	50	$-5 \rightarrow 35$
SHE	$70 \xrightarrow[09/2011]{} 58$	$-5 \rightarrow 35$
AFT	SHE and no OD	$35 \rightarrow 535$

SK-IV a new trigger was introduced, SHE, which activates, if there is no activity in the OD, the after trigger (AFT), an additional 500 μs of data (Table 3.4).

3.6 Calibration of the Detector

The calibration of the detector is of crucial importance for the event reconstruction and all the physics analysis. Items like the water properties, the PMT response or the electronics are calibrated to obtain input parameters for the Monte Carlo simulation of the detector. This way, all the physics and the detector response can be completely simulated very accurately and, therefore, the simulation can be confidently used to analyse the data.

The various calibration works carried out at SK deal with the high voltage, PMT gain, quantum efficiency, charge and timing of PMTs which are compiled in [Abe14a]. Additionally, calibrations concerning the photon tracking and the energy scale are also performed, and can be found in [Blaufuss01, Nakahata99].

References

[Einstein05] A. Einstein, *On the Electrodynamics of Moving Bodies* (1905)
[Heaviside08] O. Heaviside, *Electromagnetic Theory*, vol. 3 (Cosimo, Inc., 2008)
[Čerenkov37] P. Čerenkov, Visible radiation produced by electrons moving in a medium with velocities exceeding that of light. Phys. Rev. **52**(4), 378 (1937)
[Fukuda03] S. Fukuda, et al., The super-kamiokande detector. Nucl. Instrum. Meth. Phys. Res. Sect. A **501**(2–3), 418–462 (2003), ISSN 0168-9002, https://doi.org/10.1016/S0168-9002(03)00425-X, http://www.sciencedirect.com/science/article/pii/S016890020300425X
[Abe14a] K. Abe, et al., Calibration of the super-kamiokande detector. Nucl. Instr. Meth. Phys. Res. Sect. A. **737**, 253–272 (2014), ISSN 0168-9002, https://doi.org/10.1016/j.nima.2013.11.081, http://www.sciencedirect.com/science/article/pii/S016890021301546X

[Blaufuss01] E. Blaufuss, et al., 16n as a calibration source for super-kamiokande. Nucl. Instr. Meth. Phys. Res. Sect. A. **458**(3), 638–649 (2001), ISSN 0168-9002, https://doi.org/10.1016/S0168-9002(00)00900-1, http://www.sciencedirect.com/science/article/pii/S0168900200009001
[Nakahata99] M. Nakahata, et al., Calibration of super-kamiokande using an electron linac: The super-kamiokande collaboration. Nucl. Instr. Meth. Phys. Res. Sect. A. **421**(1–2), 113–129 (1999), ISSN 0168-9002, https://doi.org/10.1016/S0168-9002(98)01200-5, http://www.sciencedirect.com/science/article/pii/S0168900298012005

Chapter 4
The SuperK-Gd Project

The SuperK-Gd project is the proposed and approved upgrade of the Super-Kamiokande detector in order to enable it to efficiently detect thermal neutrons. The project consists in dissolving a Gd salt into SK at a concentration of 0.2%.

Once they have thermalised, the neutrons are captured by Gd, emitting a delayed 8 MeV γ cascade from its de-excitation. The detection of this cascade in coincidence with a neutrino event, will lead to the measurement of the final state neutrons produced in the neutrino interaction with very high confidence and efficiency (80%). The physics of the Gd-neutron capture is explained in Sect. 4.1.

Across this chapter, the development of the techniques for the successful realisation of SuperK-Gd are explained in Sect. 4.2. Additionally, its physics potential is briefly summarised in Sect. 4.3.

In the context of the development of the technology needed for the success of the SuperK-Gd project, the main R&D program towards SuperK-Gd, EGADS, is described in Sect. 4.4.

4.1 Gadolinium-Neutron Tagging

Gadolinium has the largest neutron capture cross-section of all stable nuclei known in nature, around 48800 barn on average, given the natural abundances of its isotopes. Therefore, it is a terrific candidate to achieve the largest neutron tagging efficiency possible in water-Čerenkov detectors.

Figure 4.1, shows a schematic diagram of the Gd-neutron capture after the inverse-β interaction of an electron antineutrino in a water-Čerenkov detector. This is the simplest process which clearly shows the different signature in the detector between, a low energy antineutrino and the analogous neutrino interaction.

After the neutron is produced, it is thermalised by its interactions with water in the detector. Once it is thermal, the neutron is captured by a Gd nucleus, resulting in an excited Gd nucleus with atomic mass one unit larger. After that, the nucleus

P. Fernández Menéndez, *Neutrino Physics in Present and Future Kamioka Water-Čerenkov Detectors with Neutron Tagging*, Springer Theses,
https://doi.org/10.1007/978-3-319-95086-0_4

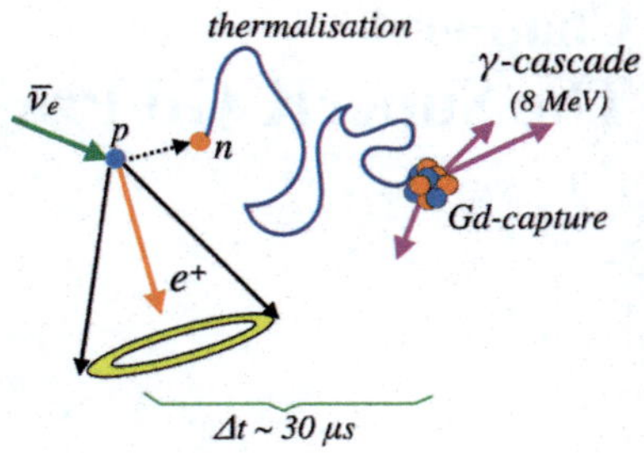

Fig. 4.1 Schematic diagram of Inverse Beta Decay (IBD) with Gd-neutron tagging

Table 4.1 List of gadolinium isotopes, natural abundances, neutron capture cross sections and total energy released in the de-excitation

Isotope	Natural abundance (%)	Cross-section (barn)	De-excitation energy (MeV)
152Gd	0.20	1050	6.25
154Gd	2.18	85.0	6.44
155Gd	**14.80**	**60700**	**8.54**
156Gd	20.47	1.71	6.36
157Gd	**15.65**	**254000**	**7.94**
158Gd	24.84	2.01	5.94
160Gd	21.86	0.765	5.64

de-excites, emitting a γ-ray cascade with total energy depending on the Gd isotope taking part in the capture.

The Gd-neutron capture occurs following the next reaction.

$$n +^X Gd \rightarrow^{X+1} Gd^* \rightarrow^{X+1} Gd + \gamma s \tag{4.1}$$

And Table 4.1 shows the relation between each isotope, its natural abundance, neutron capture cross section and emitted energy in the de-excitation.

The energy of the photons emitted after the neutron capture add up ~8 MeV, balanced by the abundances of each isotope. The photon multiplicity typically varies from three to five photons, giving the Gd-neutron capture signals a very characteristic topology, very advantageous for its reconstruction, as seen in Fig. 4.2.

The target solution in SuperK-Gd will be of 0.2% of gadolinium by mass. At this concentration, the fraction of tagged neutrons by Gd will be of 90%. This is shown in Fig. 4.3.

The mean time between the neutron production and its capture is about 30 μs. The first ~10 μs, correspond to the neutron thermalisation time, and the posterior ~20 μs, to the Gd capture and de-excitation times. This feature makes the correlation, between the neutrino prompt signal and the neutron capture, very efficient in terms of its detection.

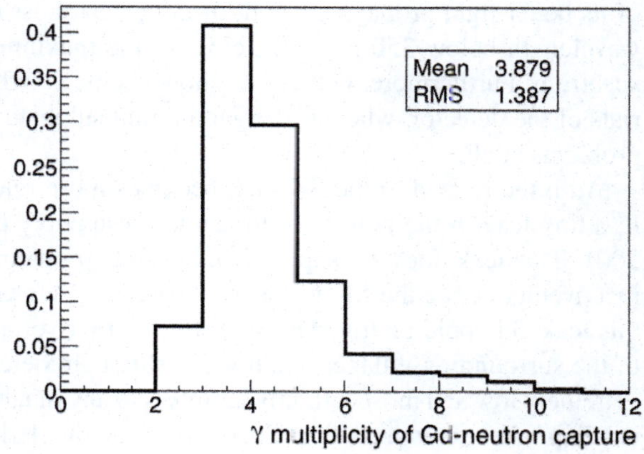

Fig. 4.2 Normalised distribution for the γ-ray multiplicity from the de-excitation of gadolinium after neutron capture

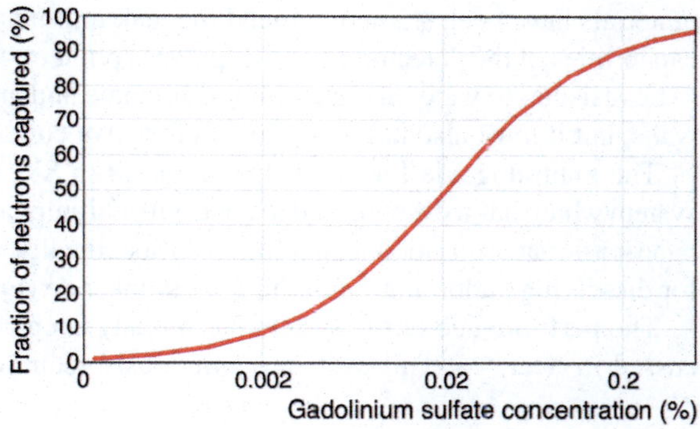

Fig. 4.3 Fraction of neutrons captured by Gd as function of its concentration

The distance between the neutrino interaction, which produces a neutron, and the position of the capture is around 2 m. This distance is mainly due to the thermalisation process in the water.

4.2 Gadolinium Addition to Super-Kamiokande

The addition of Gd into water-Čerenkov detectors is not trivial, and several considerations and studies must be done in order to assure the capability and good performance of the detector.

First thing to be considered is in which form the gadolinium should be added. Gadolinium is a metal and, therefore, does not dissolve in water, meaning that a compound of this element which is soluble in water has to be considered. Highly solvable salts of Gd were chosen as candidates from the beginning of the project. Amongst all possible salts, $Gd_2(SO_4)_3$ and $GdCl_3$ were taken in consideration due to their good properties. Finally, the latter was rejected in favour of the former because of its better light propagation and lower corrosion issues. $GdCl_3$ absorbs light with wavelength below 350 nm, very close to the maximum of the Čerenkov radiation spectrum. Furthermore, $GdCl_3$ has shown some corrosion problems in some materials of the detector, whereas the gadolinium sulphate has not shown any corrosion problems at all.

An issue related to the SK tank becomes more relevant once Gd is poured into it, a tiny leak in the bottom of the tank, presumably caused by the SK accident in 2001. This leak does not suppose any serious problem currently, but it could be an inconvenient once the SK water is Gd-doped, since Gd is a heavy metal. Through this leak Gd could be transferred to the nearby river and pollute the water supplies of the surrounding villages and towns. Although the concentration of Gd would be extremely low and most certainly harmless to any inhabitant in the area, extreme care is being taken to address this. Specially given the dark history of the zone with the itai-itai disease, caused by cadmium poisoning. In order to solve this, several sealing materials have been studied to mend the leak minimising the impact on the physics capabilities of the detector. From the physics point of view, this material has not only to be resistant to water and gadolinium sulphate and attach well to the stainless steel walls, but it must also have very low radioactive contamination.

The main upgrade for the inclusion of Gd in SK, will be the water purification system which has to be modified to cope with Gd sulphate. For this mater, the selective band-pass water filtration technology was specifically invented and developed at UCI for dissolving gadolinium into Super-Kamiokande (Fig. 4.4).

The performance of the system has already been checked at EGADS and is discussed in Sect. 4.4. This system continuously recirculates the Gd-doped ultrapure

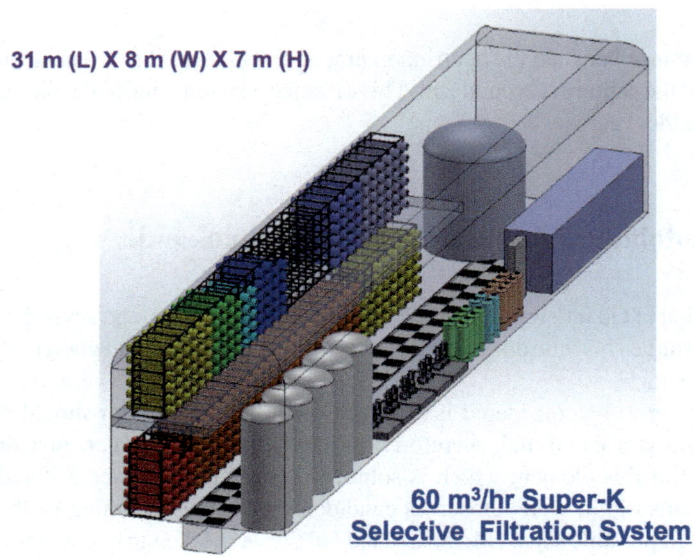

Fig. 4.4 Selective filtration system for SuperK-Gd

water, maintaining its transparency to SK ultrapure water levels and removing most of the unwanted impurities while simultaneously, and indefinitely, retains the desired levels of both the gadolinium and sulphate ions. The EGADS prototype was built specifically to show that gadolinium loading would be feasible in Super-Kamiokande and, therefore, scalability was always an important design criterion. The selective band-pass water filtration will be working together with the current water purification system of SK.

In addition to the band-pass a Gd-removal system was developed to quickly and efficiently withdraw the $Gd_2(SO_4)_3$ once the experiment is concluded. After several studies, the technique proposed is to make the Gd-doped water go through a cation ion-exchange resin as the detector is drained. This method will recover all the Gd compound avoiding the Gd to be filtered to the nearby river.

Finally, there are two important concerns for the realisation of SuperK-Gd which have been extensively addressed during the past years: the possible corrosion and radioactivity from gadolinium sulphate itself.

For the former, all materials present in SK or that will be in SuperK-Gd were soaked in 0.2% Gd-doped water and then the attenuation length change was measured independently at Kamioka Observatory with a spectrophotometer and at UCI with IDEAL (Irvine Device Evaluating Attenuation Length). All the checks showed no significant change in the water transparency.

Radioactivity contamination is one of the most relevant and remaining issues that has been and is being addressed in order to reduce its impact as much as possible in the detector performance and in the low energy physics studies. A much detailed description of the situation on this topic is treated in Sect. 5.6.

4.3 Main Physics Outcomes

The addition of Gd sulphate to SK to enable the highly efficient Gd-neutron tagging, has a huge impact in the detection and study of various physics phenomena. Next, the main physics improvements due to neutron tagging are briefly reviewed at both, low and high energy regimes.

4.3.1 Low Energy

One of the main goals for this upgrade is to be able to first detect diffuse supernova neutrino background (DSNB) antineutrinos. These are the neutrinos coming from all the past supernova explosions in the history of the universe. This measurement will provide very important information about core-collapse supernovae, like their mean energy spectrum and the star formation rate of the universe. This measurement is currently dominated by spallation background, which will be drastically reduced with neutron tagging. The details of this measurement-to-be are treated in Sect. 6.2.

Similar issues happens when dealing with antineutrinos coming from nuclear reactors in Japan and South Korea. Nowadays, this measurement is also background dominated from spallation products and solar neutrinos. Neutron tagging will enable the possibility of performing long baseline reactor antineutrino oscillation analyses with large statistics, improving the current sensitivity of the solar oscillation parameters Δm^2_{21}, θ_{12}. This potential new measurement is described in detail in Sect. 6.1.

Finally, in the situation that a supernova occurs close enough, the huge amount of neutrinos detected from it will provide much information about early stages of the core-collapse process. Although currently, SK can distinguish supernova neutrinos from antineutrinos thanks to their different cross-section angle dependence, with Gd-neutron tagging this separation will be much more effective, as will be shown in Section 6.4. Furthermore, if the supernova is less than 1 kpc away, SuperK-Gd will be able to detect the Si burning phase of the star a few (depending on the distance) days before to the core-collapse through the monotonically increasing rate of very low energy antineutrinos during this stage. The description in greater detail is given in Sect. 6.3.

4.3.2 High Energy

The advantages of Gd-neutron tagging have not been considered traditionally for high energy neutrino physics, but it improves the separation between neutrino and antineutrino in the range from 100 MeV to 10 GeV. High energy antineutrinos tend to have larger neutron multiplicities in the final state as it happens for the low energy regime. The ability to distinguish between neutrinos and antineutrinos is very useful for long baseline and atmospheric neutrinos and it is expected that the sensitivity for both, the CP violating phase and the neutrino mass hierarchy are significantly enhanced, as shown in Sect. 5.4.

In addition to the neutrino-antineutrino separation, the neutron multiplicity depends on the neutrino interaction, being able to make discern, at some level, between neutral current, deep inelastic charged current and the rest of charged current interactions. This is specially important for high energy atmospheric neutrinos, where the contamination from neutrino interacting neutral current and deep inelastic scattering becomes an issue. The classification of these events improves the sensitivity of the atmospheric oscillation analysis to the neutrino mass hierarchy. This featured is developed in Sect. 5.3.

Besides, at these energies, the neutron multiplicity can be used to improve the atmospheric and long baseline neutrino energy reconstructions. This is due to the information about the missed energy, in the production of neutral particles, that neutrons contain. The improvement in the knowledge of the neutrino energy is very useful for the atmospheric oscillation analysis, improving also the neutron mass hierarchy and CP phase sensitivities. This technique is explained in Sect. 5.5.

Finally and in addition to neutrino physics, Gd-neutron tagging is also beneficial for proton decay searches. It is possible to remove 83% of the atmospheric neutrino background, by requiring the number of neutrons in the final state to be zero. This neutron veto produces tiny detection efficiency losses, due to the fact that proton decays rarely have any neutrons in the final state.

4.4 EGADS R&D Program and Experiment

EGADS (Evaluating Gadolinium's Action on Detector Systems) is the main R&D project for testing the feasibility of Gd-doped water-čerenkov detectors and focused in the realisation of SuperK-Gd, [Magro15, Mori15]. Due to the success of its technology and its good performance, it has become the currently most sophisticated supernova antineutrino detector, belonging to the Multi-Messenger Astronomy Consortium.

During this section, the most important advances towards SuperK-Gd made at EGADS are reviewed. This is done not only in terms of the hardware needed to deal with Gd-doped water, but also for items concerning the physics itself as well, such as the light propagation or the neutron capture detection. Some of the most relevant items are the water purification system, the monitoring of the Gd concentration in the tank.

4.4.1 Detector Description

The EGADS detector is designed very similarly to Super-Kamiokande, but about 250 times smaller. The EGADS water tank contains 200 ton of ultrapure water, with 5.417 m in diameter and 4.949 m in height. It is instrumented with 240 photomultipliers, which 227 of them are similar to those in the SK detector, 151 PMTs without any cover, 16 PMTs with an FRP housing and the 60 remaining PMTs with FRP and acrylic cover, achieving a 40% photocathode coverage. As in SK, black-sheets are installed surrounding all PMTs.

For compensating the geomagnetic field, Helmholtz coils are put around the detector in the same way as SK, achieving a tiny residual magnetic field of less than 0.1 Gauss (Fig. 4.5).

4.4.2 Water Purification System

The EGADS water system can be divided into two parts which work jointly, the pre-treatment and the water circulation systems.

The pre-treatment system, shown in Fig. 4.6, contains a 15-ton tank where the gadolinium sulphate powder is dissolved rather quick by using a stirrer, and once it is completely dissolved, it passes through a series of microfilters of different diameters

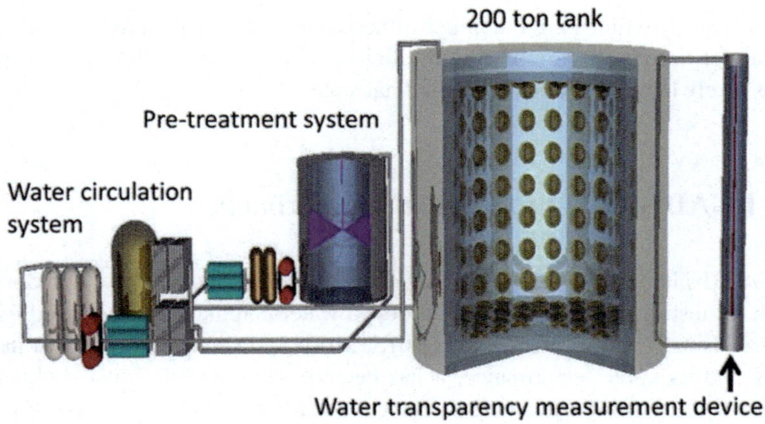

Fig. 4.5 Schematic drawing of the EGADS detector, including its water purification system and water transparency measuring device

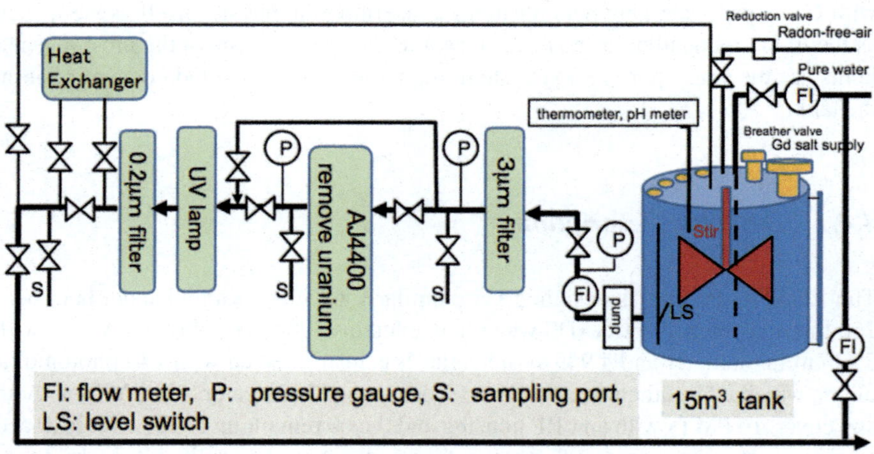

Fig. 4.6 Schematic view of the 15-ton tank and the pre-treatment system

to remove the largest dust particles, a UV lamp to kill bacteria and the resin AJ4400, specially designed to remove uranium, keeping, in its turn, all the dissolved Gd.

After the pre-treatment system, the solution is 0.2% Gd concentrated and it is pumped into the main water circulation system in Fig. 4.7. This system had to be specifically design in order to match the ultrapurity requirements like the SK water and to keep all the Gd and sulphate ions dissolved. To achieve this, the solution is chilled, to prevent the $Gd_2(SO_4)_3$ from precipitating, and passed through microfilters and ultrafiltration processes until it gets to the nanofilters, whose pore size is smaller than the Gd salt ions. This leaves pure water in part of the system which is further purified and the reunited with the previously left Gd^{3+} and $(SO_4)^{2-}$. After this,

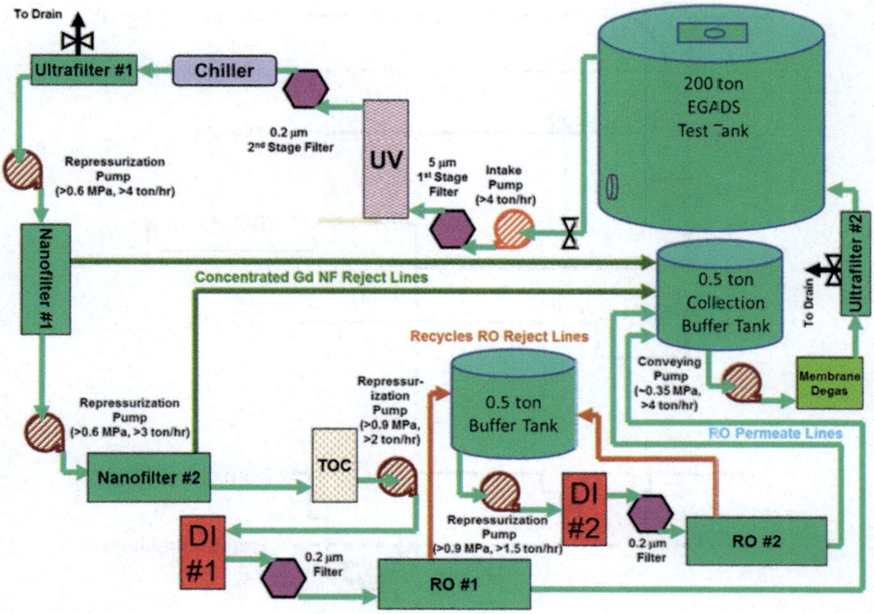

Fig. 4.7 Schematic view of the EGADS water circulation system

the Gd-doped water is injected to the 200-ton tank and continuously re-circulated through the system in order to maintain the water transparency quality.

4.4.3 Data Acquisition System

The data acquisition and electronic systems are analogous to those of SK-I to SK-III. In fact the old SK ATM electronics from those periods. The integrated charges and the time information for each PMT are recorded and digitalised by the ATM module.

The high voltage of the PMTs is set so they have the same gain and response, in a similar way to that described in Sect. 3.3.2.

Finally, in the trigger system, the signal is translated into a 200 ns width signal called $HITSUM$. If the event candidate exceeds the threshold, it is recorded and reconstructed (Fig. 4.8).

4.4.4 PMT Pre-Calibration

Before the PMT installation at EGADS, the quality of 250 PMTs was checked, of which 227 were selected to instrument EGADS. The high voltage was set in order

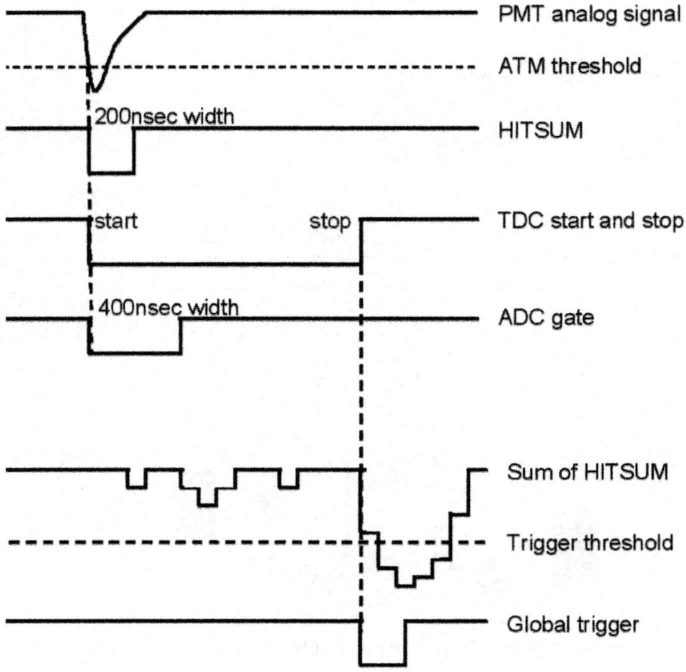

Fig. 4.8 Diagram of the trigger logic of the EGADS detector

to obtain uniform response from all PMTs, and the 1 photoelectron (p.e.) peak and dark rate of the PMTs was reevaluated.

For all this, the following setup was used at the Kamioka mine and very similar to that used for the PMTs that were to be installed in SK. A Xe lamp for measuring multiple p.e. was placed inside a black box, and connected through optical fibre to a scintillation ball inside a μ-metal box, together with an LED and the PMT to be tested. This metal box acts as shield from the geomagnetic field.

The scintillation ball is made of acrylic and is 5 cm in diameter. It contains 15 ppm of POPOP (1,4-bis(5-phenyloxazol-2-yl) benzene) as wavelength shifter and 200 ppm of MgO. Before arriving to the scintillation ball, the light from the Xe lamp was passed through an ultraviolet filter and divided to the scintillation ball and to an APD (Avalanche PhotoDiode), which monitors the intensity of the source.

In addition to the scintillation ball, an LED was used for testing the PMT sensitivity to a single photoelectron (Fig. 4.9).

The high voltage was set such that the PMT gain is 1.2×10^7. For this, seven PMTs randomly selected from the 250 PMTs were used as reference. The output charge data at four different high voltages around the nominal operational value is calculated by fitting the four data points using this formula $y = \alpha x^{\beta}$. α and β characteristic constants of each PMT, and y and x are the output charge and the high voltage.

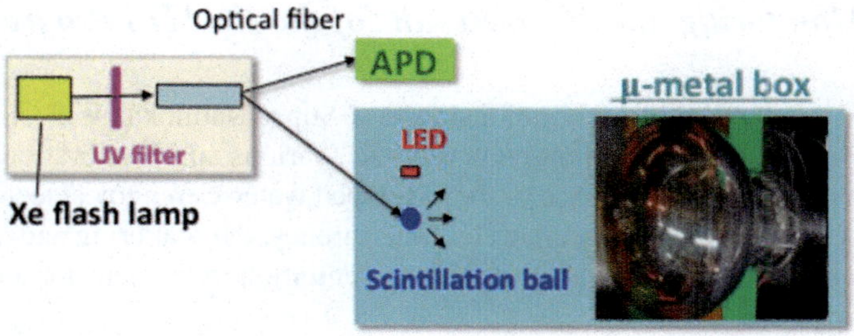

Fig. 4.9 Diagram of the pre-calibration setup for the PMTs installed at EGADS

The PMTs are selected based on the low dark rate at half p.e. and the high 1 p.e. sensitivity, being this the difference between the 1 p.e. peak and the valley between 0 p.e. and 1 p.e.

4.4.5 Detector Construction and Instrumentation

The EGADS tank was built in a new cavern specifically designed for it. Its inner faces covered by SUS304 panels. Inside the tank, everything metallic is also made out of SUS304, such as the frames supporting the PMTs, the bolts, etc. The construction and installation works were carried out with extreme care about the cleaning of every surface, being every component of the detector previously cleaned with ethanol and pure water. Further, all workers were equipped with tyvek suits, masks, cap and shoes only allowed inside the detector, and the detector was continuously flushed with fresh air preventing any dust from entering.

The PMTs were mounted on their frames outside the detector in a clean room with positive pressure to avoid the dust, as in the detector. The FRP housings and acrylic covers were also attached in this clean room after each of the PMTs was checked again, for the final tuning of the high voltage.

Once instrumented, the EGADS detector seemed to run successfully with pure water, but when adding the gadolinium, the presence of rust was detected through a large drop in the water transparency. After investigation, what was thought to be a component of stainless steel SUS304, turned out to be from other kind of stainless steel, producing the rust, already when the detector was filled with pure water. This meant that the entire detector had to be thoroughly cleaned once again, and refurbished, replacing the problematic parts by ones made of SUS304. This mishap was quickly and very efficiently solved within a month of hard work from all members of the EAGDS group.

Lastly, the detector was ready to add gadolinium, being this time the water transparency drops as expected from the addition of the gadolinium sulphate and with values very close to those of SK pure water.

4.4.6 Monitoring and Main Results of the EGADS Program

In addition to some of the calibrations done at Super-Kamiokande Sect. 7.3.14 on page 302, such as Nickel calibration or auto-Xe, various other calibrations ought to be done to prove the performance of the Gd-doped water-Čerenkov detectors. These measurements have to do with the water transparency, the scattering and absorption properties of the Gd-doped water, the Gd concentration in the detector and the Gd-neutron captures themselves.

In order to monitor the transparency of the EGADS water, the UDEAL (Underground Device Evaluating Attenuation Length) device was installed nearby the 200-ton tank. UDEAL is composed by long vertical pipe of 8.6 m instrumented with an array of seven lasers on the top, covering the main range of the Čerenkov radiation spectrum (337, 375, 405, 445, 473, 532 and 595 nm) (Fig. 4.10).

The attenuation length is then measured for all these wavelengths by measuring the light from the lasers after their light go through the column of water filling the pipe beneath. For better accuracy in the behaviour along the detector, the measurements are done for various water levels and with water from three different sampling points in the detector. This measurement is one of the most important in the R&D program, because it shows the great performance and stability of the water system, and the

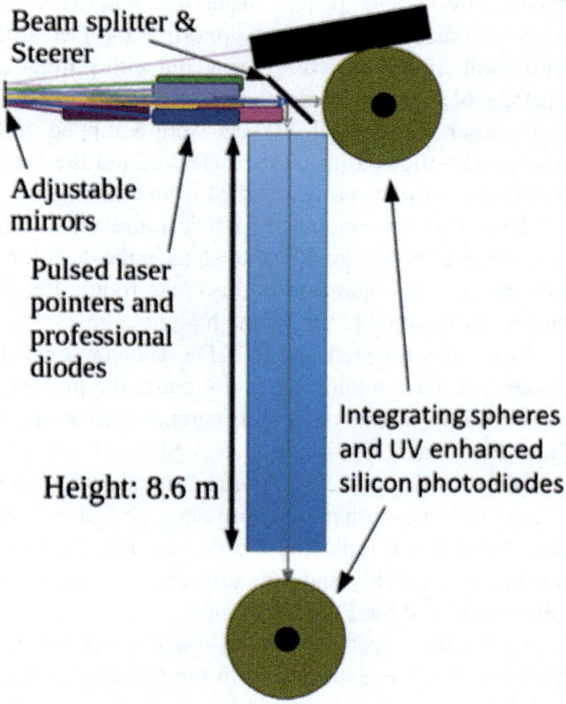

Fig. 4.10 Schematic drawing of the UDEAL device for measuring the EGADS water transparency

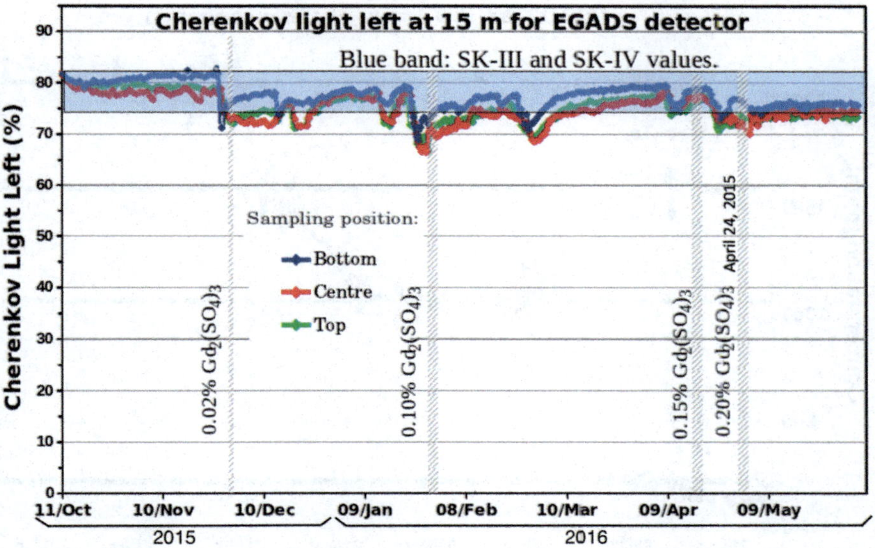

Fig. 4.11 Čerenkov light fraction remaining at 15 m of traveling through EGADS tank water for increasing concentrations of gadolinium sulphate. The green, red, and blue lines correspond to water from top, centre and bottom of the tank, and the pale blue line is the range of ultrapure water transparencies during SK-III and SK-IV

very little effect that the Gd salt has in the water transparency as compared with the usual SK values, as seen in Fig. 4.11.

For the measurement of the Gd concentration and its uniformity in the detector tank, water from the three UDEAL sampling points was extracted and measured with an Atomic Absorption Spectrometer (AAS). Figure 4.12 shows the agreement in the concentration of the three regions of the detector and with the expected concentration.

To further investigate and learn more about the effect of the added gadolinium salt in the light propagation two independent instruments measured the fraction of light absorbed and scattered, by the Gd-doped water. This is a new parameter that must be introduce in the Monte Carlo simulations. Both devices were measuring Rayleigh scattering, one introducing a laser injector inside the EAGDS tank, and the other one, called SAADI and developed at UCI, which compares the scattering of Gd with other compounds of known scattering properties. The former provides the most solid result. The measurements were done for three different concentrations of $Gd_2(SO_4)_3$. By comparing the fraction of light scattered with and without a reflector in the beam axis, the scattering coefficient due to the Gd-doped water can be extracted. The results indicate that the amount of light loss due to Gd is mostly due to absorption and that scattering contribution is $\lesssim 10\%$.

Finally, for checking and proving the capabilities of Gd-doped water-Čerenkov detectors, a device for mimicking an inverse β decay was used. It consists in an americium and beryllium (AmBe) source, which is placed inside the EGADS

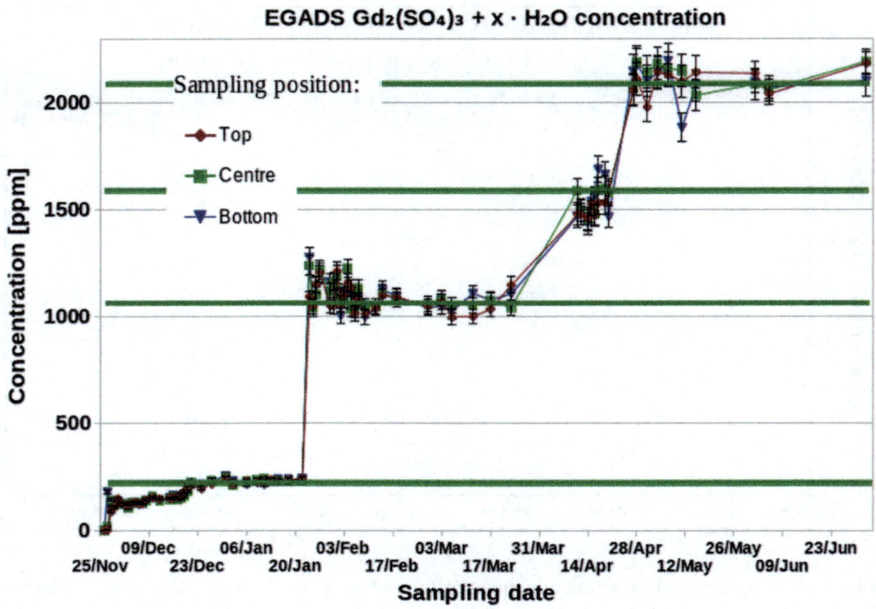

Fig. 4.12 $Gd_2(SO_4)_3$ concentration in the EGADS 200-ton tank measured with an AAS for top (red), centre (green) and bottom (blue) regions. The expected concentration is represented by the green horizontal lines

detector. The AmBe calibration device is made out of a core of Am and Be covered by a BGO ($Bi_4Ge_3O_{12}$) crystal which emits scintillation light when hit by γ rays. In the core of the AmBe source, a 4.43 MeV photon and a neutron are produced by the next reactions.

$$^{241}Am \rightarrow \alpha + ^{237}Np$$

$$\searrow$$

$$^{9}Be + \alpha \rightarrow ^{12}C^{*} + \boxed{n}$$

$$\searrow$$

$$^{12}C^{*} \rightarrow ^{12}C + \boxed{\gamma (4.43\ \mathrm{MeV})} \quad (4.2)$$

The scintillation light emulates the prompt neutrino signal and it is used as trigger for searching the delayed neutron capture by Gd in the tank. The measurements were done for the three different Gd concentrations as in the Rayleigh calibration. The data show very good agreement with the Monte Carlo simulations as can be seen from the distributions of the capture time and gamma cascade reconstructed energy in Figs. 4.13 and 4.14.

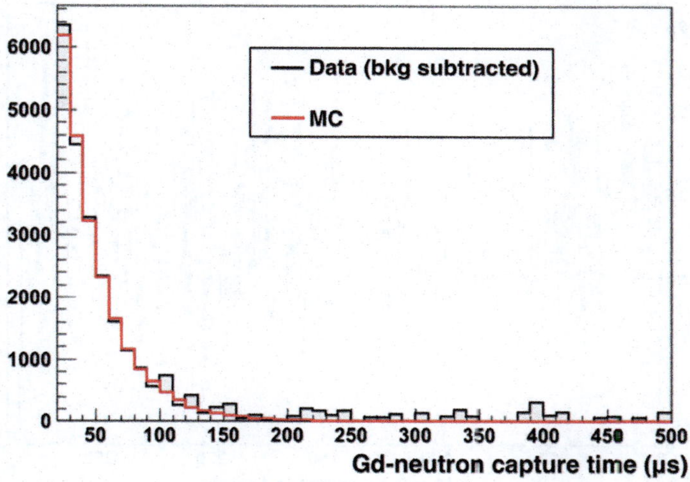

Fig. 4.13 Data (black) and Monte Carlo (red) distributions for the time difference between prompt and delayed Gd-neutron capture signals for data after background is subtracted data (right)

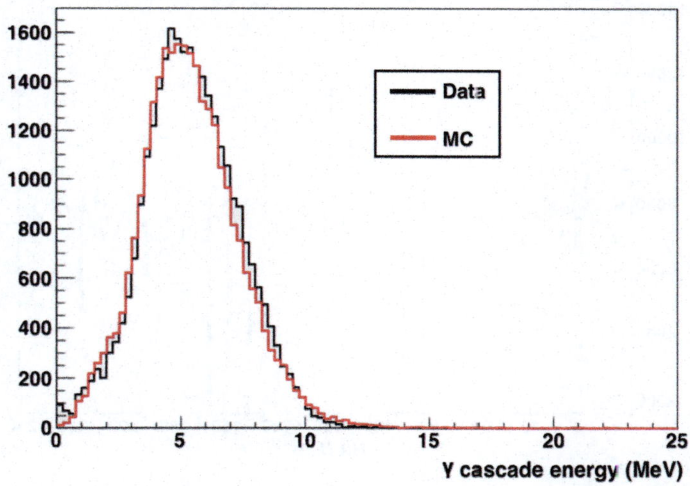

Fig. 4.14 Data (black) and Monte Carlo (red) distribution for the reconstructed γ cascade energy from the Gd de-excitation after the neutron capture

In addition to the on-site calibrations, the performance of the detector and status of its components are checked every 2 h remotely by all the EGADS members. For the data quality check, a series of basic plots are verified as well. Figures 4.15, 4.16, 4.17 and 4.18 show some of the most characteristic distributions.

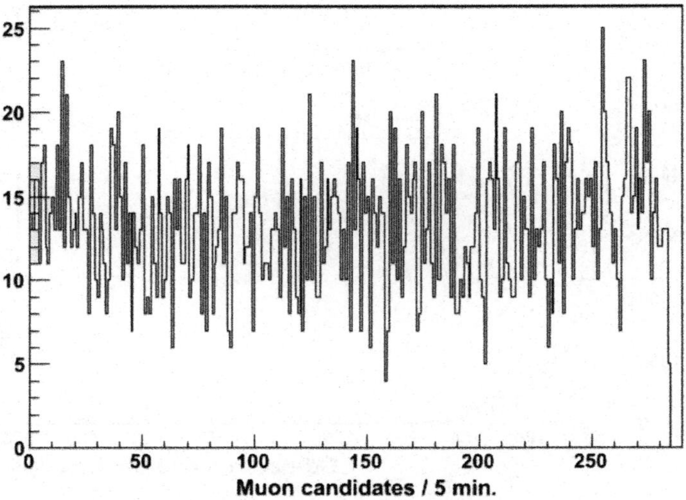

Fig. 4.15 Cosmic muon rate. Number of cosmic muon candidates per five minutes

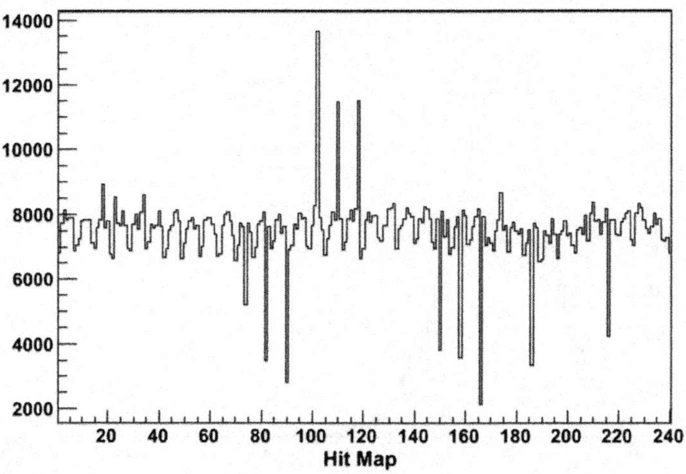

Fig. 4.16 Hit rate for each PMT. The narrow peaks correspond to R&D PMTs for Hyper-K

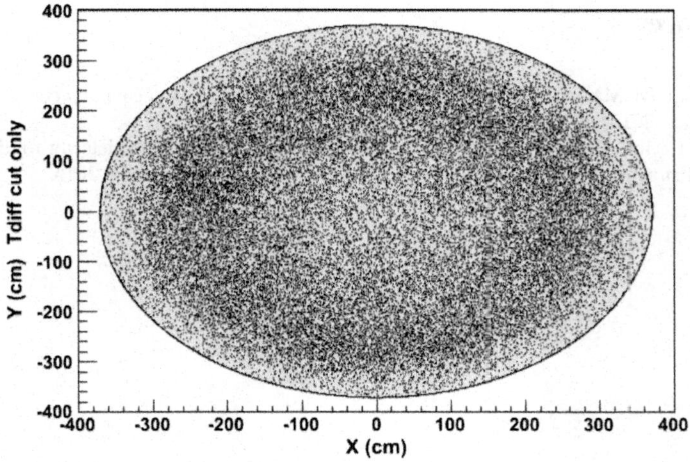

Fig. 4.17 Scatter plot of X and Y coordinates of reconstructed events more than 1 ms ago from the previous event (Tdiff cut)

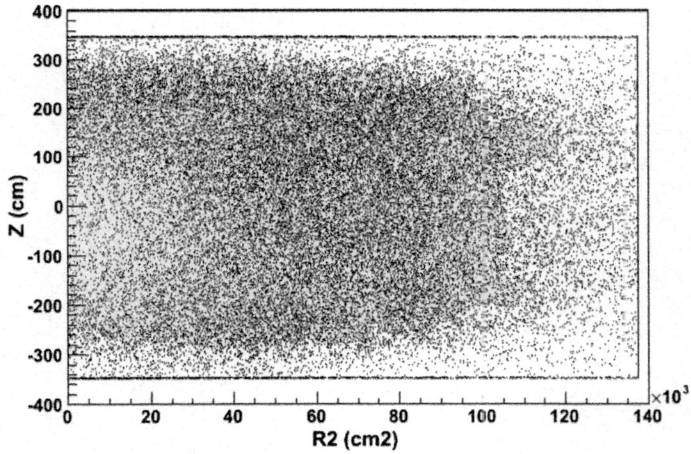

Fig. 4.18 Scatter plot of Z and R^2 coordinates of reconstructed events more than 1 ms ago from the previous event (Tdiff cut)

References

[Magro15] L.M. Magro, Egads approaching gadzooks!, in *EPJ Web of Conferences*, vol. 95. (EDP Sciences, 2015), p. 04041

[Mori15] T. Mori, Development of a gadolinium-doped water cherenkov detector for the observation of supernova relic neutrinos. Ph.D. thesis, University of Okayama, 2015

Chapter 5
On Relevant Items for SuperK-Gd Physics

In this chapter, key aspects of Gd-neutron tagging in a water-Čerenkov detectors are described, studied and developed using the SuperK-Gd project.

First, the neutron production mechanisms from neutrino interactions are reviewed in Sect. 5.1. Then, in Sect. 5.2, and based on the previous, the detector response to the Gd-tagged neutron signals is analysed in order to build the neutron reconstruction algorithm, which will play the key role in the present chapter for various physics analyses.

The neutron reconstructed information is then used to develop new tools which will improve the atmospheric and long baseline neutrino oscillation analyses, these are the separation between neutral and charged current interactions, the neutrino-antineutrino distinction and the neutron corrections to the reconstructed neutrino energy, all of them described in Sects. 5.3–5.5 respectively.

At the end of this chapter, in Sect. 5.6, the radioactive contamination that the Gd salt could bring along is studied in detail, from the measuring method to the new backgrounds produced by itself.

5.1 Neutron Production in Neutrino-Nucleus Interactions

Since this work strongly relies on the study of the different applications of the neutron production from neutrino interactions in water-Čerenkov detectors, it is worth to have a section explaining some of the theory of the processes involved, as well as the remaining issues in this, yet novel, topic. This, strongly depends on the neutrino energy, therefore, in this overview the low and high energy regimes will be differentiated (Fig. 5.1).

At low energies, up to 100 MeV, only electron antineutrinos are able to produce a charged lepton (positron) accompanied with neutrons. This is the inverse-β decay, seen in Sect. 2.4. This feature provides a very simple and efficient way of discerning antineutrinos from neutrinos at low energies, once neutrons can be tagged.

P. Fernández Menéndez, *Neutrino Physics in Present and Future Kamioka Water-Čerenkov Detectors with Neutron Tagging*, Springer Theses,
https://doi.org/10.1007/978-3-319-95086-0_5

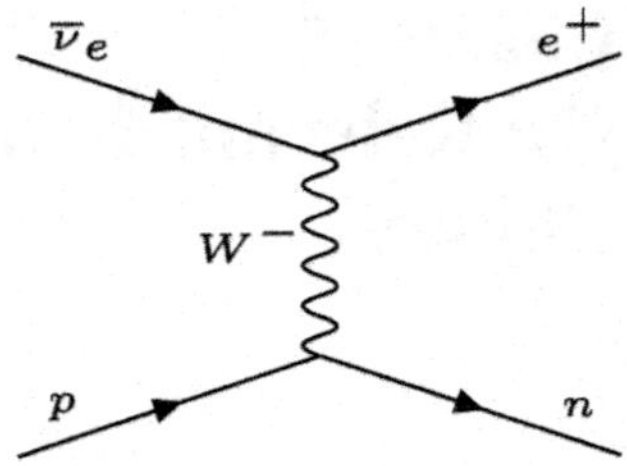

Fig. 5.1 Feynman diagram of Inverse Beta Decay (IBD) process

$$\overline{\nu}_e + p \rightarrow e^+ + n \tag{5.1}$$

This is the fundamental motivation for the development of neutron tagging in water-Čerenkov detectors. In SuperK-Gd, this reaction has as target, the first measurement of DSNB and the antineutrinos coming from the Si burning phase of a star.

At higher energies, from 100 MeV on, the neutrino interactions are more complex and varied, meaning that there will be various neutron production mechanisms. In fact, here, the neutron production not only depends on the neutrino interaction, but also on the secondary interactions inside the nucleus. This makes the neutron multiplicity larger than in the low energy regime and, thus, more difficult to model and compute. Because of this, the neutrino-antineutrino differences, in terms of the neutron multiplicity, are drastically dimmed as compared to the low energy regime.

The overall neutrino cross-section at these energies can be divided into three well defined categories for both charged and neutral current interactions. These are the elastic and quasi-elastic scatterings, the resonant and coherent meson productions and the deep inelastic scattering. In this thesis though, and motivated by the atmospheric neutrino oscillations analysis, they will be treated in three groups CC (containing charged current quasi-elastic and meson production scatterings), CC DIS and NC (containing all the neutral current interactions). The latter is defined in this manner, because all neutral current interactions share the lack of charged leptons in the final state.

The CCQE scattering dominates the cross-section below 1 GeV, making its understanding crucial for accelerator neutrinos and the lower spectrum of atmospheric neutrinos.

$$\nu_l + n \rightarrow l^- + p, \, \overline{\nu}_l + p \rightarrow l^+ + n \tag{5.2}$$

For energies above 1 GeV CC-RES interactions dominate, mainly to the resonance of Δ baryons. For higher energies, CC DIS interactions dominate (Fig. 5.2).

These interaction will happen mainly on the oxygen nuclei of water. The secondary interactions within the nuclear media are able to knock out free nucleons from the nucleus, more specifically neutrons. Depending on the interaction, the neutrino energy fraction left inside the nucleus changes, and so does the neutron production due to the secondary interactions (Fig. 5.3).

The neutrino-nucleon cross sections are modified by the transport of particles in the surrounding nuclear medium and the binding energy of the target nucleons. The

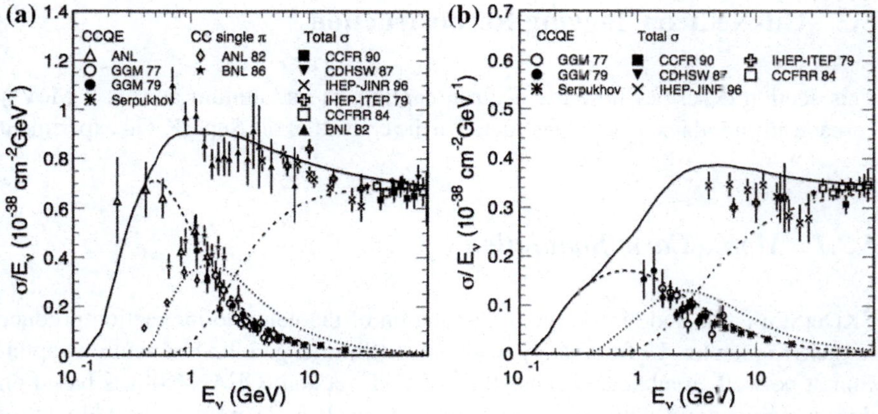

Fig. 5.2 Charge current interaction cross sections for **a** neutrinos and **b** antineutrinos. The dashed lines are calculated quasi-elastic, dotted lines are single-meson, dash-dotted are for deep-inelastic scatterings and dots represent the current experimental measurements

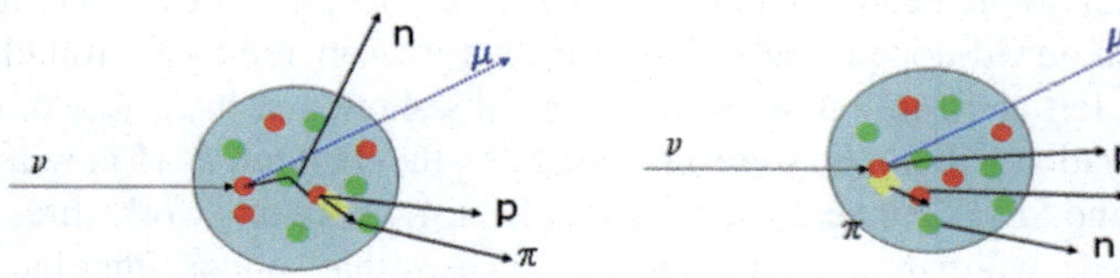

Fig. 5.3 Schematic diagrams involving a typical FSI process (left) and for a reaction where a pion is produced then absorbed in the same nucleus. These explains that the large neutron multiplicities are due to the secondary interactions inside the nucleus

nucleon knockout cross-section for the neutrino-nucleon interaction inside a nucleus can be calculated following Eq. (5.3).

$$\mathrm{d}\sigma_{tot}^{\nu_l A \to l' N X} = \int\limits_{\text{nucleus}} \mathrm{d}^3 r \int \frac{\mathrm{d}^3 p}{(2\pi)^3} \Theta(p_F(r) - p) \frac{k \cdot p}{k^0 \cdot p^0} \mathrm{d}\sigma_{tot}^{med} P_{\mathrm{PB}}(\mathbf{r}, \mathbf{p}) M_N(\mathbf{r}, \mathbf{p}) \tag{5.3}$$

This expression takes into account the Fermi momentum (p_F), the kinematics correction inside the nucleus ($\frac{k \cdot p}{k^0 \cdot p^0}$), the interaction cross-section in the nuclear media ($\mathrm{d}\sigma_{tot}^{med}$), de Pauli-blocking factor (P_{PB}), and the multiplicity of the final state with a given number of free nucleons ($M_N(\mathbf{r}, \mathbf{p})$). Basically, Eq. (5.3) tells that the neutron multiplicity from secondary interactions is due to the energy fraction transferred to the nucleus, from which, some nucleons may be have enough energy to overcome their binding energy.

5.2 Gd-Neutron Tagging Reconstruction

This section describes how the neutron capture on gadolinium and its 8 MeV γ cascade are simulated and reconstructed in the context of the SuperK-Gd experiment.

5.2.1 Monte Carlo Simulation

SKDetSim is the code in charge of simulation of the neutrino interaction products in the SK detector. Unfortunately, with SKDetSim the gadolinium neutron capture cannot be well simulated, because the GEANT version (GEANT3) it is based on, differs significantly with the latest experimental data. Therefore, the Gd-neutron capture information vectors were simulated independently from SKDetSim with GEATN4 and then, introduced into the detector's simulator to reproduced the signal at SuperK-Gd.

In addition to the neutron captures, Gd also has the effect of worsening the light transmission on Gd-doped water. This effect was considered to be around 15% loss in the light left at 15 m, but more recent results show that the effect of Gd in the light propagation will not be so severe. Actually, the light left at 15 m will be within the SK-III and SK-IV range, as discussed in Sect. 4.4.6. In this work, three scenarios are presented, two extreme and conservative cases that suppose that the 15% loss in the light left at 15 m is all due to Rayleigh scattering or only absorption of the gadolinium. The third case considers the transparency of pure water, assuming that there is no significant transparency loss. For this, SKDetSim was tuned to reproduce the three water transparency cases.

In Fig. 5.4, it is shown how the number of hits within 20 ns surrounding the Gd capture, where the differences between the three cases is most noticeable. The distribution for the pure water peaked at larger values than those with 15% light loss at 15 m.

Between the two extreme scenarios it is possible to discern different behaviours. For the all scattering case, the number of hits tends to be larger than for the all absorption one. This is because in the all absorption scenario, more photons are absorbed resulting in fewer hits in the detector, whereas in the scattering case, the photons are more dispersed. This behaviour is very similar to that of the energy reconstruction because it relies mainly on the number of hits recorded.

These differences are propagated to the reconstructed vertex and direction. Both extreme cases will have poorer resolution as compared to the pure water because less light is collected by the detector. Of these two the scenarios, that with vertex and direction poorer resolution should be the all scattering one. This is due to the fact that in this case, photons loose information of the point they are coming from, meaning that the time of flight (ToF) subtraction will leave these photons outside the time window used by the BONSAI (Branch Optimisation Navigating Successive Annealing Iterations) fitter, [Smy07]. This behaviour can be seen in Fig. 5.5, which

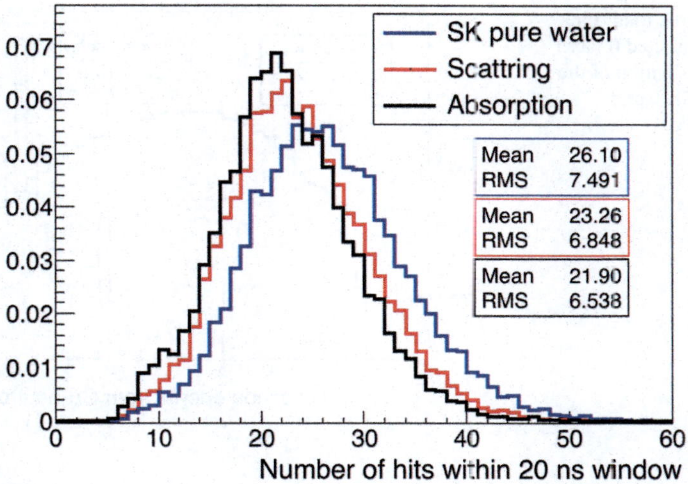

Fig. 5.4 Number hit PMTs within 20 ns time window surrounding the Gd-neutron capture for SK pure water (blue), and the extreme cases where the 15% loss in the light left at 15 m due to Gd is all because of scattering (red) or absorption (black)

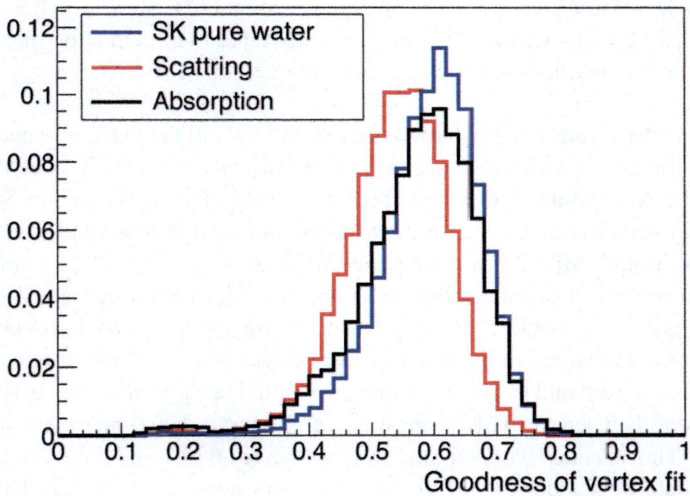

Fig. 5.5 Goodness of BONSAI vertex fit of the Gd-neutron capture for SK pure water (blue), and the extreme scenarios where the 15% loss in the light left at 15 m due to Gd is all because of scattering (red) or absorption (black). Larger values of the goodness represent better reliability of the vertex reconstruction

shows the goodness for the vertex fit. For the fit of the direction, the effect was not so noticeable, being the goodness of the fit ($DirKS$) very similar in all three cases.

The latest measurements from UDEAL, Fig. 4.11, show that such a large drop, as 15% in the water transparency, is not seen anymore. In fact, the decrease in the

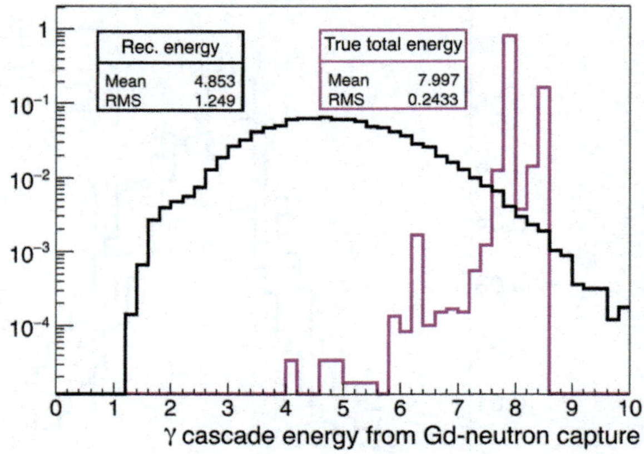

Fig. 5.6 True total (pink) and reconstructed (black) energy distribution of the Gd signal for SK water transparency

light left at 15 m will not be larger than 8% with respect to current SK's average pure water values.

The last issue concerning the Monte Carlo simulation of the Gd-neutron capture concerns the detector background. This has great importance for such low energy signals as the 8 MeV γ cascade. Despite being a signal of 8 MeV, which is a sizeable energy, it is shared by various photons, typically three to five, making that the mean value of the reconstructed energy around 4.8 MeV, close to SK's energy threshold (Fig. 5.6).

It is, therefore, important to have a reliable estimate of the detector's background. For that purpose it is not enough to consider the dark rate of the PMTs, but to take into account the rest of the backgrounds present in the detector as well. In Sect. 5.2.1, it is shown the fake candidates selected after running the first candidate selection algorithm through Monte Carlo simulated SK dark noise, of 5.7 kHz, and through SK-IV noise data. The latter consists in the hit information of SK-IV data after removing the signal from the event. This shows that the actual background of SK is quite larger than that assuming the dark rate of the PMTs in terms of the number of hits. For that reason and to better estimate the actual background, data was taken for various periods when the T2K beam was not in operation and no other event in SK was seen. This is called T2K dummy trigger data (Figs. 5.7 and 5.8).

These data were added to the full length of each simulated event as the background of the detector, and in substitution of the random dark noise. This technique enables a better description of the detector and the development of a much more reliable Gd-neutron capture signal detection algorithm. Figure 5.9 shows the number of hits within 20 ns time window for fake candidate selection as in Sect. 5.2.1, but for T2K dummy trigger data.

SK-IV data and T2K dummy trigger data fake candidate selection show a very similar distribution, corroborating the validity of the usage of T2K dummy trigger data as the background of the detector.

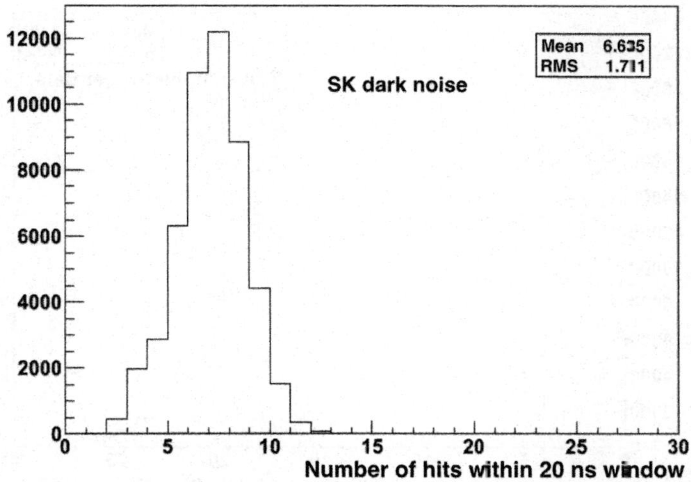

Fig. 5.7 Number of hits within 20 ns time window surrounding the neutron tag candidate for the simulated dark noise with no neutron tagging

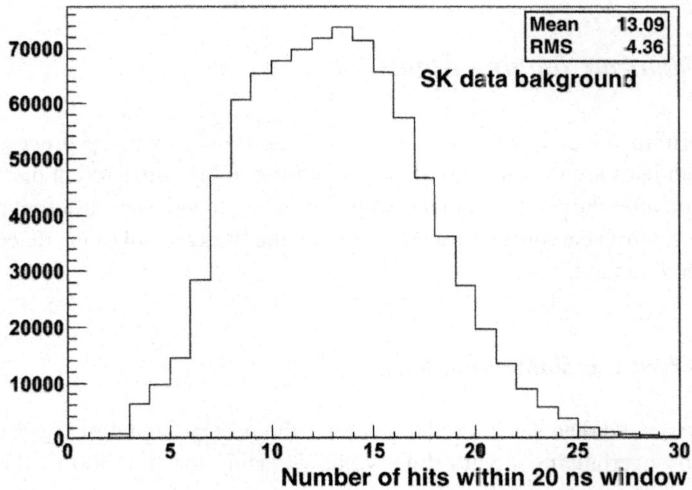

Fig. 5.8 Number of hits within 20 ns time window surrounding the neutron tag candidate for the SK data with no neutron tagging

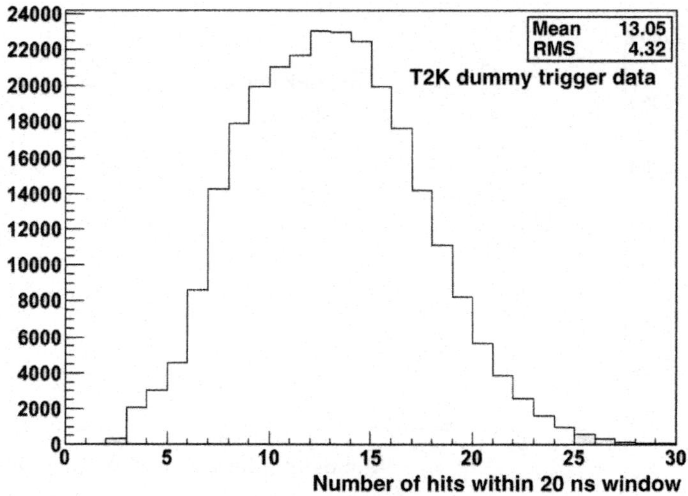

Fig. 5.9 Number of hits within 20 ns time window surrounding the neutron tag candidate for the T2K dummy trigger data with no neutron tagging

5.2.2 Neutron Tagging Algorithm

The algorithm for detecting the Gd-neutron captures a two-step process is used. First, candidates are selected based on the number of hits clustered in the hit timing distribution after the prompt signal is triggered. The second, uses different properties of the Gd-neutron captures to discern this from the background of the detector using a likelihood method.

5.2.2.1 First Candidate Selection

For the first candidate selection the hits after the prompt signal of the neutrino are scanned by overlapping 1.3 μs time windows, being the first 500 ns the same of the previous time window. The hit timing information of the event is ToF subtracted assuming the distance from PMTs to the neutrino signal vertex, which is a good first order approximation. Inside these 1.3 μs time windows, the maximum number of events within a 50 and 200 ns time windows are computed (Fig. 5.10).

Here, the mentioned ToF subtraction is crucial due to the multiple photon emission of the gadolinium, having each of them its own direction and path to the walls of the detector.

The criteria for the candidate selection are rather loose, but effective in identifying the true neutron captures and keeping the fake candidate selection rather low for each event. The requirements are chosen for the maximum number of hits in the 1.3 μs window within 50 ns (N_{50}) and 200 ns (N_{200}), being a candidate selected if Eq. (5.4) is satisfied.

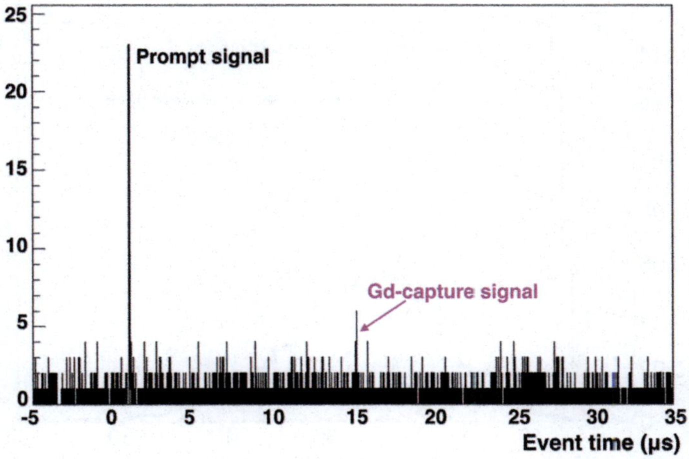

Fig. 5.10 Hit timing distribution of PMTs showing the prompt and Gd-capture delayed signals

Table 5.1 Efficiencies and remaining background for the first candidate selection of the Gd-neutron capture signals and for the three water transparency cases considered

	Gd-neutron capture			Background		
	SK water	All scat.	All abs.	SK water	All scat.	All abs.
Candidate selection	99.3%	99.0%	98.0%	2.75 event^{-1}	2.60event^{-1}	2.75event^{-1}

$$N_{200} \geq 20, \; N_{50} \geq 11 \tag{5.4}$$

These selection requirements have been optimised for the case where the water transparency of SuperK-Gd has similar values as that of SK. Although the first selection could have been adapted to other water transparencies, the same cuts were done, so they give an estimate of the performance of the selection for different water properties. The three water transparencies explored are those shown in Figs. 5.4 and 5.5. Table 5.1 shows the efficiencies of the first candidate selection algorithm and the fake capture candidates per neutrino event due to the detector's background. The latter correspond to the Gd-neutron tagging detector background. The length of the event is assumed to be 535 μs corresponding to the SHE+AFT trigger, shown in Table 3.4.

Once a candidate has been selected, a new time window for the candidate is defined. This contains the PMT hits contributing to the signal in the centre of the window.

This delayed event is treated then as a low energy signal and reconstructed with the BONSAI fitter. Several of the variables fitted will contribute to the construction of the likelihood distribution, explained in the next section, to further discern true Gd-neutron captures from the detector background.

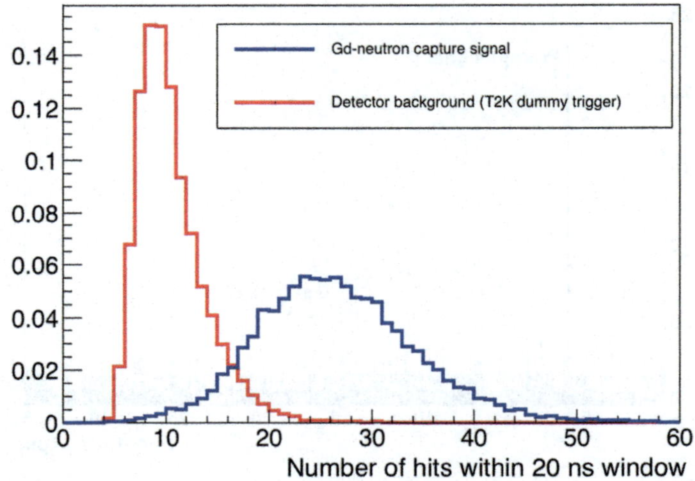

Fig. 5.11 Number of hits within 20 ns time window surrounding the neutron tag candidate for the true Gd-neutron signal (blue) and the remaining background from the first selection (red)

5.2.2.2 Gd-Neutron Capture Selection Likelihood

For discriminating between the actual neutron capture signal and fake first candidates from background, a likelihood function in terms of several variables is constructed. These variables take into account the detector response to the physics of the Gd capture. The variables for the construction of the likelihood distribution are described next and shown for the SK water transparency case. For the other two water scenarios these variables show little deviations, which will be shown at the end of this section in the likelihood distributions and efficiencies for all scenarios (Figs. 5.11, 5.12, 5.13, 5.14, 5.15, 5.16, 5.17, 5.18, 5.19, 5.20 and 5.21).

- N_{20}: The number of hits in 20 ns around the delayed signal candidate.
  For true Gd-neutron captures the values tend to be ~2.5 times larger than for the fake candidates, which also show a much narrower distribution due to the broad spectra of the Gd γ cascade.
- Nbf: The the number of hits used by BONSAI, the low energy fitter.
  In this case, the Gd-neutron captures show a narrow distribution near 150 hits, whereas for the background due to the randomness of the PMT hit distribution the distribution is much wider and with smaller values.
- *Signal significance*: $\dfrac{N_{eff} - 2/11 \cdot (N_{1300} - N_{200})}{\sqrt{\frac{2}{13} N_{1300}}}$, is the signal significance is based on the number of effective hits in 200 ns window with respect to the average number of hits outside that window. The effective number of hits is computed taking into account the poor charge resolution at these energies, angular dependences of the PMT acceptance and that for energies below 20 MeV, approximately 1 p.e. is produced per hit.

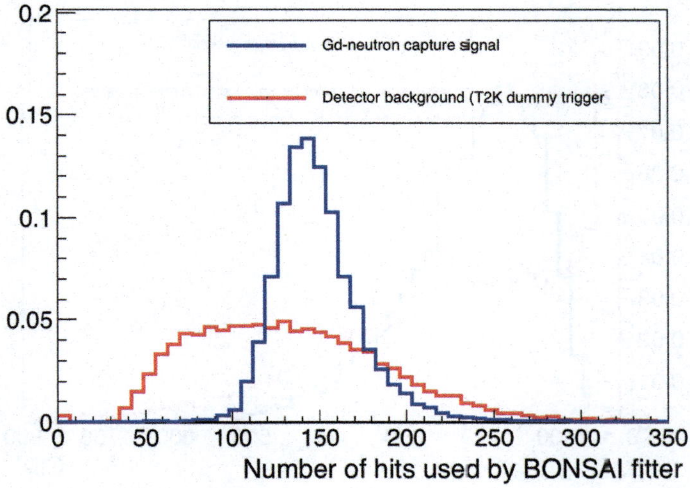

Fig. 5.12 Number of hits used by BONSAI for fitting the signal for the true Gd-neutron signal (blue) and the remaining background from the first selection (red)

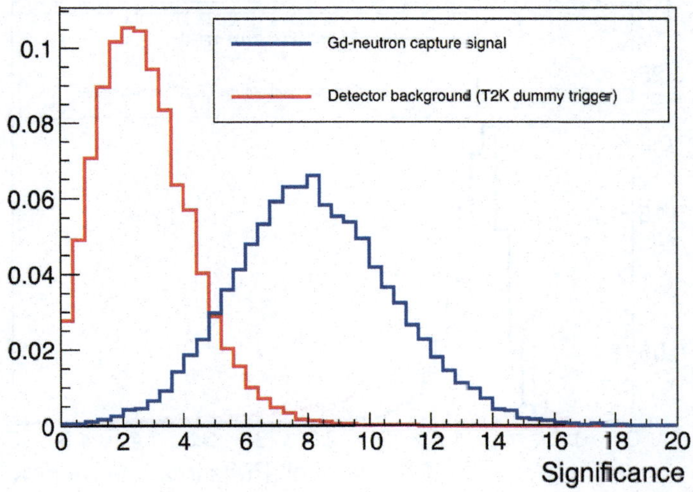

Fig. 5.13 Signal significance of the the candidate for the true Gd-neutron signal (blue) and the remaining background from the first selection (red)

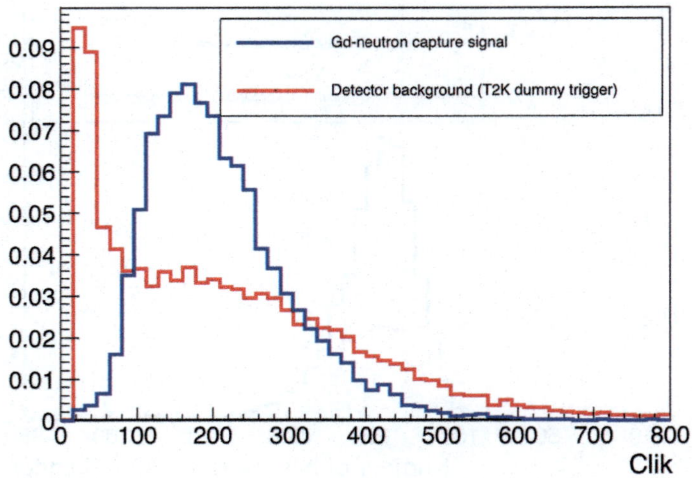

Fig. 5.14 Distribution of the *clik* variable for the true Gd-neutron signal (blue) and the remaining background from the first selection (red)

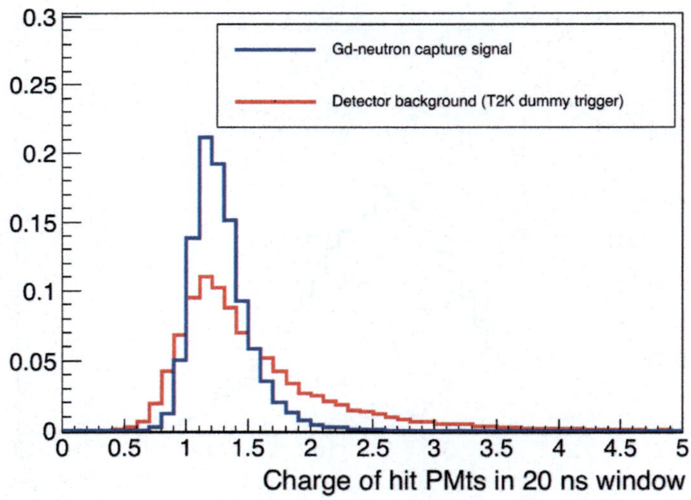

Fig. 5.15 Sum of charge of hit PMTs within 20 ns window for the true Gd-neutron signal (blue) and the remaining background from the first selection (red)

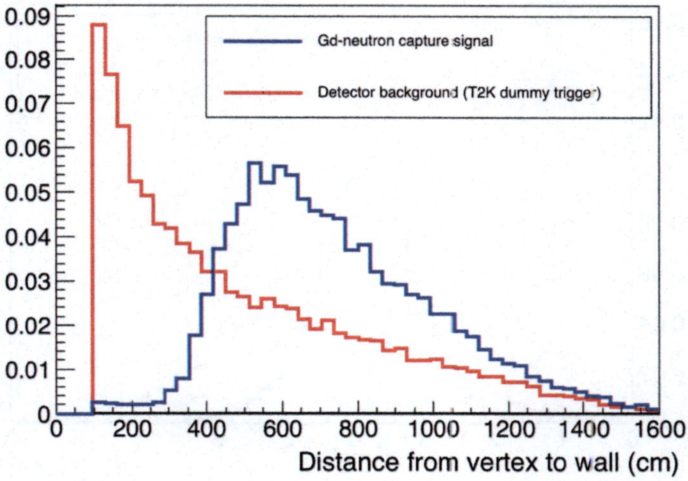

Fig. 5.16 Distance from candidate vertex to nearest wall for the true Gd-neutron signal (blue) and the remaining background from the first selection (red)

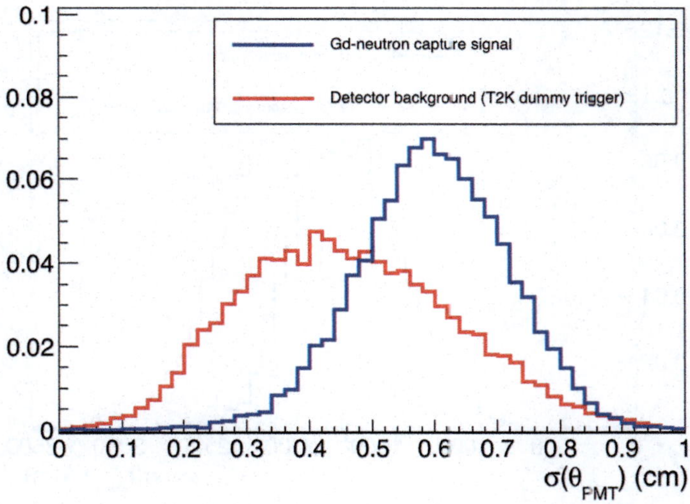

Fig. 5.17 RMS of the angle between the vector of delayed vertex and each hit PMT and the reconstructed direction of the delayed signal for the true Gd-neutron signal (blue) and the remaining background from the first selection (red)

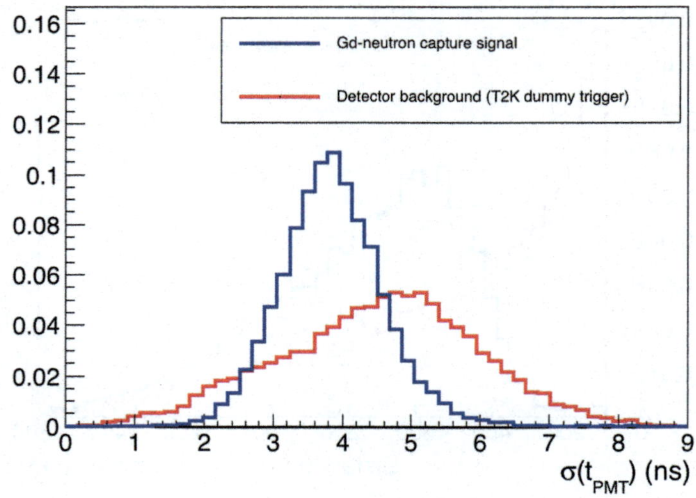

Fig. 5.18 RMS of the distance distribution from each it PMT to the reconstructed vertex for the true Gd-neutron signal (blue) and the remaining background from the first selection (red)

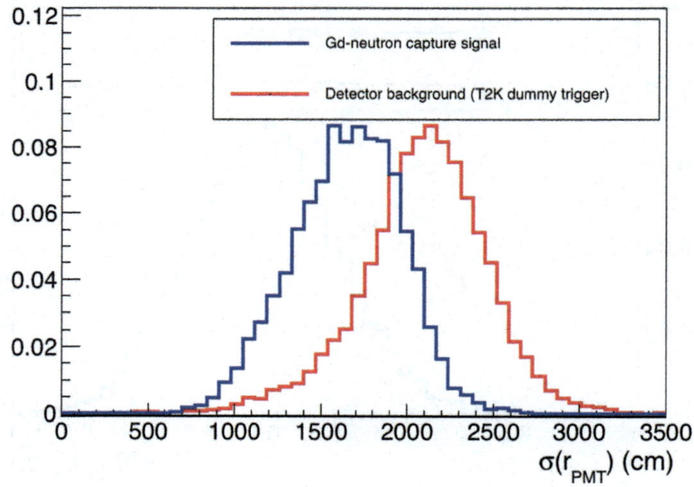

Fig. 5.19 RMS of the time difference from each hit PMT to the time of the candidate for the true Gd-neutron signal (blue) and the remaining background from the first selection (red)

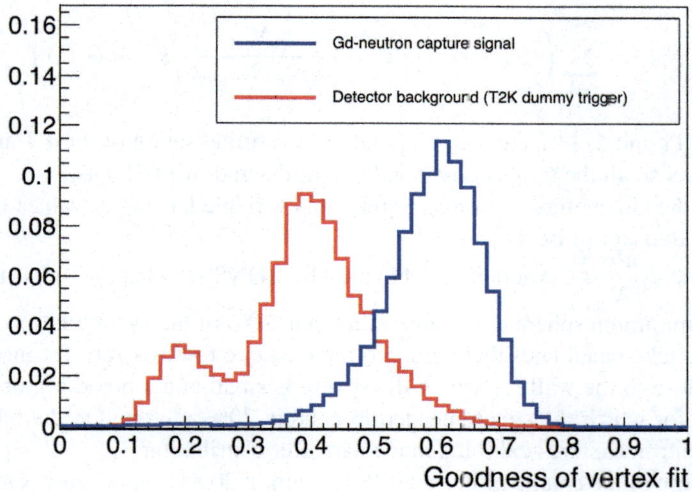

Fig. 5.20 Goodness of the reconstructed candidate vertex by BONSAI fitter for the true Gd-neutron signal (blue) and the remaining background from the first selection (red)

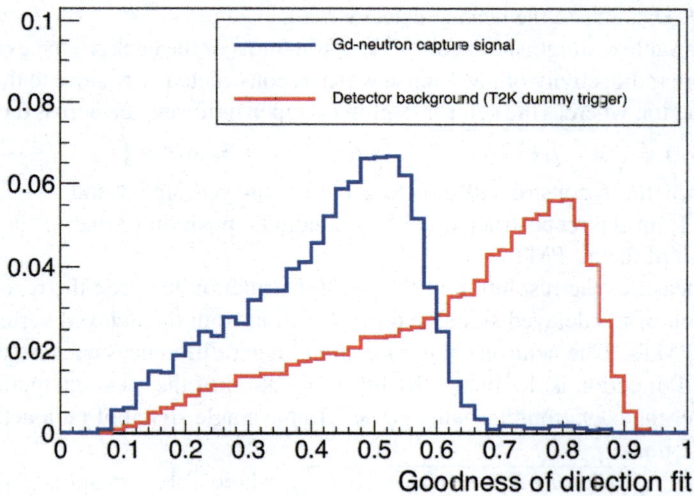

Fig. 5.21 Goodness of the reconstructed candidate direction by BONSAI fitter for the true Gd-neutron signal (blue) and the remaining background from the first selection (red)

$$N_{eff} = \sum_{i=1}^{N_{200}} \left((X_i + \epsilon_{tail} - \epsilon_{dark}) \frac{N_{all}}{N_{alive} S(\theta_i, \phi_i)} e^{\frac{r_i}{\lambda_{run}}} G_i(t) \right) \tag{5.5}$$

The $2/11$ and $2/13$ factors correspond to the normalisation of the 1.1 and 1.3 µs windows so all the terms correspond to a time window of 200 ns.
Being the Gd capture a stronger signal, this variable has larger values for it than for the detector noise.

- $clik = N_{20}\frac{nbsr0}{N_{eff}}$: Another variable used by BONSAI, where $nbsr0$ is the radius of the minimum sphere containing more than 20% of the event hits.
For the fake candidates there are two regions, one towards zero for those events very close to the wall for which the sphere is small and a broad region at larger values for which the sphere needed to contain 20% is large. On the other hand, true neutron captures exhibit a much narrower distribution.
- Q_{20}: Sum of the charge of those hit PMTs within 20 ns time window surrounding the delayed signal.
As in the case of Nbf, the true neutron capture distribution is narrower than that for the background, meaning that the charge is similar for all neutron captures.
- $dWall$: Distance to the nearest detector wall.
This variable distinguishes neutron tags from most of the background events coming from radioactivity or PMT noise which reconstructed very close to the walls of the detector, whereas the neutron captures happen uniformly all across the detector.
- $\sigma(\theta_{PMT}) = \sqrt{\langle\theta_{PMT}\rangle^2 - \langle\theta_{PMT}^2\rangle}$, where $\theta_{PMT} = \arccos\left(\vec{u}_\gamma \cdot \frac{\vec{r}_{PMT}-\vec{r}_\gamma}{|\vec{r}_{PMT}-\vec{r}_\gamma|}\right)$, and being $\vec{u}_\gamma$ the reconstructed direction of the delayed signal and $\vec{r}_\gamma - \vec{r}_{PMT}$ the vector from the reconstructed delayed candidate position to that of the hit PMT for each of the hit PMTs.
This measures the resolution of the angle distribution between the reconstructed direction of the delayed signal and the direction from the delayed vertex to each of the PMTs. True neutron captures exhibit bigger differences because the reconstructed direction is dominated by hit information of the most energetic photon, however the background usually comes from a single signal of radioactive decay or PMT noise.
- $\sigma(t_{PMT})$: $\sqrt{\langle|t_\gamma - ToF|\rangle^2 - \langle(t_\gamma - ToF)^2\rangle}$, where ToF corresponds to the time of flight from the reconstructed delayed vertex to each of the hit PMTs.
Here it is shown that once ToF is subtracted the time of the hit PMTs is within 4 ns, but for the background events their differences tend to be larger.
- $\sigma(r_{PMT}) = \sqrt{\langle|\vec{r}_\gamma - \vec{r}_{PMT}|\rangle^2 - \langle(\vec{r}_\gamma - \vec{r}_{PMT})^2\rangle}$, where $\vec{r}_\gamma$ is the position of the reconstructed vertex of the delayed signal and $\vec{r}_{PMT}$ is the position of each hit PMT.
This variable shows the resolution of the distances between the delayed signal and the hit PMTs. Fake candidates have larger values on average than actual neutron captures because the signal of the latter comes from the same point, where background events may come from noise of different PMTs and the vertex reconstruction is not so good.

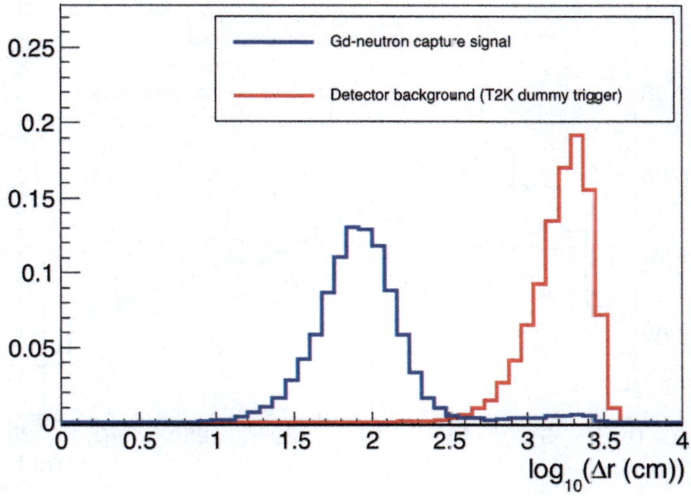

Fig. 5.22 Decimal logarithm of the distance between prompt and delayed candidate vertices for the true Gd-neutron signal (blue) and the remaining background from the first selection (red)

- *goodness*: This is goodness of the BONSAI fitted vertex of the delayed signal. Larger values mean better quality and reliability of the reconstruction, indicating that actual Gd-captures are better reconstructed than the background.
- $dirKS2$: This is goodness of the fit direction of the delayed signal. Larger values of this variable indicate worse direction reconstruction, confirming that also for the direction reconstruction the Gd-neutron captures perform better than the background.
- $\Delta r = \sqrt{(x_\nu - x_\gamma)^2 + (y_\nu - y_\gamma)^2 + (z_\nu - z_\gamma)^2}$: Spatial difference in cm between prompt and delayed signal.
  For the distribution the decimal logarithm was taken to better see the behaviour and typical values. Gd-neutron captures tend to occur at order 1 m from the prompt signal, whereas the background events are randomly distributed in the detector independently from the prompt vertex. This is the most relevant variable for the likelihood computation.
- $dt = t_\gamma - t_\nu$: Time difference in μs between prompt and delayed signal.
  This quantity shows the different time distribution with respect to the neutrino signal. It is flat distributed for fake candidates. For true Gd-neutron captures, on the other hand, the distribution is the convolution of an increasing behaviour corresponding to the amount of neutrons thermalised until 10 μs and a decaying exponential with mean lifetime of $\sim$20 μs corresponding to the Gd capture time of the neutrons. For times larger than 100 μs it is very unlikely for the Gd-neutron captures to happen (Figs. 5.22 and 5.23).

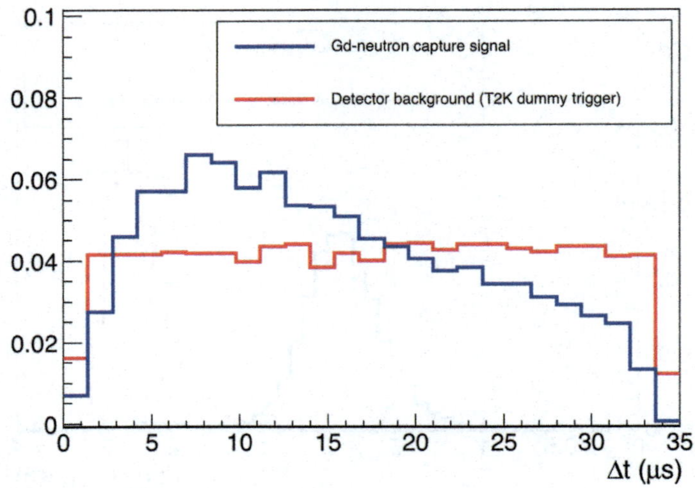

Fig. 5.23 Time distribution of neutron capture candidates from the prompt signal for the true Gd-neutron signal (blue) and the remaining background from the first selection (red)

Figure 5.24 shows the linear correlation coefficients of all the previous variables for the construction of the likelihood distribution, showing that the majority of them are completely uncorrelated. The number of hits in 20 ns window and the signal significance are the most correlated variables because they both use the number of hits within certain time windows. Other variables show a moderate correlation, due to the similarities between timing and spacing of hit PMTs. These correlations were checked not to have any worsening in the performance of the separation.

The likelihoods for the three water transparencies are shown in Figs. 5.25, 5.26 and 5.27.

- SK water transparency
- 15% transparency loss due to Rayleigh scattering
- 15% transparency loss due to absorption.

The cut point was chosen to be the same for all three cases like in the candidate first selection. This point was chosen so the purity of the remaining sample of Gd-neutron tags was a higher as possible for the SK water transparency case. Table 5.2 shows the efficiencies of selecting actual neutron tags and background events after the cut is applied.

The method shows a great efficiency in selecting the Gd-neutron captures and with great confidence, given the low contamination from background events for the three cases. However, a slightly decreasing trend in the Gd-neutron captures efficiency is seen going from SK pure water to all absorption passing through all scattering.

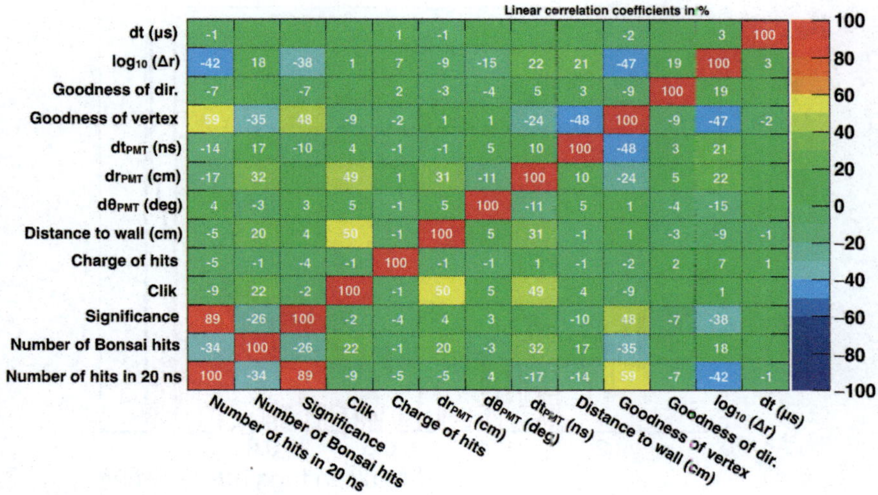

Fig. 5.24 Matrix of the linear correlation coefficients of the variables used in the identification as Gd-neutron capture of the delayed candidate signals

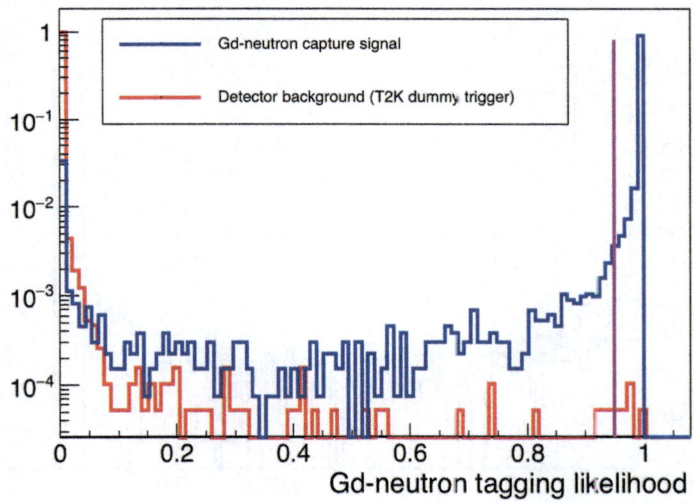

Fig. 5.25 Gd-neutron capture likelihood distribution for actual signal (blue) and detector background (red), assuming that gadolinium sulphate does not have any effect on the water transparency. The pink vertical line shows the selection point at 0.95

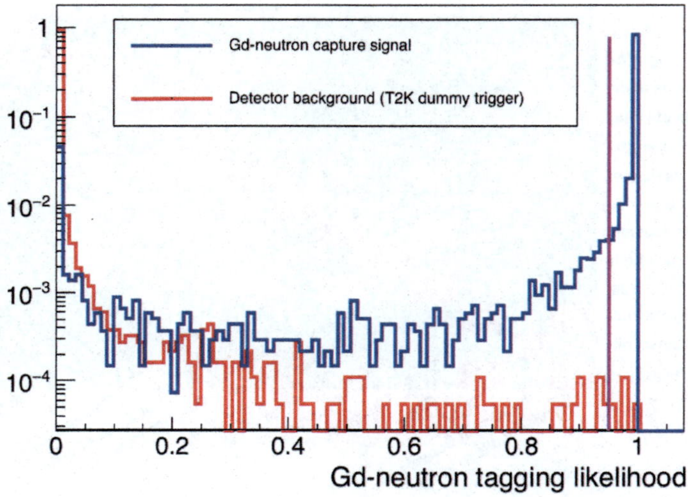

Fig. 5.26 Gd-neutron capture likelihood distribution for actual signal (blue) and detector background (red), assuming 15% water transparency loss due to additional Rayleigh scattering component of gadolinium sulphate. The pink vertical line shows the selection point at 0.95

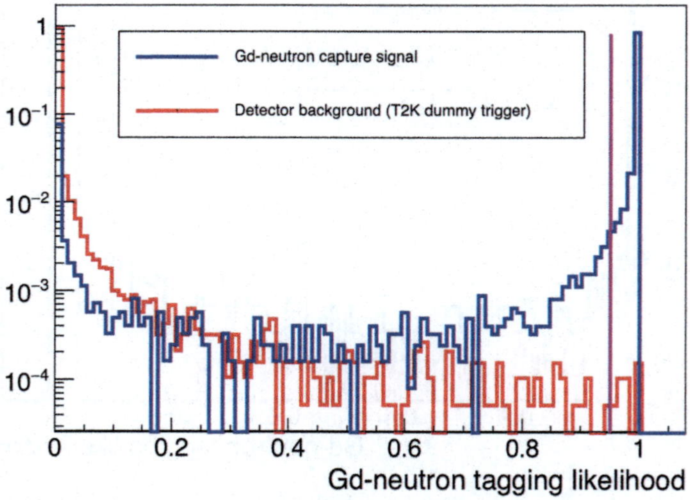

Fig. 5.27 Gd-neutron capture likelihood distribution for actual signal (blue) and detector background (red), assuming 15% water transparency loss due to additional absorption component of gadolinium sulphate. The pink vertical line shows the selection point at 0.95

Table 5.2 Efficiencies and remaining background of the likelihood distribution for the Gd-neutron tagging selection and for the three water transparency cases

	Gd-neutron capture			Background		
	SK water	All scat.	All abs.	SK water	All scat.	All abs.
Likelihood selection (%)	93.3	89.8	87.7	0.009	0.011	0.018

Table 5.3 Overall efficiencies and backgrounds of the Gd-neutron tagging selection methods for the case for the SK water transparency case

	Gd-neutron capture (%)	Background
Candidate selection	99.3	2.72 event^{-1}
Likelihood selection	93.3	0.009%
Total efficiency	92.7	2.4×10^{-4} event^{-1}

Table 5.4 Overall efficiencies and backgrounds of the Gd-neutron tagging selection methods for the case for the 15% water transparency loss due to Rayleigh scattering

	Gd-neutron capture (%)	Background
Candidate selection	99.0	2.63 event^{-1}
Likelihood selection	89.8	0.011%
Total efficiency	88.9	2.9×10^{-4} event^{-1}

Table 5.5 Overall efficiencies and backgrounds of the Gd-neutron tagging selection methods for the case for the 15% water transparency loss due to absorption

	Gd-neutron capture (%)	Background
Candidate selection	98.0	2.72 event^{-1}
Likelihood selection	87.7	0.018%
Total efficiency	86.0	4.9×10^{-4} event^{-1}

5.2.3 Overall Gd-Neutron Tagging Efficiency

The summary of both selection steps is shown next in terms of their efficiencies and the total efficiencies for all water transparency cases (Tables 5.3, 5.4 and 5.5).

Given the latest EGADS results, only a ∼8% loss and mainly due to absorption, it is safe to know that the expected total efficiency for Gd-neutron tags will be approximately the average of the SK pure water and all absorption cases, around 89%. The systematic error in the efficiency is taken from the difference in the efficiency of these two extreme cases, which is of the order of 5% of the averaged efficiency. With this assumption not only the uncertainty in the water transparency effect is considered, but also any other error concerning the simulation, as discussed in [Mori15].

For a 0.2% concentration of gadolinium sulphate in water the total amount of neutrons captured by Gd will be 90% and the rest will be captured by hydrogen. This gives a total neutron tagging efficiency of ~80%±4%, this values will be taken as the expected efficiency and associated systematic error for the physics analyses coming in the following sections. The 5% error of the efficiency is taken as the systematic error because the water transparency in which most of the variables rely, is the largest uncertainty in this method. This will be taken as the systematic error associated to the Gd-neutron tagging for the atmospheric neutrino oscillation analysis in Sect. 6.6.

5.3 Classification of Charged Current and Neutral Current Interactions

In this section a new method for distinguishing between neutral current (NC), charged current deep inelastic scattering (CC DIS) and the charged current (CC) electron neutrino interactions, is presented. This method is based on the neutron multiplicity along with other properties of the neutrino interaction final state. This feature is only considered for the MultiRing MultiGeV e-like sample of the atmospheric neutrino data. This sample contains events with more than one reconstructed ring, which the most energetic ring is e-like and visible energy greater than 1.33 GeV. It is the most contaminated sample by NC and $\overset{(-)}{\nu}_{\mu}$. The latter interact mostly via CC DIS interactions, motivating the grouping of the events in the three sets of interactions. The contents of these sample is shown in Table 5.6.

For completeness, Table 5.7 shows all the variety of interactions that CC DIS and NC englobe.

Table 5.6 Relative contents of each interacting mode for the MultiRing e-like atmospheric neutrino sample

Interaction mode	MultiRing e-like (%)
CC non-DIS ν_e	21.0
CC non-DIS $\overline{\nu}_e$	7.3
CC DIS ν_e	18.7
CC DIS $\overline{\nu}_e$	3.4
CC non-DIS ν_μ	4.8
CC non-DIS $\overline{\nu}_\mu$	1.1
CC DIS ν_μ	19.8
CC DIS $\overline{\nu}_\mu$	1.8
NC ν	22.1

Table 5.7 Set of CCDIS and NC interactions for atmospheric neutrinos that will be try to discern from the rest of CC interactions

Interaction	Name
$\nu + N/P \rightarrow l + N/P +$ mesons	CC deep inelastic (JET set)
$\overline{\nu} + N/P \rightarrow \overline{l} + N/P +$ mesons	
$\nu + N \rightarrow \nu + N + \pi^0$	NC1π (from Δ resonance)
$\overline{\nu} + N \rightarrow \overline{\nu} + N + \pi^0$	
$\nu + P \rightarrow \nu + P + \pi^0$	
$\overline{\nu} + P \rightarrow \overline{\nu} + P + \pi^0$	
$\nu + N \rightarrow \nu + P + \pi^-$	
$\overline{\nu} + N \rightarrow \overline{\nu} + P + \pi^-$	
$\nu + P \rightarrow \nu + N + \pi^+$	
$\overline{\nu} + P \rightarrow \overline{\nu} + N + \pi^+$	
$\nu + O^{16} \rightarrow \nu + O^{16} + \pi^0$	NC1π
$\overline{\nu} + O^{16} \rightarrow \overline{\nu} + O^{16} + \pi^0$	
$\nu + N/P \rightarrow \nu + N/P +$ multiπ	NCmultiπ
$\overline{\nu} + N/P \rightarrow \overline{\nu} + N/P +$ multiπ	
$\nu + N \rightarrow \nu + N + \eta^0$	NC1η (from Δ resonance)
$\overline{\nu} + N \rightarrow \overline{\nu} + N + \eta^0$	
$\nu + P \rightarrow \nu + P + \eta^0$	NC1η (from Δ resonance)
$\overline{\nu} + P \rightarrow \overline{\nu} + P + \eta^0$	
$\nu + N \rightarrow \nu + \Lambda + \kappa^0$	NC1κ (from Δ resonance)
$\overline{\nu} + N \rightarrow \overline{\nu} + \Lambda + \kappa^0$	
$\nu + P \rightarrow \nu + \Lambda + \kappa^+$	NC1κ (from Δ resonance)
$\overline{\nu} + P \rightarrow \overline{\nu} + \Lambda + \kappa^+$	
$\nu + N/P \rightarrow \nu + N/P +$ mesons	NC deep inelastic (JET set)
$\overline{\nu} + N/P \rightarrow \overline{\nu} + N/P +$ mesons	
$\nu + N/P \rightarrow \nu + N/P +$ mesons	NC deep inelastic (JET set)
$\overline{\nu} + N/P \rightarrow \overline{\nu} + N/P +$ mesons	
$\nu + P \rightarrow \nu + P$	Elastic
$\overline{\nu} + P \rightarrow \overline{\nu} + P$	
$\nu + N \rightarrow \nu + N$	
$\overline{\nu} + N \rightarrow \overline{\nu} + N$	

This separation will become important for neutrino-antineutrino separation and the neutron correction of the reconstructed energy of the events in this sample. The last two techniques will be developed in Sects. 5.4 and 5.5 respectively.

Given the energy regime of this sample, being able to make a more pure sample of electron neutrinos and antineutrinos, will have an significant impact on the sensitivity to the neutrino mass ordering from the atmospheric oscillation analysis.

For this discrimination, the number of tagged neutrons is crucial because their production depends on the interaction mode and also on the secondary interactions of mesons in the nuclear media, as discussed in Sect. 5.1.

The method consists in the construction of a likelihood distribution for each of the interaction groups NC, CC DIS and CC non-DIS using a neural network method, [Hoecker07]. The variables used are described and motivated next.

- Neutron multiplicity: Neutrinos interacting via neutral current interactions leave a larger fraction energy in the target nucleus than those interacting charged current. This makes the typical neutron production of NC events larger than CC ones. The CC DIS events have a intermediate behaviour, a large part of the neutrino energy is transferred to the nucleus, although not as much as in the NC case because of the charged lepton produced in these interactions. The three groups of interactions have then, different typical neutron multiplicities.
- Additional particle production: For these separations the number of rings and the likelihood of the event to have more than one ring are used. They show again that in NC and CCDIS events the neutrino energy is largely transferred to the nucleus producing, typically, more particles than CC $\overset{(-)}{\nu}_e$, similarly to the neutron production. This means that the number of rings tends to be larger for NC and CCDIS.
  In this context, the number of electrons from muon decays and the distance from the prompt vertex to the furthest of them have also a significant importance in the separation. This becomes specially relevant for discerning $\overset{(-)}{\nu}_\mu$ from $\overset{(-)}{\nu}_e$ as the associated charged lepton to the former are μs which will usually decays inside the detector. On the other hand, NC interactions produce more decay-e than CC $\overset{(-)}{\nu}_e$, but its distribution is peaked at zero, being the distance to the furthest decay-e in between CC $\overset{(-)}{\nu}_e$ and CCDIS $\overset{(-)}{\nu}_\mu$ because, although more decay-e are produced none of them is coming from a μ.
  Finally, the sum of particle ID likelihoods of each ring, this is used for deciding whether the ring is e-like from μ-like. This variable shows which species, e-like or μ-like rings, are dominant for each event. Here the differences are small but still they help the discrimination, specially for CCDIS $\overset{(-)}{\nu}_\mu$ for which it takes larger values, i.e. more μ-like, than for the other species.
- Kinematics: Two variables having to do with the kinematics of the different event types are defined. The pseudorapidity (η) and the energy fraction of the most energetic ring. Both aim to distinguish NC from CC interactions due to the more isotropic behaviour of the final products of the former, i.e. energy is more equally distributed among all the event reconstructed rings. The definition of pseudorapidity has been adapted from collider physics taking by reference the total direction of the event, therefore this variable will show how separated are all the rings between them weighted by their momenta.

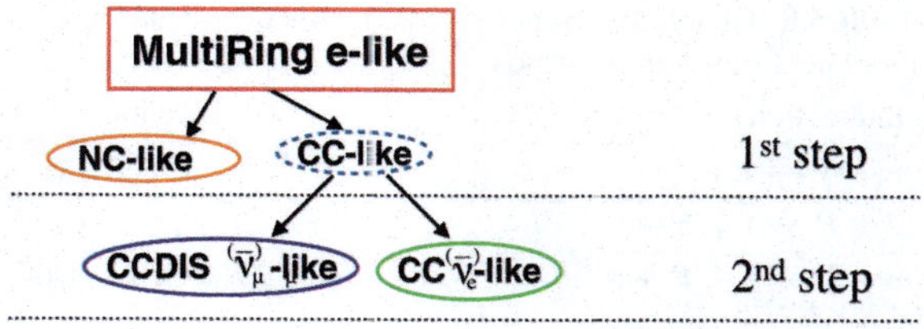

Fig. 5.28 Scheme of the two-step process in MultiGeV MultiRing e-like sample to separate NC interactions and ν_μ from CC ν_e events

$$\eta = \frac{1}{E_{vis}} \sum_{j=1}^{nring} \vec{u} \cdot \vec{p}_j \tag{5.6}$$

where $E_v is$ is the reconstructed visible energy of the event, $\vec{u}$ the direction of the event and $\vec{p}_j$ the 3-momentum of each ring.

These variables are used in a two-step process neural network to discard first the NC from CC DIS and CC events. Secondly muon neutrino and antineutrino interactions, which mainly CC DIS in this sample are separated from CC events. Figure 5.28 shows the procedure schematically.

This method is fully developed and applied for Gd-neutron tagging and H-neutron tagging in Sects. 6.6.2 and 7.2.1. It is also developed for the Hyper-K analysis with neutron tagging in Sect. 6.8.

5.4 Neutrino-Antineutrino Separation

For atmospheric neutrinos, the main advantage of neutron tagging up to now was the improvement in the distinction of neutrinos and antineutrinos. The latter tend to produce more neutrons in the final state that the former, being the most discerning variable for the distinction. In this section, a new method for the $\nu - \overline{\nu}$ separation are shown and explain.

The separation is only possible for charged current interactions. The neutrino charged current interactions at atmospheric and long baseline energy ranges are listed in Table 5.8.

Unfortunately, as explained previously, the secondary interactions with the nuclear media produce more neutrons independently of the neutrino or antineutrino character. This makes the separation dimmer than with the inverse-β interaction of low energy antineutrinos. For that reason, dedicated methods are preformed and optimised for the separation in each atmospheric and long baseline sample.

The $\nu - \overline{\nu}$ separation methods are developed in Sects. 6.6.3, 6.7.4 and 7.2.2, for atmospheric and long baseline Gd-tagging and for atmospheric H-tagging respectively, are explained. It is also implemented for the respective analyses, atmospheric oscillation analysis with Gd-tagging in Sect. 6.6.5, T2K analysis with Gd-tagging in Sect. 6.7.6 and atmospheric oscillation analysis with H-tagging in Sect. 7.3. This

Table 5.8 Charged Current ν interactions showing the tendency of antineutrinos to usually produce more neutrons than neutrinos

Interaction	Name
$\nu + N \rightarrow l + P$	CCQE
$\overline{\nu} + P \rightarrow \overline{l} + N$	
$\nu + P \rightarrow l + P + \pi^{+}$	CC1π (from Δ resonance)
$\overline{\nu} + N \rightarrow \overline{l} + N + \pi^{-}$	
$\nu + N \rightarrow l + P + \pi^{0}$	
$\overline{\nu} + P \rightarrow \overline{l} + N + \pi^{0}$	
$\nu + N \rightarrow l + N + \pi^{+}$	
$\overline{\nu} + P \rightarrow \overline{l} + P + \pi^{-}$	
$\nu + O^{16} \rightarrow l + O^{16} + \pi^{+}$	CC1π
$\overline{\nu} + O^{16} \rightarrow \overline{l} + O^{16} + \pi^{-}$	
$\nu + N/P \rightarrow l + N/P + \text{multi}\pi$	CCmultiπ
$\overline{\nu} + N/P \rightarrow \overline{l} + N/P + \text{multi}\pi$	
$\nu + N \rightarrow l + P + \eta^{0}$	CC1η (from Δ resonance)
$\overline{\nu} + P \rightarrow \overline{l} + N + \eta^{0}$	
$\nu + N \rightarrow l + \Lambda + \kappa^{0}$	CC1κ (from Δ resonance)
$\overline{\nu} + P \rightarrow \overline{l} + \Lambda + \kappa^{0}$	
$\nu + N/P \rightarrow l + N/P + \text{mesons}$	
$\overline{\nu} + N/P \rightarrow \overline{l} + N/P + \text{mesons}$	CC deep inelastic (JET set)

separation is also implemented, although not shown explicitly, and applied in the atmospheric and long baseline analyses at Hyper-K in Sect. 6.8.

Depending on the analysis and the neutrino sample, the separation will be done by making cuts or constructing a likelihood distribution with several variables in addition to the neutron multiplicity. These variables take into account all the differentiating features of neutrinos and antineutrinos. Next, all the variables that will be used are described and explained.

- Number of Gd-tagged neutrons: As mentioned before, this variable is the main motivation for the implementation of this separation. The different behaviour of neutrinos and antineutrinos comes from the primary interaction of charged current events. At $10^{-1} - 10^{2}$ GeV energy range the charged current interactions are dominant and they are more likely to produce a larger neutron multiplicity in the antineutrino case.
- Number of μ decay electrons: This variable accounts for the charged pions production difference between neutrinos and antineutrinos. Due to charge balance of the interaction, neutrinos are more likely to produce π^{+} and antineutrinos π^{-}. The formers decay mostly into μ^{+}s and these into e^{+}, which are delayed from the primary interaction and tagged as decay-e. On the other hand, the laters, are quickly absorbed by the water in the detector (mainly by ^{16}O), hardly ever being able

to decay. This, together with the neutron multiplicity are the most discriminating variable from the neutrino and antineutrino separation.

- Time to furthest decay electron (MultiRing samples only): This shows that if a decay-e comes from a π^- or π^+. If the decay happens at late times it means that the parent particle cannot be absorbed by water and, most likely, this would correspond to a π^+. On the other hand if the decay is very close in time to the primary interaction it will not have had enough time to be absorbed even if it corresponds to a π^-. Therefore, it is expected that ν have larger times than $\overline{\nu}$.
- Distance to furthest decay electron (MultiRing samples only): This has the same motivation as the previous variable, it simply translates time to distance.
- Sum of particle ID likelihood (MultiRing samples only): This variable takes into account that some of the produced charged particles in the interaction will not produce enough Cherenkov light to reconstruct a ring, but they contribute to other rings. In that sense, νs are more likely to produce charged hadrons, meaning that these events may have mis-reconstructed rings, diluting the ring pattern.
- Number of rings (MultiRing samples only): Similar to the previous case, CC ν interactions tend to produced more charged hadrons than the analogous for CC $\overline{\nu}$, and this contributes to the larger number of Cherenkov rings produced and detected.
- Energy fraction of most energetic ring (MultiRing samples only): Variable which takes into account the differences in the CC cross-section between neutrinos and antineutrinos. Neutrinos tend to have less fraction of the momentum in its most energetic ring.
- Pseudorapidity (MultiRing samples only): It is the same variable as defined in Eq. (5.6). This variable also accounts for the different expressions of the cross-section for neutrinos and antineutrinos resulting in different angle and energy distributions.
- Angle between neutrino beam and charged lepton scattering angle (T2K samples only): This variable is only possible for long baseline neutrinos where the direction of the neutrino is known. The scattering angle dependence is different for neutrino and antineutrino interactions, providing additional and very important information as compared to the atmospheric neutrino analysis.

For the case of atmospheric neutrinos, the methods can be divided into two different strategies. In the SubGeV and single ring MultiGeV samples, simple cuts are done on the two most relevant variables, the number of neutrons and the number of decay electrons. On the contrary, in the MultiRing samples, the neutron multiplicities grow significantly making even dimmer the distinction, therefore, a similar method to the NC-CC separation is shared by these samples. It consists in the computation of a likelihood distribution via a neural network with all the variables described before. After the NC-CC and neutrino-antineutrino separations done for the atmospheric fully contained events, the samples have the scheme shown in Fig. 5.29.

For the T2K analysis, the neutrino-antineutrino separation is done using just the number of neutrons and the cosine of the scattering angle. In this case, a likelihood distribution is constructed in favour of simple cuts due to the continuous distribution of the scattering angle.

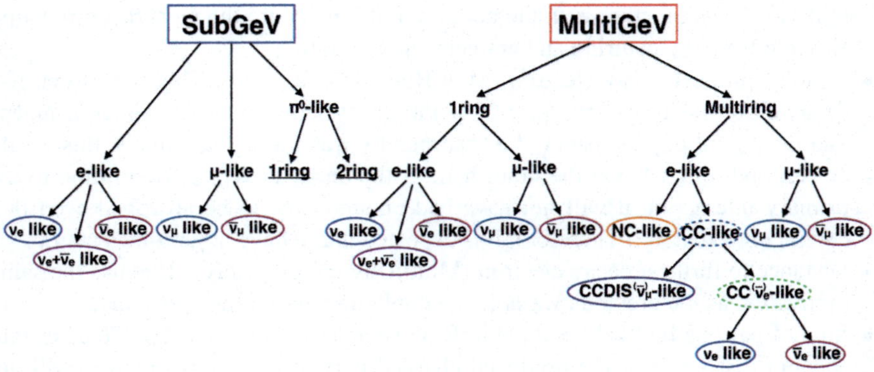

Fig. 5.29 Scheme of the new sample definition for atmospheric neutrinos based on the neutron tagging event selection

5.5 Neutron-Corrected Reconstructed Neutrino Energy

The neutrino energy reconstruction is very important for any water-Čerenkov neutrino detector, specially in the atmospheric oscillation analysis. Neutrino oscillation probabilities change very rapidly with energy and different energy regions are sensitive to particular oscillation parameters. For this reason, it is crucial to have a reliable and accurate energy reconstruction.

Neutrino energy is reconstructed as the so called visible energy, and defined as the sum of the energy of each ring produced. Therefore, it does not take into account the energy transferred to the nucleus and transformed into neutral hadrons ($\pi, \kappa, \eta \ldots$) produced in the interaction.

The more energetic is the incoming neutrino the more energy fraction is lost due to neutral hadron production. These usually interact inside the nuclear media, which translates into the increasing possibility of neutron production. This conclusion is extracted directly from Eq. (5.3), which shows that the neutron multiplicity encloses information about the neutrino energy transferred to the nucleus and, therefore, about the fraction of the neutrino energy potentially invisible to the detector.

Figure 5.30 shows an example of the fraction of invisible energy as a function of the neutrons tagged by gadolinium, exhibiting a clear correlation between both quantities.

This dependency gives the opportunity for improving the reconstructed visible energy by applying a neutron correction function to all the events in a particular sample. The monotonically increasing behaviour is shared by all samples, but with changes in the parameters of the fitting function, depending on the neutrino energy and flavour. For each sample, the correction function is obtained by fitting the distribution shown in Fig. 5.30 to a polynomial function. The correction is applied by using the expression in Eq. (5.7).

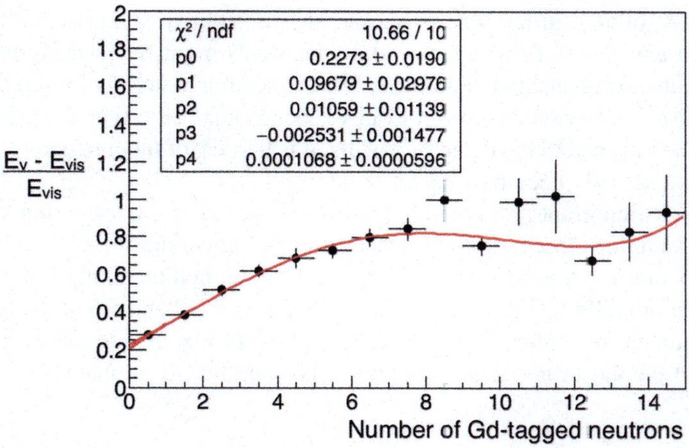

Fig. 5.30 Plot showing the dependence of the quotient of the non-reconstructed energy and the visible energy with the number of Gd-tagged neutrons for MultiRing $\overline{\nu}_e$-like atmospheric neutrino Monte Carlo

$$E_{rec}^{Gd} = E_{vis}(1 + f(\text{Gd-neutrons})) \tag{5.7}$$

where f is the corresponding sample correction function and E_{rec}^{Gd} is the corrected reconstructed neutrino energy.

This is a very novel, but certainly interesting feature. In this thesis, this technique will be developed for Gd-neutron tagging at Sects. 6.6.4 and 6.7.5 to be implemented for the atmospheric neutrino oscillation and the T2K oscillation analyses respectively. It has been also done for the Hyper-K analyses in Sect. 6.8, although it is now explicitly shown to reduce the extension of this dissertation. In Sect. 7.2.3, this method is implemented for current SK data with H-neutron tagging and applied to the atmospheric oscillation analysis shown in Sect. 7.3.

5.6 Radioactive Contamination

The inclusion in SuperK-Gd of additional radioactive contamination could be potentially harmful for the future of the project. If the radioactive contamination is too large, the capability of SuperK-Gd to discover DSNB neutrinos could be ruined and the current low energy solar analysis of SK would be impossible. In addition, if the radioactive contamination is huge, the signals coming from radioactive decays could saturate the detector. Therefore, the measurement and monitoring of the radioactive contamination of new materials in SuperK-Gd is crucial.

Hence, an exhaustive campaign of radioactivity measurements has been and is being carried on for the pursuit of SuperK-Gd. Of all of the sources from which radioactivity might enter the detector, the most dangerous is the Gd compound itself.

This will be diluted in the whole volume and radioactive signals could be reconstructed at any point of the detector and thus, undistinguishable from the true low energy neutrino signals. Other sources, like the sealing materials for repairing SK's leak, could also contribute to the radioactive background. However, since these materials will be attached to the detector's walls, it is harder for them to enter the fiducial volume and actually become a background.

The most important and potentially harmful radioactive decays that will occur, are those from the three radioactive decay chains, shown in Fig. 5.31.

The upcoming sections present the measuring method using high purity germanium detectors (HPGe), Sect. 5.6.1, and the in Sects. 5.6.2 and 5.6.4, the main background sources are explained as well as the latest latest techniques developed for the removal of the most radioactive contaminants from the Gd sulphate.

5.6.1 Measurement of Radioactive Contamination in Materials with HPGe Germanium Detectors: Gadolinium Sulphate

The principal method for measuring the radioactive contamination of $Gd_2(SO_4)_3$ are HPGe detectors. High precision Ge detectors are able to measure a broad range of energies and, therefore, the majority of the characteristic photon peaks from the isotopes of the three radioactive decay chains. Based on the intensity of the peak, the detection efficiency of the detector and the time of exposure, the amount of a certain isotope can be calculated, [Vidmar09].

Several batches of the Gd compound have been bought to different companies in order to measure and compare its radioactive contamination to choose the cleanest and purest for the detector.

The majority of the measurements have been done at Canfranc Underground Laboratory (LSC), in Spain. In addition, some complementary measurements are done at the Kamioka Observatory and at Boulby mine (Fig. 5.32).

For each decay chain, various energy peaks were analysed in order to obtain the global radioactivity of each chain. The typical photons for each decay chain are shown in Fig. 5.33.

Next, all the measurements done are shown in Tables 5.9 and 5.10. The $Gd_2(SO_4)_3$ batches measured can be divided into two groups. Table 5.9 shows the first batches received, which contain a sizeable and fluctuating radioactive contamination. Furthermore, the decay chains in these batches were far from equilibrium.

In Table 5.10, the group of the latest batches measured is shown, presenting a much lower radioactive contamination. The key point for this improvement in purity was the official approval of SuperK-Gd in 2015, and the collaboration with the providing companies to improve their production and purification methods. The limits in the radioactive contamination shown are estimated according to [Heisel09].

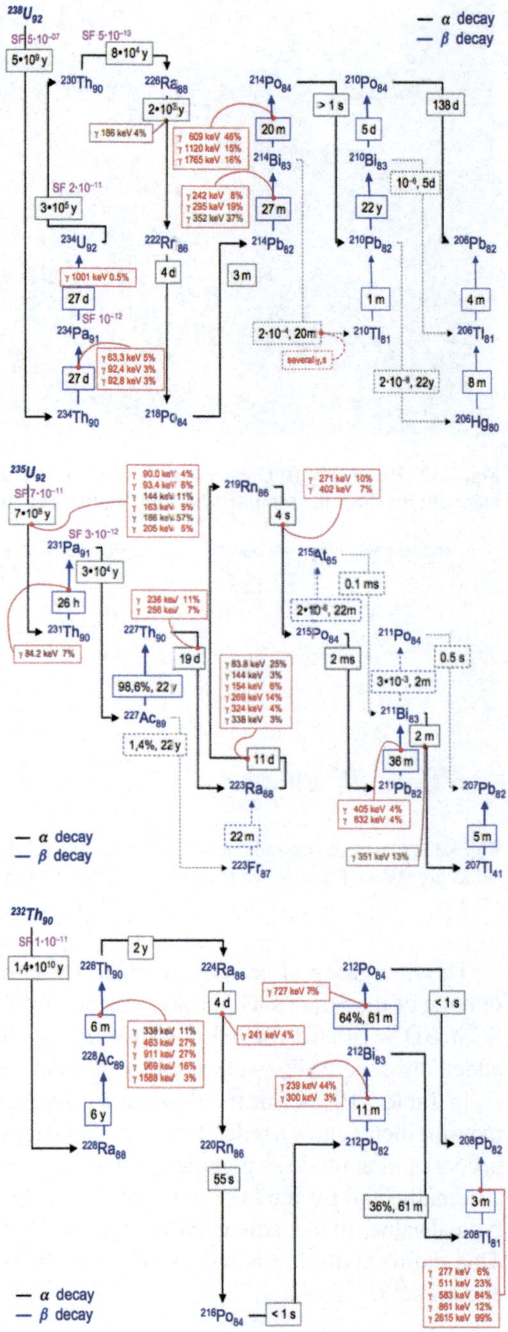

Fig. 5.31 From top to bottom, ^{238}U, ^{235}U and ^{232}th radioactive decay chains. Black lines between isotopes indicate α-decays, blue lines β-decays and pink writing spontaneous fissions with their half-life. Half-life for α and β-decays is written over the lines between isotopes. In addition, the energy and probability of the most characteristic photons are shown in red next to the corresponding isotopes

Fig. 5.32 Pictures of HPGe detectors in LSC (left) and of two of the HPGe detectors containing a marinelli in which the sample is kept during the measurement

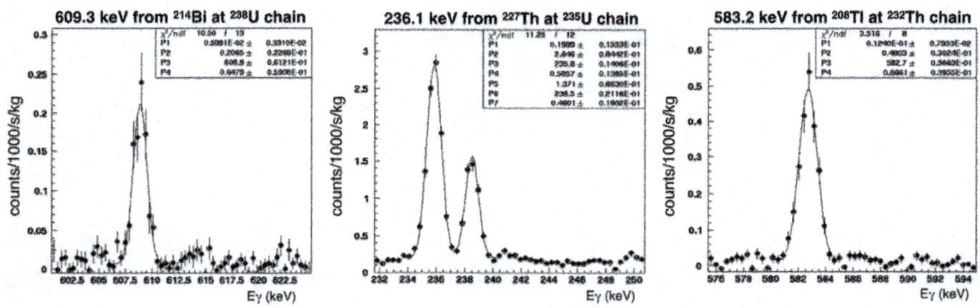

Fig. 5.33 Data from the radioactivity measurement of the first $Gd_2(SO_4)_3$ batch. Most characteristic peaks are shown for isotopes from the three decay chains

The first group of batches, in Table 5.9, correspond to the earlier days of development of the SuperK-Gd project. Some of them were used at EGADS, GSF-1008-SFM-BD was used for the non-instrumented detector and GSF-1308-SFM-1 was added once the PMTs were installed in EGADS.

In Table 5.10, a clear trend towards cleaner gadolinium sulphate is seen. The last measurements show a reduction of the radioactivity contamination of more than three orders of magnitude as compared with those from the first group.

Finally, and for the sake of simplicity in Sects. 5.6.2–5.6.4, Table 5.11 shows the typical values of the radioactive contamination for the cleanest samples in Table 5.10. This radioactivity levels will be used for the background estimation in Sects. 5.6.4 and 6.2–6.5.

Table 5.9 Radioactivity contamination measurements with HPGe at Canfranc corresponding to the first group of batches. Units are mBq per kg of $Gd_2(SO_4)_3$

Chain	^{238}U		^{232}th		^{235}U		Others		
Sub-Chain	^{238}U	^{226}Ra	^{228}Ra	^{228}th	^{235}U	^{227}Ac	^{40}K	^{138}La	^{176}Lu
GSF-0904-SFM-1	51 ± 21	8 ± 1	11 ± 2	28 ± 3	<32	214 ± 10	29 ± 5	8 ± 1	80 ± 8
GSF-1008-SFM-BD	<33	2.8 ± 0.6	270 ± 16	86 ± 5	<32	1700 ± 20	12 ± 3	<	21 ± 2
GSF-1208-BEJ-1	292 ± 67	74 ± 2	1099 ± 10	504 ± 6	<112	2956 ± 30	101 ± 10	683 ± 15	566 ± 6
GSF-1302-CHS-1	74 ± 28	13 ± 1	205 ± 6	127 ± 3	<25	1423 ± 21	60 ± 7	3 ± 1	12 ± 1
GSF-1303-BEJ-1	242 ± 60	13 ± 2	21 ± 3	374 ± 6	<25	175 ± 42	18 ± 8	42 ± 3	8 ± 2
GSF-1308-SFM-1	22 ± 20	1.3 ± 0.5	4.0 ± 0.7	117 ± 2	<2.6	231 ± 6	7 ± 3	3.5 ± 0.5	25 ± 1
GSF-1412-SFM-1	<76	<1.4	2 ± 1	29 ± 2	<1.8	190 ± 6	<5	23 ± 1	2.5 ± 0.6
GSF-1307-TAI-1	47 ± 26	5 ± 1	14 ± 2	13 ± 1	<12	<6	3 ± 2	<1	1.6 ± 0.3
GSF-1307-TAI-2	73 ± 27	6 ± 1	3 ± 1	411 ± 5	<30	<18	8 ± 4	<2	<2

Table 5.10 Radioactivity contamination measurements with HPGe at Canfranc corresponding to the second group of batches. Units are mBq per kg of $Gd_2(SO_4)_3$

Chain	^{238}U		^{232}th		^{235}U		Others		
Sub-Chain	^{238}U	^{226}Ra	^{228}Ra	^{228}th	^{235}U	^{227}Ac	^{40}K	^{138}La	^{176}Lu
GSF-1412-SFM-1	<76	<1.4	2 ± 1	29 ± 2	<1.8	190 ± 6	<5	23 ± 1	2.5 ± 0.6
GSF-1508-KJD	<37	<0.8	<1.1	2.0 ± 0.5	<0.6	11 ± 4	<3	<0.6	2.9 ± 0.2
GSF-1512-NYC-1	<139	<2.1	2.8 ± 1.9	1.8 ± 0.9	<2.4	<10	<14	<1.9	<1.6
GSF-1604-NYC-160303	<20	<0.64	<0.67	0.5 ± 0.2	<0.7	<2.3	<1.6	<0.3	<0.4
GSF-1604-NYC-160311	<59	<0.7	3.2 ± 1.0	<1.4	<1.2	<4.1	<2.7	<0.2	<0.7
GSF-1604-SHT-1	<25	<0.6	<0.7	0.9 ± 0.3	<3.1	<6.1	<2.1	<0.5	0.4 ± 0.3
GSF-1611-SHT-003	<13	<0.3	<0.3	<0.4	<0.6	<1.9	<1.8	<0.3	0.4 ± 0.1

5.6.2 ^{238}U Spontaneous Fission

Spontaneous fission (SF) is the process in which a heavy nucleus splits in various pieces, usually two with half weight of the original, plus various neutrons, [Vogt11].

Concerning the backgrounds for low energy neutrino physics, the spontaneous fission from ^{238}U is the most relevant. In this process, photons of varying energy and various neutrons may be produced. In some cases, a high energy (a few MeV) photon and a neutron are emitted, producing the very same signature in the detector as an inverse-β antineutrino reaction. Therefore, this is an irreducible background for DSNB, reactor, and supernova early warning measurements.

The number of events expected from this process is estimated for the standard radioactive contamination in Table 5.11.

For the computation of this background, the following features of this spontaneous fission are considered.

- Spontaneous fission (SF) likelihood: Only 5×10^{-7} SF occur per ^{238}U decay [Popeko80].
- Photon energy: This process may produce high energy γs of several MeV. The likelihood P at this high energy regime depends on the photon energy [Sobel73].

$$P(E) = 0.69 \cdot e^{-\frac{E(MeV)}{1.41}} \tag{5.8}$$

  Meaning that between any given energies the amount (N) of photons in that energy range is the following.

$$N(E_{min} \geq E_{\gamma}(MeV) \leq E_{max}) = e^{-\frac{E_{min}}{1.4}} - e^{-\frac{E_{max}}{1.4}} \tag{5.9}$$

  In order for the ^{238}U photons to have enough energy to be detected as electrons, they need to be 0.511 MeV more energy, in order to account for the electron mass that is added in the reconstruction software of low energy events in SK.
- Neutron multiplicity: To match the inverse-β interactions, only one neutron has to be produced, therefore, only those SFs with neutron multiplicity equal to one are relevant for this background. According to [Ethvignot05, Popeko75], the amount of SFs emitting just a neutron is 28%. There seems to have a weak dependence with the energy of the emitted γ, but it is ignored in this calculation.

Table 5.11 Standard radioactive contamination values of the gadolinium sulphate

Radioactive chain	Part of the chain	mBq/kg
^{238}U	^{238}U	50
	^{226}Ra	5
^{232}Th	^{228}Ra	10
	^{228}Th	100
^{235}U	^{235}U	32
	^{227}Ac / ^{227}Th	300

After all this is taken in consideration, it is possible to compute the amount of these decays that will occur in the SK fiducial volume (SKFV) per unit time, energy range and for the standard amount of ^{238}U (50 mBq/kg). It is done in Eq. (5.10), where the 80% neutron-tagging efficiency of Gd-doped SK at 0.2% concentration is taken into account from Sect. 5.2.

$$N_{SF}(1\gamma + 1Gd\text{-}tagged\ n) = 21.75\left(e^{-\frac{E_{min}}{1.4}} - e^{-\frac{E_{max}}{1.4}}\right)\frac{1}{day \cdot SKFV} \tag{5.10}$$

It will be used for estimating the background to the reactor antineutrinos in Sect. 6.1, to DSNB in Sect. 6.2 and to pre-supernova antineutrinos in Sect. 6.3.

5.6.3 Radioactivity-Induced Neutron Production

If the amount of neutrons produced due to the radioactive contamination is large, there will be many neutron captures with no prompt signal. These could be mistaken for neutrino signals or even saturate the detector if the contamination is too large.

To estimate this source, the number of neutrons produced by radioactivity is computed. For this, it is assumed a the radioactive contamination in Table 5.11. The computation is done using the SOURCES-4C code, [Wilson02].

The obtained spectrum of the neutrons produced by the radioactive contamination levels shown in the table, split into the different contributing processes Fig. 5.34.

The total amount of neutrons obtained is 5.084×10^{-13} neutrons/s/cm^3. The translation to neutrons/day/SKFV is calculated in Eq. (5.11).

$$N_{rad}^{neutrons} = 988.3\frac{neutrons}{day \cdot SKFV} \tag{5.11}$$

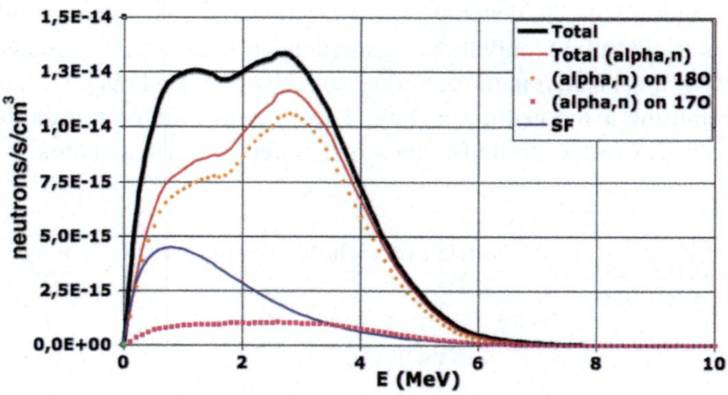

Fig. 5.34 Neutron production rates and spectrum assuming the standard radioactive contamination of gadolinium sulphate

The amount of neutrons produced by the standard radioactivity levels is sizeable value that will, certainly, have an important impact on the low energy analyses.

As it can be seen in Fig. 5.34, the main contribution to the neutron production are (α, n) reactions. However, the neutron multiplicity for this process is estimated to be two. This fact would neglect this background for low energy solar neutrinos. Going deeply into the (α, n) reactions in the detector, they are mediated by both isotopes of oxygen ^{18}O and ^{17}O. Equation (5.12) shows the neutron production mechanism of (α, n) on ^{18}O.

$$\begin{aligned} ^{18}O + \alpha(E_{kin} \approx 6\,\text{MeV}) \rightarrow^{22} Ne^{*}(E_{exc} \approx 15\,\text{MeV}) \rightarrow \\ \rightarrow^{20} Ne + n(E_{kin} \approx 2.2\,\text{MeV}) + n(E_{kin} \approx 2.5\,\text{MeV}) \end{aligned} \tag{5.12}$$

Because the multiplicity of neutron coming from SF are 72% of the times larger than one, they are considered negligible compared to the number of neutrons from (α, n) reactions. Therefore, and assuming the 80% Gd-tagging efficiency, the amount of events with a single detected neutron is reduced as shown in Eq. (5.13).

$$N_{rad}^{neutrons} = 32\% \cdot 988.3 \frac{neutrons}{day \cdot FV} = 316.3 \frac{single\ neutrons}{day \cdot SKFV} \tag{5.13}$$

Due to the spectrum of the Gd-neutron captures, in Fig. 5.35, this background may not only affect the low energy neutrino analysis, Sect. 6.5, but also, the analyses involving antineutrinos . In this case, the high amount of neutrons could produce

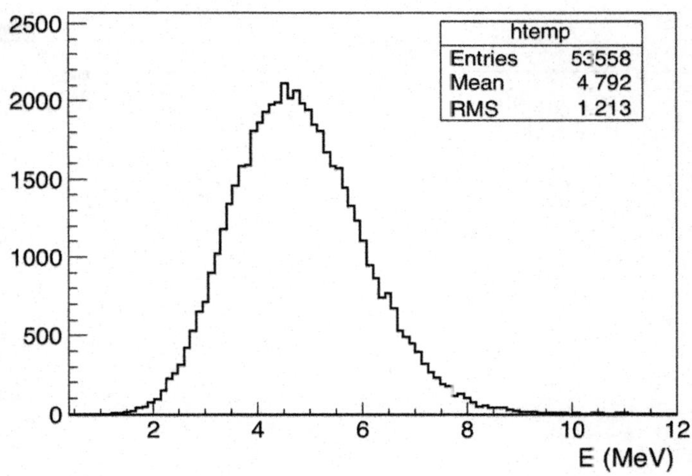

Fig. 5.35 Reconstructed energy of the 8 MeV γ cascade from the Gd-neutron capture

an accidental signal between a low energy neutrino and a capture of a neutron from radioactivity. This will have the same imprint in the detector as a low energy electron antineutrino. This background will be estimated for reactor antineutrinos, DSNB and pre-supernova antineutrinos in Sects. 6.1–6.3.

5.6.4 β-Rays from Radioactivity

Radioactivity contamination also brings with β decays as source of background for low energy neutrino physics. The main contributions are from the ^{208}Tl ($Q_\beta = 5.00$ MeV), ^{212}Bi ($Q_\beta = 2.25$ MeV) and ^{214}Bi ($Q_\beta = 3.27$ MeV) isotopes, those with higher Q_β-value of the three radioactive chains.

This source of background, as it happens in Sects. 5.6.3, may be confused with the prompt signal of a low energy neutrino or as a fake neutron-capture candidate, if it coincides with a solar neutrino candidate. Since the corresponding background estimations depend highly on the analysis, they are done specifically for each case in Sects. 6.1 to 6.3 and 6.5.

The reconstructed energy spectrum for the three decays is shown in Figs. 5.36, 5.37 and 5.38.

^{208}Tl β-decay is the most potentially harmful one due to its large Q_β-value. Hence, for the suppression of this background source it is needed to reduced the levels of ^{228}Ra and ^{228}th from the Th decay chain.

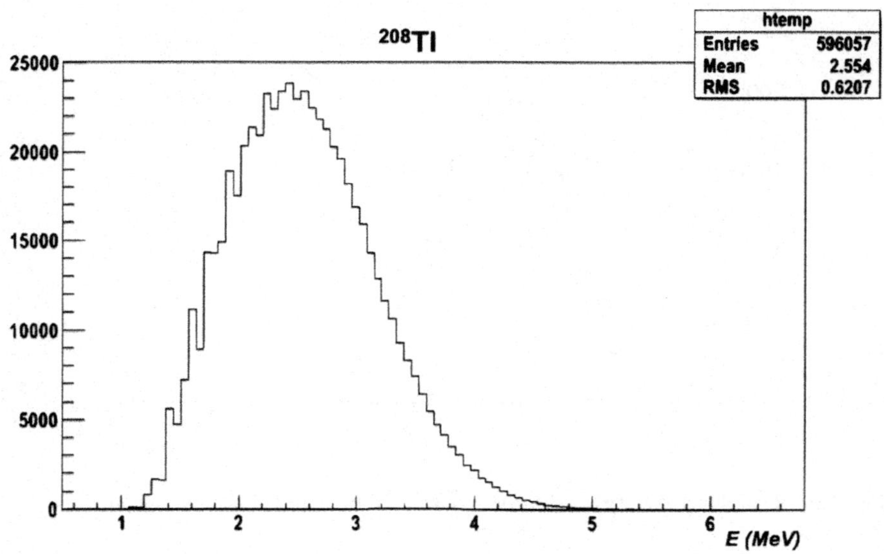

Fig. 5.36 Reconstructed energy spectrum of ^{208}Tl β-decay

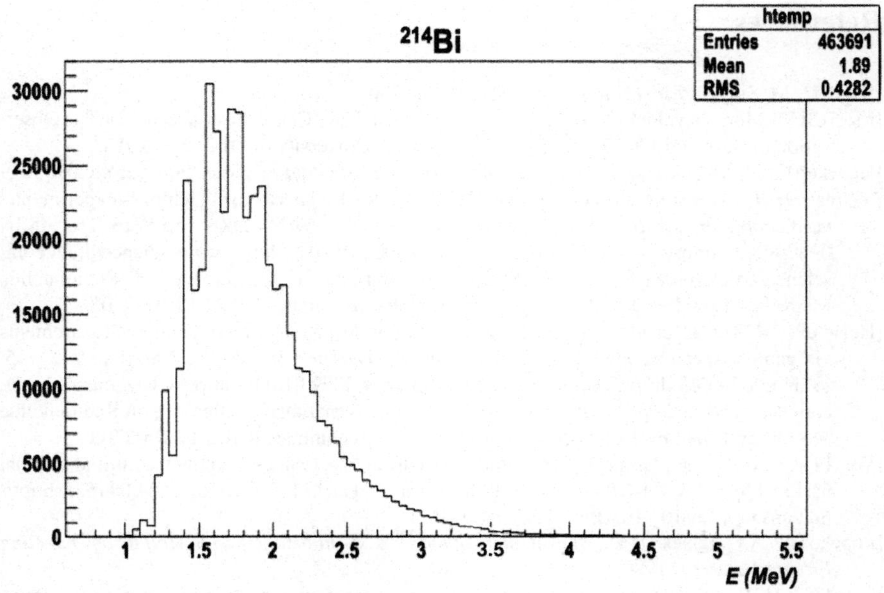

Fig. 5.37 Reconstructed energy spectrum of ^{214}Bi β-decay

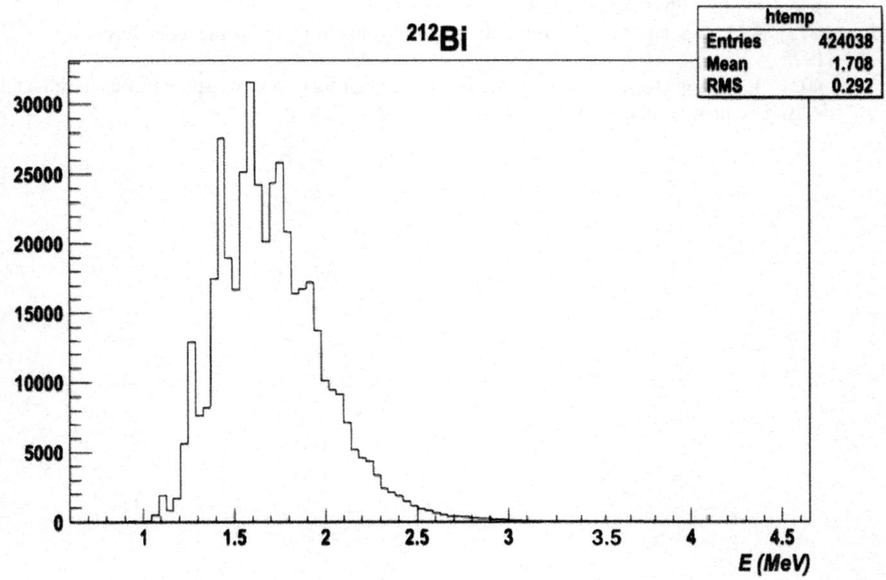

Fig. 5.38 Reconstructed energy spectrum of ^{212}Bi β-decay

References

[Smy07] M. Smy, in *Proceedings of the 30th ICRC*, 2007

[Mori15] T. Mori, Development of A Gadolinium-doped Water Cherenkov Detector for The Observation of Supernova Relic Neutrinos. Ph.D. thesis, University of Okayama, 2015

[Hoecker07] A. Hoecker et al.,*Tmva-toolkit for Multivariate data Analysis*, 2007. arXiv:0703039

[Vidmar09] T. Vidmar et al., Analysis of HPGe spectra by spectrum matching—experimental verification, in *Applied Radiation and Isotopes*, vol 67(5), 2009, pp. 716–718. ISSN 0969-8043. https://doi.org/10.1016/j.apradiso.2009.01.016. http://www.sciencedirect.com/science/article/pii/S0969804309000220. 5th International Conference on Radionuclide Metrology-Low-Level Radioactivity Measurement Techniques ICRM-LLRMT'08

[Heisel09] M. Heisel et al., Statistical analysis of low-level material screening measurements via gamma-spectroscopy, *Applied Radiation and Isotopes*, vol. 67(5), 2009, pp. 741–745. ISSN 0969-8043. https://doi.org/10.1016/j.apradiso.2009.01.028. http://www.sciencedirect.com/science/article/pii/S0969804309000281. 5th International Conference on Radionuclide Metrology-Low-Level Radioactivity Measurement Techniques ICRM-LLRMT'08

[Vogt11] R. Vogt et al., Event-by-event study of neutron observables in spontaneous and thermal fission. Phys. Rev. C **84**, 044621 (2011). https://doi.org/10.1103/PhysRevC.84.044621. http://link.aps.org/doi/10.1103/PhysRevC.84.044621

[Popeko80] A.G. Popeko et al..*Measurement of the* ^{238}U *spontaneous-fission Halflife by Detecting Prompt Neutrons*, 1980

[Sobel73] H.W. Sobel et al., High-energy gamma rays from spontaneous fission of ^{238}U. Phys. Rev. C **7**(4) (1973)

[Ethvignot05] T. Ethvignot et al., Neutron multiplicity in the fission of ^{238}U and ^{235}U with neutrons up to 200 MeV. Phys. Rev. Lett. **94**(052701) (2005)

[Popeko75] A.G. Popeko et al., Multiplicity of prompt neutrons in spontaneous fission of ^{238}U (1975)

[Wilson02] W. Wilson et al., Sources 4C: a code for calculating (alpha, n), spontaneous fission and delayed neutron sources and spectra. Am. Nucl. Soc. (2002)

Chapter 6
SuperK-Gd Physics Potential

In this section, all the measurements and potential searches that SuperK-Gd would be able to performed, are addressed. This section takes into account all the advantages of the 80% efficiency neutron tagging technique using gadolinium, but also the drawbacks that the inclusion of radioactive contamination could have in these measurements. For the latter the typical values in Tables 5.9 and 5.10 are taken in order to make the detectability predictions.

6.1 Reactor Neutrinos

The detection of reactor neutrinos is, after the DSNB and pre-supernova neutrino searches, the most important measurements of SuperK-Gd. Nuclear reactors produce a huge amount of antineutrinos due to their nuclear reactions.

Currently, in SK these antineutrinos are masked by solar neutrinos and spallation products, but enabling an efficient neutron tagging technique, these low energy antineutrinos can be identified through their inverse-β decay.

Since the nuclear program of Japan is moving towards the reactivation of most of its nuclear power plants, in this section, the full operation of all reactors will be assumed. Japanese reactors are the main producers of reactor antineutrinos arriving to SK, but their not the only ones. South Korean reactors are close enough and ought to be taken into account. The relation of nuclear plants considered with their operating power and distance to the Kamioka Observatory is shown in Table 6.1 and extracted from [Murayama02].

The electron antineutrino oscillation probabilities can be assumed to be through vacuum and can be approximated by the following expression.

P. Fernández Menéndez, *Neutrino Physics in Present and Future Kamioka Water-Čerenkov Detectors with Neutron Tagging*, Springer Theses,
https://doi.org/10.1007/978-3-319-95086-0_6

$$P(\overline{\nu}_e \rightarrow \overline{\nu}_e) = 1 - \sin^2 2\theta_{12} \sin^2 \left(1.27 \Delta m^2_{12} \frac{L_{reactor}}{E_\nu}\right) \tag{6.1}$$

The expected spectrum is shown in Fig. 6.1, from which it can be seen the oscillation peaks, assuming 3 MeV energy threshold, the most important one is that around 5 MeV and the valley from the disappearance regime at 3 MeV.

The expected rate of reactor antineutrinos detected by SuperK-Gd is of the order of 30 events/day in the 22.5 kton fiducial volume. Several studies have been made on the impact of this new measurement that SuperK-Gd would bring along, with high sensitivity to the solar oscillation parameters. The main backgrounds for this measurement are products from spallation due to cosmic muons, which some of them leave the same signature as a low energy inverse β decay, such as carbon isotopes or ^{16}N, [Li14]. As for the backgrounds introduced by the radioactive contamination of the gadolinium sulphate, the signals mimicking the inverse-β reaction are the ^{238}U spontaneous fission and accidental coincidences of solar neutrino candidates with radioactivity-induced neutrons and β-decays. Assuming the standard radioactive contamination levels at Table 5.11, the amount of events from ^{238}U SF are computed following Eq. (5.10), with $E_{min} = 3$ MeV (SK low energy threshold) and $E_{max} = 10$ MeV.

$$N_{SF}(1\gamma + 1Gd\text{-}tagged\ n) = 2.6 \frac{1}{day \times SKFV} \tag{6.2}$$

Table 6.1 List of nuclear reactor power plants from which SK is able to detect antineutrinos

Reactor	Power (GW)	Distance (km)
Kashiwazaki	24.6	160
Ohi	13.7	180
Takahama	10.2	191
Hamaoka	10.6	213
Tsuruga	4.5	139
Mihama	4.9	145
Fukushima 1	14.2	344
Fukushima 2	13.2	344
Tokai	3.3	295
Shimane	3.8	414
Ikata	6.0	561
Genkai	6.7	755
Onagawa	4.1	430
Tomari	3.3	784
Sendai	5.3	824
Sendai	1.6	81

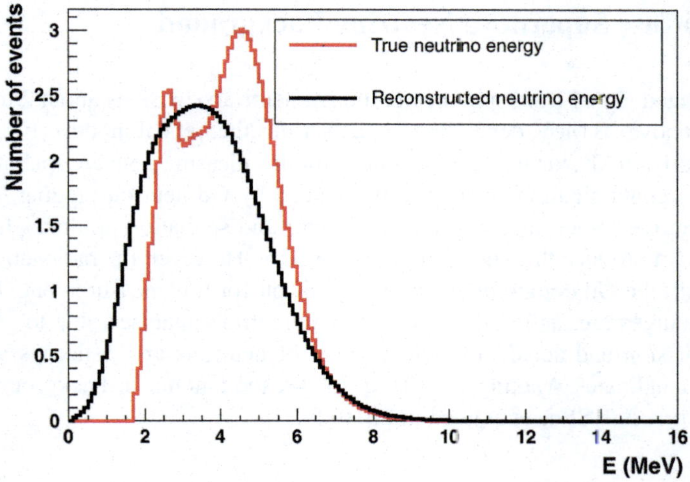

Fig. 6.1 Reactor neutrino spectrum at Kamioka for true (red) and reconstructed (black) antineutrino energy

On the other hand, the background estimation from the two possible accidental coincidences, with neutrons and with β-decays, is calculated in Eq. (6.4) and Sect. 5.2.2 respectively. In these computations, it is considered that the amount of solar neutrinos in the final sample is around 200 events/day/SKFV, according to [Abe16c], and that every event is 535 μs long (SHE+AFT, seen in Table 3.4). For the accidental coincidences with neutrons coming from radioactivity Eq. (5.13) is considered.

$$N^{neutron}_{accidental} = 8.0 \times 10^{-6} \frac{1}{day \times SKFV} \tag{6.3}$$

The amount of β-decays passing the selection methods for the Gd-neutron tagging explained in Sect. 5.2.2, are 20.4% for ^{208}Tl, 1.7% for ^{214}Bi and 0.2% for ^{212}Bi. Assuming the standard radioactivity levels, the final amount of accidental coincidences is computed, analogously to Eq. (6.4), in Sect. 5.2.2.

$$N^{\beta}_{accidental} = 9.2 \times 10^{-2} \frac{1}{day \times SKFV} \tag{6.4}$$

The largest contaminations come from the ^{238}U spontaneous fission. However, two important items have to be taken into account. The first one is that the resin AJ4400, already tested at EGADS, is able to reduce 100 times the amount of uranium present in the Gd salt, becoming the amount of this background also reduced by a factor 100. Furthermore, the radioactivity contamination of the latest Gd batches, in Table 5.10, is two to three orders of magnitude smaller. This means that, for these Gd batches, the background from radioactive contamination will be reduced by the same factor, being negligible compared to the reactor neutrino signal.

6.2 Diffuse Supernova Neutrino Background

The expected signal from supernova relic antineutrinos in SK is about 2 to 5 events per year above 10 MeV, depending on the various theoretical models (Fig. 6.2).

Currently, in SK, the main backgrounds for this measurement are spallation products. This contamination is drastically reduced by Gd-neutron tagging, as DSNB antineutrinos interact inverse-β in the detector and spallation products, beyond 10 MeV, do not produce this signature in the detector. However, the radioactivity introduced with the Gd compound may be significant for this measurement. The background sources are, as in all low energy antineutrino analyses, due to ^{238}U spontaneous fission and accidental coincidences of neutrons and β-decays with solar neutrino candidates. Making use of Eq. (5.10), the amount of background signals from the uranium SF is estimated in Eq. (6.5).

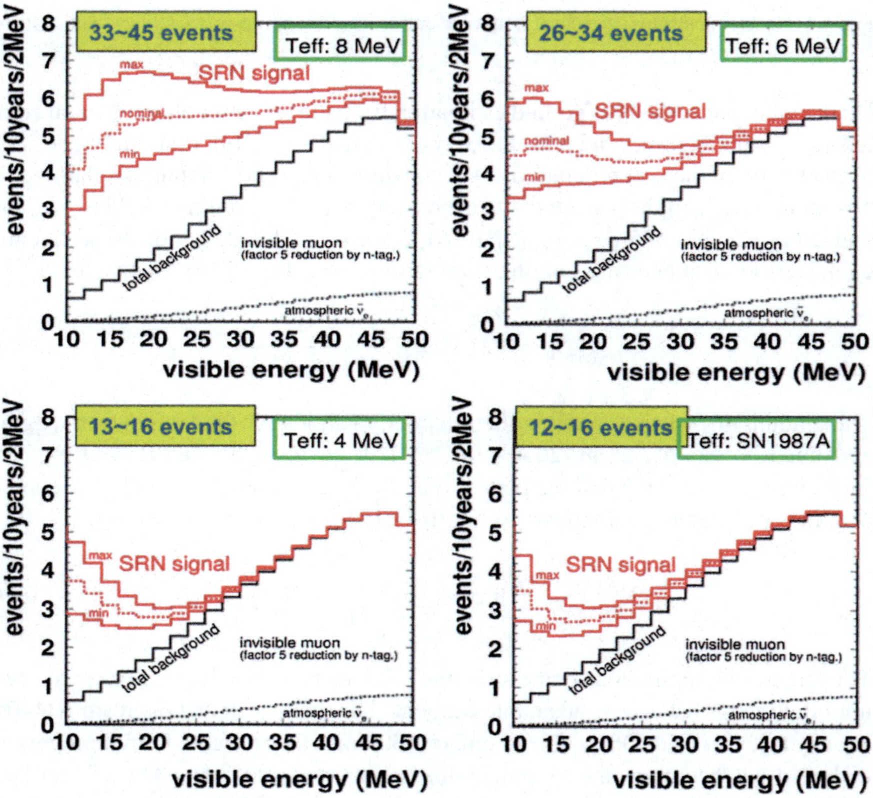

Fig. 6.2 Expected number of events and energy spectrum for DSNB assuming different effective neutrino temperatures (T_{eff}) for 10 years of SuperK-Gd data taking

$$N_{SF}(1\gamma + 1Gd\text{-}tagged\ n) = 4.4\frac{1}{year \times SKFV} \tag{6.5}$$

For the estimate of the amount of accidental coincidences, it assumed that ~4 events will pass the solar cuts per day and above 10 MeV, [Abe16c]. With the considerations, for the neutrons and βs from radioactivity, made in Sect. 6.1, the background signals are calculated in Eqs. (6.6) and (6.7).

$$N^{neutron}_{accidental} = 2.9 \times 10^{-3}\frac{1}{year \times SKFV} \tag{6.6}$$

$$N^{\beta}_{accidental} = 0.3\frac{1}{year \times SKFV} \tag{6.7}$$

For the DSNB measurement, as in the reactor neutrino case, the principal background comes from the ^{238}U SF. But again, the AJ440 resin performance for removing uranium and the lower radioactive contamination of the latest Gd batches, suggest that this background source will become, at least, reduced by a factor 100. In addition, the much lower radioactive contamination from the last Gd batches measured, indicate that the background from ^{238}U SF will be at least two orders of magnitude smaller than the estimate in Eq. (6.5), using the standard radioactive contamination tables. In this sense, the DSNB measurement is expected to have negligible background, crucial for the future of SuperK-Gd.

6.3 Silicon Burning Stage

This section deals with the detectability of pre-supernova neutrinos. During silicon burning phase, the luminosity is about eight orders of magnitude less than the peak of core-collapse supernovae, but while this peak lasts just a few milliseconds, the Si burning phase takes a few days. This measurement would be of enormous importance, because it helps understanding of the stellar evolution mechanisms at this stage, few moments prior to the core-collapse. In addition, it can be used as early warning for optical observatories to the upcoming supernova.

Given the low energy spectrum of the neutrinos from this origin, a tiny fraction of the antineutrinos could be detected through their inverse-β reaction in SK. According to [Odrzywolek07], the number of events above the inverse-β threshold that could be detected by SuperK-Gd is 796 assuming a very close supernova, such as Betelgeuse at 0.2 kpc away. For this study, a 3 MeV kinetic energy threshold for the positron is considered, which would lead to ~187 events in the 24 h previous to the core-collapse. Assuming the same efficiency of 11% used in the solar analysis for this low energy positron and 80% efficiency in the neutron tagging, the amount of events detected would be 16.4.

In this case, the main source of background are reactor antineutrinos. Backgrounds due to Gd radioactive contamination completely negligible. The background due to the ^{238}U SF is estimated in Eq. (6.8).

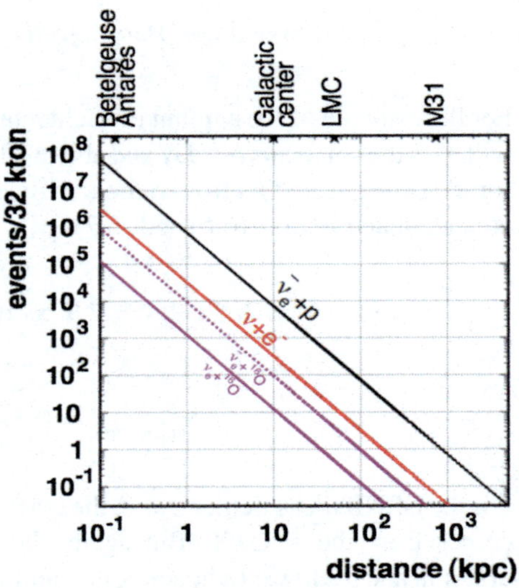

Fig. 6.3 Supernova neutrino and antineutrino total fluxes as function of the distance to the collapsing star

$$N_{SF}(1\gamma + 1Gd\text{-}tagged\ n) = 0.15 \frac{1}{day \times SKFV} \tag{6.8}$$

For the case of the accidental coincidence backgrounds, it is enough refer to the corresponding calculations made for the reactor antineutrinos in Eqs. (6.4) and (6.4).

The significance for the pre-supernova stage measurement, assuming a star 0.2 kpc away, would be of $16.4/\sqrt{16.4+30} = 2.4\sigma$. The actual significance for particular case of Betelgeuse may vary because of the uncertainty in its distance to the Earth.

6.4 Supernova Burst

If a supernova happens within the Milky Way, it will produce an enormous amount of neutrinos and antineutrinos that will be detected by SK. Although, electron neutrinos can be distinguished from electron antineutrinos via their scattering angle differences, in SuperK-Gd this separation will be largely improved due to the Gd-neutron tagging.

These neutrinos and antineutrinos will have energies ranging from 10 up to 30 MeV and the signal will last for a few seconds, [Totani98b]. This is shown in Fig. 6.4.

This measurement, will provide much information about early stages of the core-collapse process, its spectrum and time profile, yielding to more detailed picture of the whole core-collapse process, extracting the neutrino and antineutrino fluxes independently.

The backgrounds for this measurement are completely negligible given the huge signal in such narrow time window, shown in Fig. 6.3.

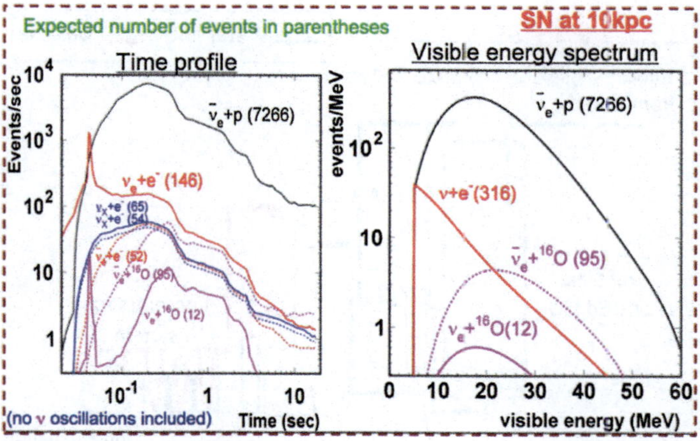

Fig. 6.4 Supernova neutrino and antineutrino fluxes as function of time (left) and energy (right) for a supernova occurring 2 kpc away from the Earth

6.5 Solar Neutrinos

In order to maintain the current low energy solar analysis of SK, [Abe16c], it is crucial to keep the low energy radioactive backgrounds as low as possible. The new backgrounds for this analysis that SuperK-Gd will need to deal with, are those coming from the Gd radioactive contamination, namely, the signals from β-decays and from the Gd-capture of neutrons.

According to [Abe16c], the amount of solar events per day in SK fiducial volume is around 200. This value will be used to quantify the size of the backgrounds next. All the SK solar selection criteria are run over the Gd-neutron capture and ^{208}Tl, ^{214}Bi and ^{212}Bi spectra. The remaining fraction of this events are 40, 0.19, 0.01 and $<2 \times 10^4\%$.

With this, the total amount of background signals capable of contaminating the solar neutrino sample, due to neutron production from radioactivity, is calculated in Eq. (6.9). This estimate is based on Eq. (5.13) and for the standard radioactivity levels.

$$N^{neutron} = 126.5 \frac{1}{day \times SKFV} \tag{6.9}$$

Based on the survival fraction of β-decays to the solar cuts and the standard radioactive contamination of Gd, the amount of this contamination is done in Eq. (6.10).

$$N^{\beta} = 2.6 \times 10^5 \frac{1}{day \times SKFV} \tag{6.10}$$

Both background contributions are huge, being that related to the β-decays around three orders of magnitude larger than the expected ~200 solar neutrino candidates

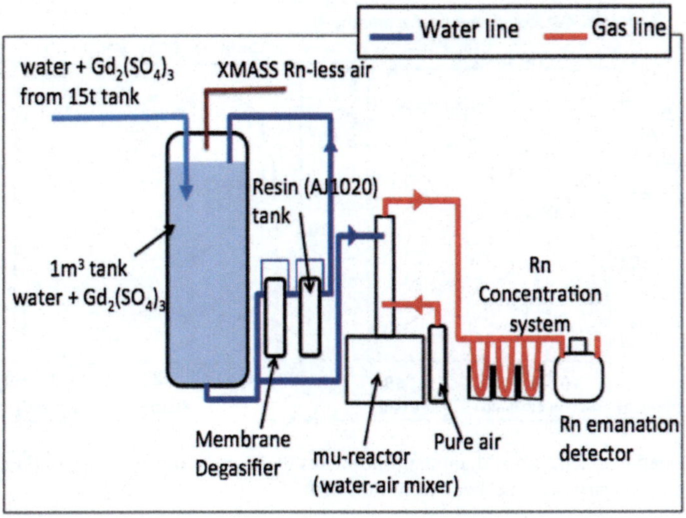

Fig. 6.5 Scheme of the radium-removal system developed at EGADS

per day in SK. This contribution is mainly due to the large Q_β-value of the ^{208}Tl β-decay, from the ^{232}Th decay chain.

Nevertheless, the size of this background is drastically reduced by the cleanness if the radioactivity levels from Table 5.10 are considered. These samples show a notable reduction of the contamination levels from the ^{232}Th decay chain. This fact, basically lowers each of the radioactive background from β-decays a factor 100 or 1000, depending on the gadolinium sulphate batch, as compared with the estimates for the standard radioactivity levels shown in Eq. (6.10).

Furthermore, a system for reducing this background is being developed at EGADS. The system consists in making the dissolved gadolinium passing through a specially designed resin (AJ1020Gd), which removes radium keeping the $Gd_2(SO_4)_3$ concentration. ^{228}Ra is one of the longest-lived parent isotopes of ^{208}Tl and, therefore, the origin of the origin of its β-decays. In Fig. 6.5, a diagram of the Ra-removal system setup at EGADS is shown.

In addition to ^{228}Ra, ^{228}Th is the other longest-lived parent isotope of the ^{208}Tl, therefore, it is crucial to remove thorium from the Gd compound too. Several strategies to achieve this are under investigation, such as a special kind of filtrating paper, that can remove up to 98% of the thorium, and a PH shock to the Gd solution to make the thorium precipitate so it could be easily removed. However, the case of the ^{228}Th is not so severe because its half-life is 1.9 years. This means that once ^{228}Ra is removed, the radioactivity background will be reduced by half after 2 years.

In conclusion, although this is a serious remaining issue, the high purity of the last Gd batches and the ongoing efforts in the development of new methods and techniques to reduce it, suggest that the low energy solar neutrino analysis will also be able to be done in SuperK-Gd.

6.6 Atmospheric Neutrinos

In this section, the atmospheric neutrino analysis is developed for including Gd-neutron tagging. In Sect. 6.6.1.2, the basic features of the atmospheric analysis in SK are reviewed. These are the simulation of atmospheric neutrinos in Sect. 6.6.1.2 and the event reduction and reconstruction in Sects. 6.6.1.2 and 6.6.1.3. These will be shared by the future SuperK-Gd.

In Sects. 6.6.2–6.6.4 the CC-NC and ν-$\overline{\nu}$ separations, and the neutron-corrected neutrino energy are implemented for the atmospheric analysis in SuperK-Gd. Lastly, in Sect. 6.6.5, a global atmospheric oscillation sensitivity analysis is done, including the previous tools and comparing the obtained sensitivities to CP violating phase, neutrino mass hierarchy and θ_{23} to the current SK standard analysis.

6.6.1 Monte Carlo Simulation, Event Classification and Reconstruction in Super-Kamiokande

Generating a prediction for the number of atmospheric neutrino interactions observed in Super-Kamiokande is not a simple task. Results depend heavily on detector knowledge, flux and particle interaction theoretical models, and reconstruction within the detector. The whole process, from production, in the atmosphere, until detector signal, is simulated and the expected output event rate distributions as function of direction and energy.

The Monte Carlo simulation is divided in two stages, firstly the initial neutrino flux and interaction in water is done, and the outgoing particles effects on the detector are simulated based on the calibration data. Of these two, the one which carries the largest uncertainties is the atmospheric flux modelling, since there are not many input data helping to constrain their predictions.

Concerning the data management, events pass through a series of cuts in order to reduce as much as possible the background. Once these selection criteria are applied, the number of rings, the particle identification, direction, vertex position and momentum are computed for the remaining events. In this section, the SK-IV event reduction and reconstruction are briefly reviewed. The reader may want to refer to [Pik12, Irvine14], where this is extensively described.

6.6.1.1 Atmospheric Neutrino Simulation

Atmospheric Neutrino Flux Simulation

There are several models for the atmospheric neutrinos flux model, for this analysis is based on the HKKM flux of 2011, overlapping with the Volkava flux for events above 10 TeV.

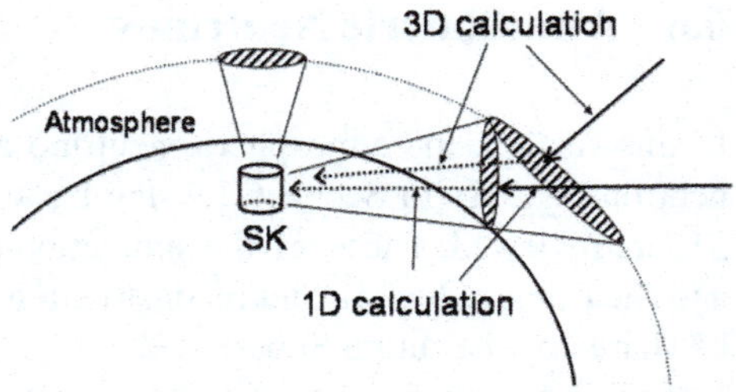

Fig. 6.6 Three-dimensional calculation of the neutrino flux in Honda flux

The atmospheric neutrino flux is a convolution of the spectrum at the highest altitudes in atmosphere which yields to neutrino production. By using a 3-dimensional tracking for the cosmic rays tracking, then secondary particles are produced (Fig. 6.6).

The neutrino flux simulation also considers the geomagnetic effects and solar activity modulation. The models are done based on high altitude experiments, such as AMS (Alpha Magnetic Spectrometer), and balloon-borne experiments. Most of the incoming cosmic rays are hydrogen ions interacting with the higher layers of the atmosphere, and this are deviated populating the flux towards the west direction (Fig. 6.7).

In addition, geomagnetic effect are also considered, which being slightly higher at Kamioka area than the average produces an asymmetry between upward and downward directions for low energy cosmic rays, as seen in Fig. 6.8.

The second part of the simulation, deals with the modelling of hadronic showers induced by the cosmic rays. Two models are used here depending on the energy region, for energies below 32 GeV the JAM model is used based on the HARP experiment data, and for energies higher than 32 GeV the DPMJET-III model is used (Fig. 6.9).

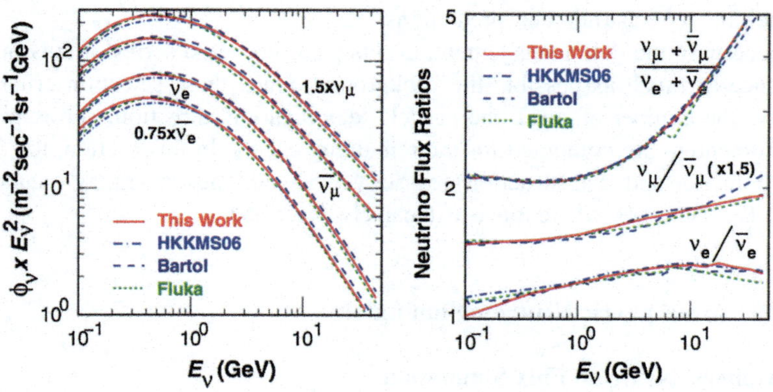

Fig. 6.7 Atmospheric neutrino fluxes (left) averaged over all directions and neutrino flux ratios (right) for different models, Bartol flux (dashed dark blue), Fluka (dotted green), HKK06 flux (dashed light blue) and HKK11 (solid red)

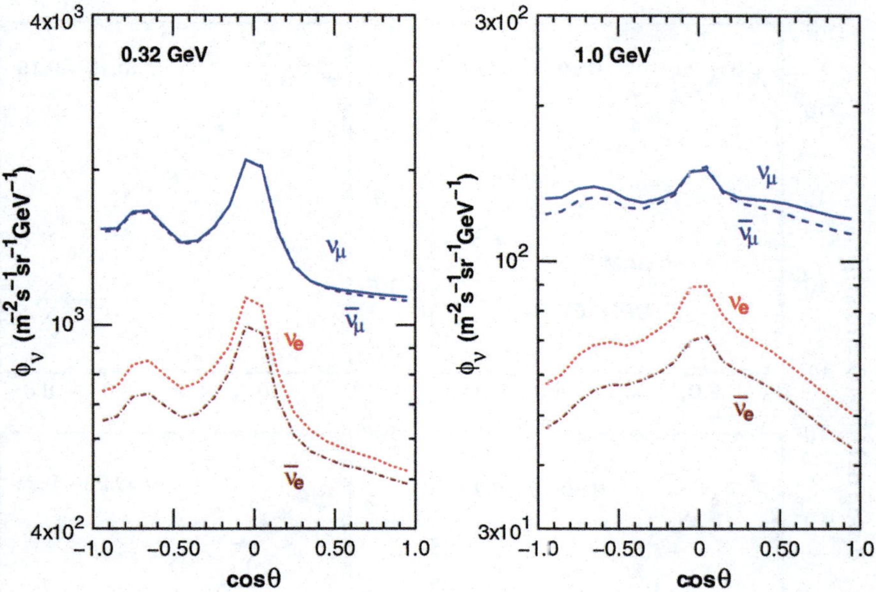

Fig. 6.8 Zenith angle distribution for neutrino flux incident on SK for two different energies and for all neutrino species

Neutrino Interaction Simulation

Neutrinos interact with nucleons and electrons in the water and the surrounding rock of the SK detector. For this simulation, the NEUT libraries are used along with other models, like NUANCE, for cross checks. The interactions simulated are listed on Tables 5.7 and 5.8.

The simulation of quasi-elastic scattering (QE) by NEUT is done differently for free and bound nucleons, corresponding to the hydrogen and oxygen nuclei respectively. For bound nucleons, the whole nucleus is considered, approximating it to a relativistic Fermi gas and taking into account the Pauli blocking effect for momenta larger than the Fermi momentum, $p_F = 225$ MeV/c (Fig. 6.10).

In single meson and coherent π production interactions, the simulation is based on Rein and Sehgal's model (Fig. 6.11).

In addition to resonance production, neutrinos can also coherently produce single pion final states. In this case, the neutrino coherently scatters from the entire nucleus, transferring negligible energy to the target, which produce no nuclear recoil and a distinctly forward-scattered pion, compared to their resonance-mediated counterparts (Fig. 6.12).

Neutrino deep inelastic scattering has long been used to validate the Standard Model and probe nucleon structure. Over the years, experiments have measured cross sections, electroweak parameters, coupling constants, nucleon structure functions, and scaling variables using such processes. In deep inelastic scattering, the neutrino scatters off a quark in the nucleon via the exchange of a virtual W or Z boson

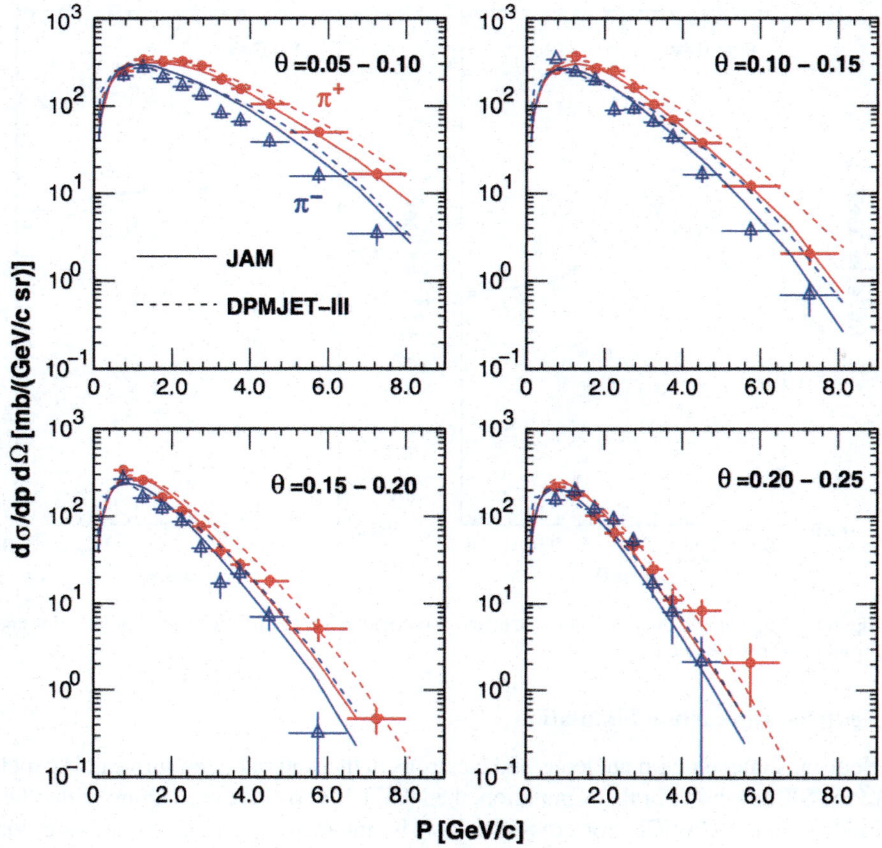

Fig. 6.9 Comparison of interaction models JAM (solid lines) and DPMJET-III (dashed lines) and data from HARP (points), for π^+ in red and π^- in blue

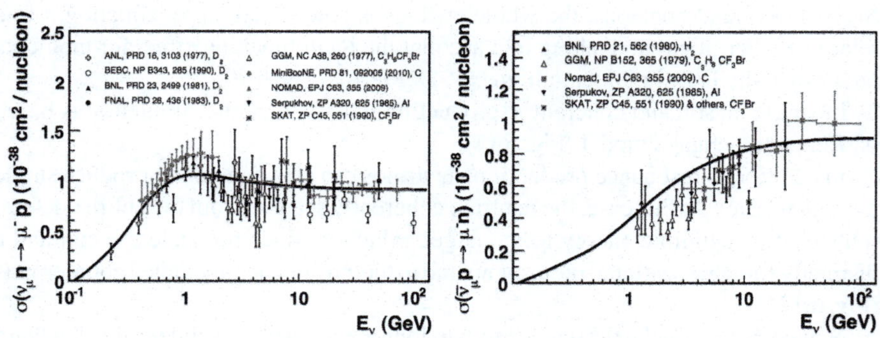

Fig. 6.10 Muon neutrino (left) and antineutrino (right) charged current quasi-elastic cross sections compared to experimental data from various experiments

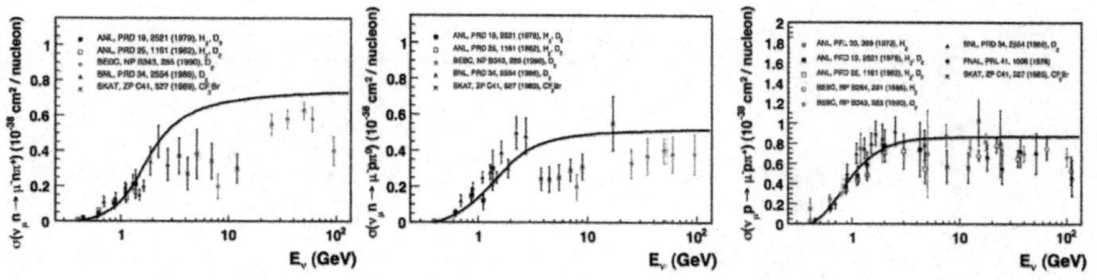

Fig. 6.11 Muon neutrino charged current single pion production cross section for π^- (left), π^0 (centre) and π^+ (right) compared to experimental data from various experiments

Fig. 6.12 Simulation (solid line) and measurements of coherent pion production cross sections from a variety of nuclear targets and for both NC and CC

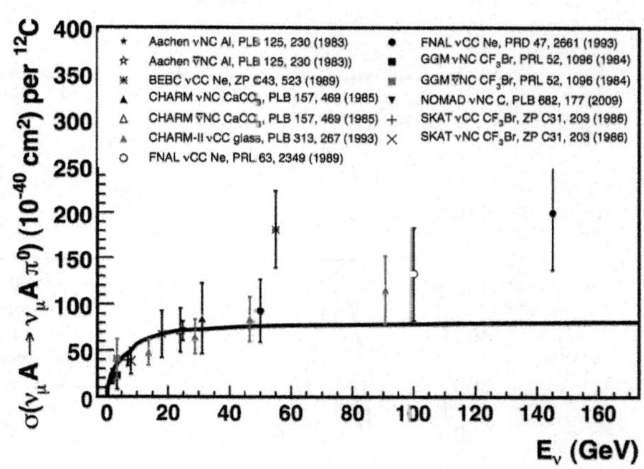

producing a lepton and a hadronic system in the final state. The neutrino cross section agrees well with the simulation for the total cross section where the DIS interactions dominate as seen in Fig. 6.13 for high energies, $\sim 10^2$ GeV.

Finally, nuclear effects are also considered in the simulation. They are crucial for secondary interactions, which play a very important role in the analysis presented in this section. These secondary interactions occur inside the nuclear media of the oxygen nuclei. This is simulated for π, κ and η interacting within the nucleus using a cascade model.

For interactions with pions of low momentum pions (less than 500 MeV/c), the Salcedo model is used after tuning using experimental data. For momentums larger than 500 MeV/c, nucleons are treated as free particles, being able to used data from π-^{12}C scattering. Similar procedure is done for other mesons, as kaons and etas.

Detector Simulation

After the neutrino interaction simulation, the produced particles are introduced into the SK detector simulation, SKDetSim. This simulation englobes the particle tracks, the Čerenkov photon generation and the PMT response to these photons (Fig. 6.14).

This simulation is based on CERN's GEANT3 package. Photons, electrons and muons are simulated with the default tools of GEANT3 and the hadron interactions by using the FLUKA (energies larger than 10 GeV), HETC (energies smaller than 10

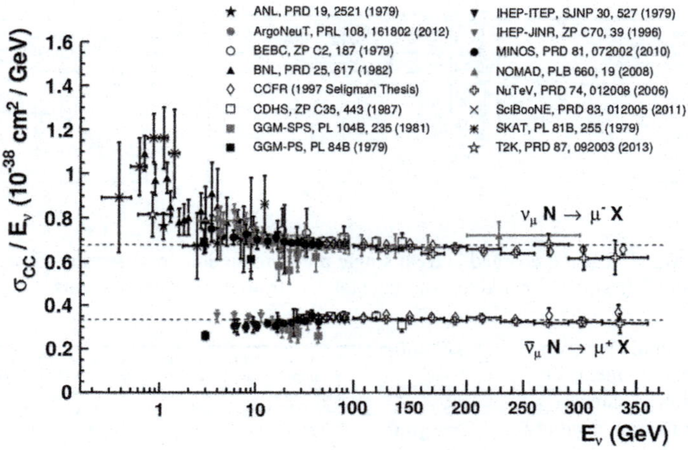

Fig. 6.13 Simulation (solid line) and measurements for inclusive total neutrino cross section of neutrinos and antineutrinos

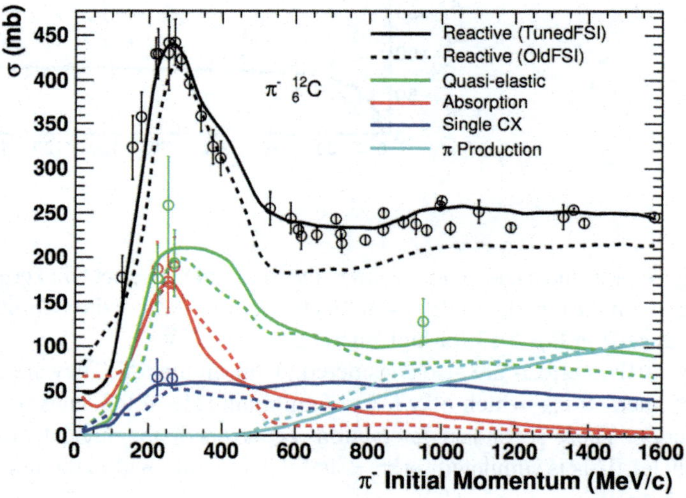

Fig. 6.14 Cross section of π^--^{12}C scattering with respect to the π^- momentum. Solid lines are the NEUT calculation before tuning, dashed lines are NEUT calculation after tuning, compared to the experimental points

GeV), MICAP (for neutrons) and NEUT cascade model (for low momentum pions). The particles produce Čerenkov photons, whose wavelength and group velocity is computed for each case depending on the refractive index and absorption and scattering properties of the SK water through which they travel.

Once this radiation gets to the PMT phototubes, its charge and timing is considered as well as the calibration data, such as gain of PMTs and quantum efficiency.

6.6.1.2 Atmospheric Data Reduction

There are specifically selection criteria for atmospheric neutrinos. SK's high energy trigger fires a little more than hundred times a day, but most of these events correspond to cosmic muons, high energy spallation events or PMT flashers. Therefore it becomes necessary to pass them through several reduction steps to reduce non-neutrino events.

There are different reduction procedures depending for each of the three event topologies, that is Fully Contained (FC), Partially Contained (PC) and Upward-going Muons (UpMu) (Fig. 6.15).

FC events correspond to those neutrinos which interact in the ID and the outgoing particles do not travel beyond the ID wall, being the produced photons captured by the detector's PMTs. Furthermore, FC events are classified into SubGeV, $E_{vis} < 1.33$ GeV, MultiGeV $E_{vis} \geq 1.33$ GeV and MultiRing, if more than one ring is reconstructed. Each of these samples is divided into e-like and μ-like, depending on the pattern of its most energetic ring. Additionally, for SubGeV events there are two more samples, SingleRing π^0 and π0-like, which contain events with the signature of the π^0 decay into two photons.

PC consists in those events in which the neutrino vertex is inside the ID but some of the produced particles in the interaction have enough energy to cross the ID walls and leave a signal in the OD. These are mainly high energy muon events since electrons transfer their energy very quickly to the surrounding water. Finally, UpMu events are those muons which enter the detector from below, indicating that the come from a muon neutrino which has interacted with the rock beneath the detector. These events are only upward-going because downward-going events of this kind could not be distinguished from cosmic muons. Furthermore, UpMu events are split into stopping-muons, meaning the incoming μ deposits all its energy in the ID, and through-going muons, which travel across the whole detector, these have the largest energy of all events SK is able to detect (Fig. 6.16).

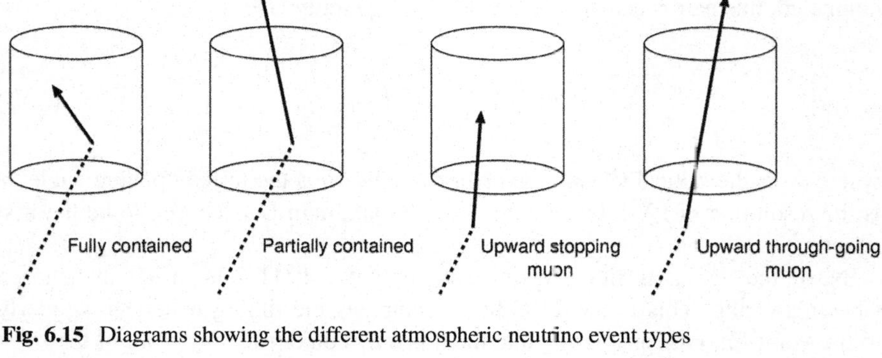

Fig. 6.15 Diagrams showing the different atmospheric neutrino event types

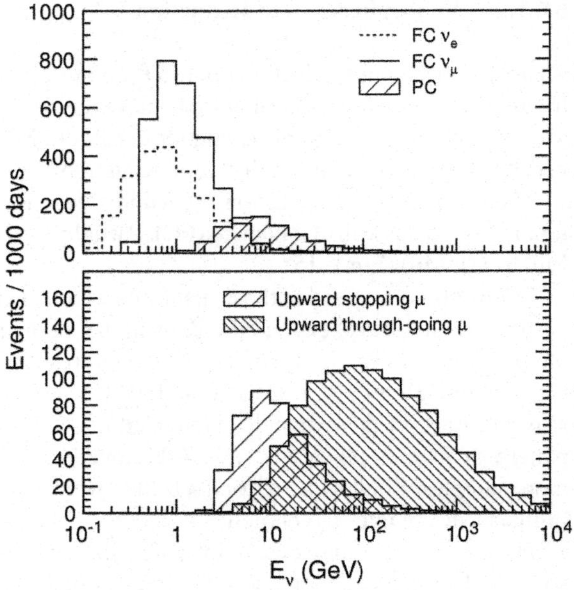

Fig. 6.16 Reconstructed energy spectrum for each sample after reduction

6.6.1.3 Atmospheric Event Reconstruction

Vertex and Direction Reconstruction

This reconstruction uses the timing information as in the case of low energy, but this time in three steps. The first step is a simple fit called Point-Fit, where the vertex is estimated by assuming that all photons are emitted from a point-like source. After subtracting the ToF, a distribution of the residual PMT hit time is constructed and the reconstructed vertex is that which maximises distribution.

In the second step, the direction and the outer edge of the dominant ring are computed, this pair is tested by calculating the quantity $G(\theta_t)$.

$$G(\theta_t) = \frac{e^{-\frac{(\theta_t-\theta_C)^2}{2\sigma^2}}}{\sin\theta_t}\int_0^{\theta_t} q(\theta)\mathrm{d}\theta \tag{6.11}$$

where θ_C is the critical Cherenkov opening angle, θ_t is the tested opening angle, σ is the resolution of PMT hits around θ_t. The maximum G is chosen to be the first ring fit.

Next, the vertex is fitted again using only the PMT hits inside the chosen Cherenkov ring. This is the TDC-fit. It computes the timing residuals, similarly to the Point-Fit, but considering that the hits may come from any point of the track (Figs 6.17 and 6.18).

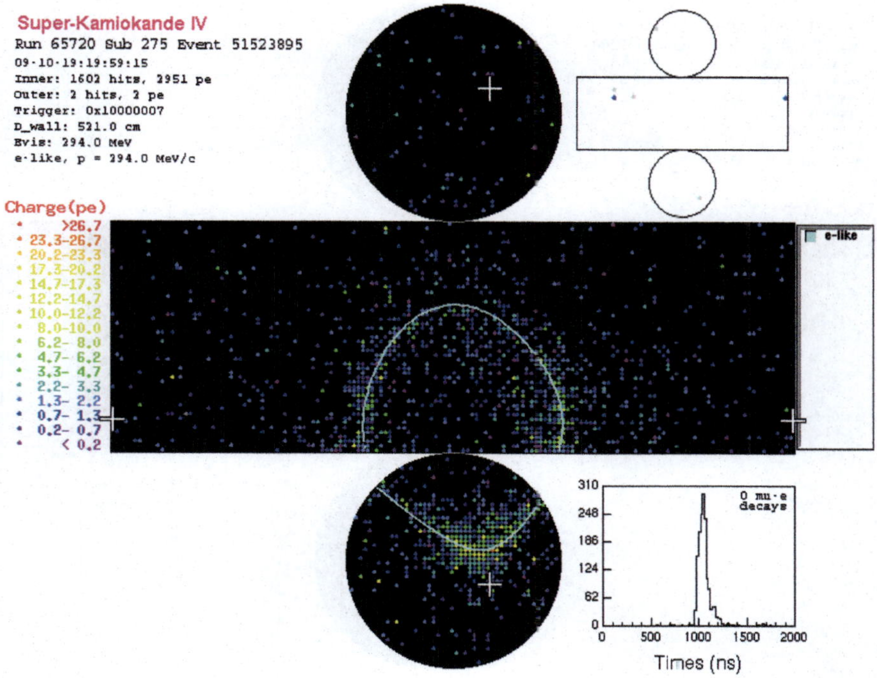

Fig. 6.17 Unfolded SK detector showing a reconstructed e-like ring

Ring Counting

Once the initial Čerenkov ring and the vertex are determined, other ring candidates are searched via a Hough transform method, technique used to extract a certain shape from an image, and illustrated in Fig. 6.19.

A likelihood distribution is used to discern if the secondary ring candidate is consistent with an actual ring. This process is repeated up to a ring multiplicity of five (Figs. 6.20 and 6.21).

For those events with just one ring the vertex fit is further optimised with the expected light pattern, given that the ring is either e-like or μ-like, this is the MS-Fit.

Particle ID Reconstruction

The fitted Čerenkov rings are the classified as e-like or μ-like by the Particle Identification (PID) likelihood. Electrons and γ rays produce diffuse ring patterns due to the electromagnetic showers, whereas muons and charged pions produce rings with much sharper edges, as previously seen in Figs. 6.17 and 6.18. In addition, the latter ones have may have smaller Čerenkov angles if they do not achieve the ultra-relativistic regime. These differences are taken into account by the PID.

The e-like or μ-like categorisation is done by comparing the observed charge pattern with the expected cases for electrons and muons and obtaining for which the maximum likelihood is achieved. The expected charge for a given PMT assuming the ring is produced by an electron has the following form.

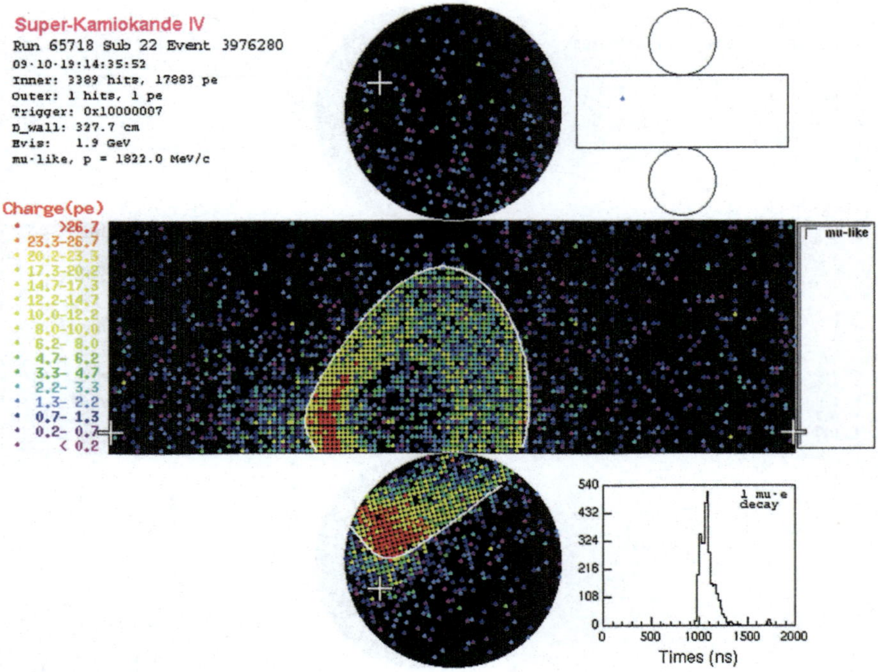

Fig. 6.18 Unfolded SK detector showing a reconstructed μ-like ring

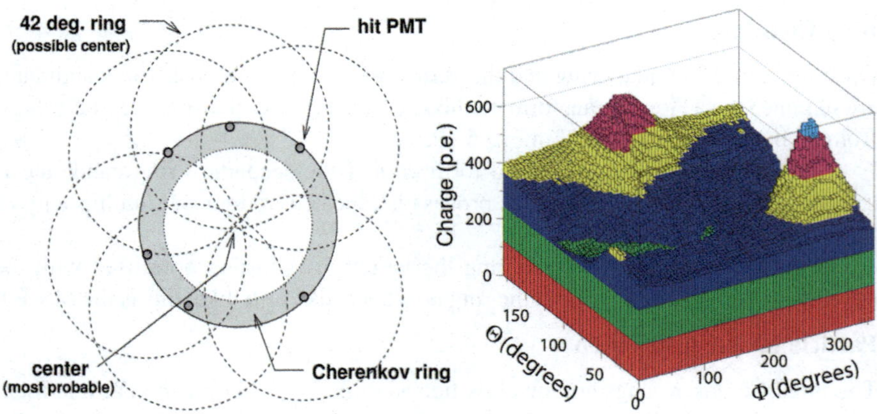

Fig. 6.19 Two dimensional drawing of ring candidate searching based on the PMTs with maximum charge in the left, and in the right, charge distribution as function of zenith and azimuthal angles in the detector, being those peaks with higher charges also the most likely rings

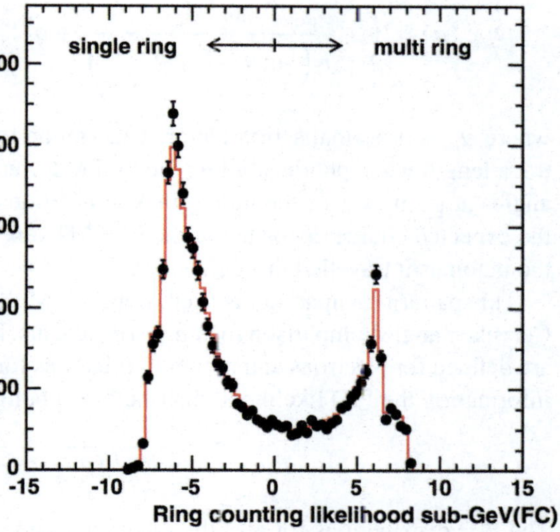

Fig. 6.20 Ring counting likelihood for the SubGeV events for Monte Carlo (red) and SK-IV data (black dots)

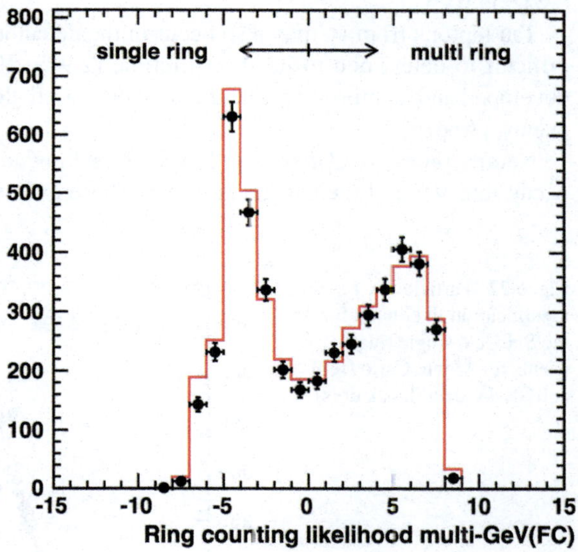

Fig. 6.21 Ring counting likelihood for the MultiGeV events for Monte Carlo (red) and SK-IV data (black dots)

$$q_i^{exp}(e) = \alpha_e Q^{exp}(p_e, \theta_i^i)\left(\frac{16.9\ \mathrm{m}}{r_i}\right)^{3/2} e^{-\frac{r_i}{\lambda(run)}} f(\theta_i) + q_i^{scatt} \tag{6.12}$$

where α_e is a normalisation factor, Q^{exp} is the expected charge distribution for an electron as a function of Cherenkov ring opening angle and momentum, r_i is the distance from the vertex to the ith hit PMT, $\lambda(run)$ is the water attenuation length, $f(\theta_i)$ the acceptance as function of incident angle and q_i^{scatt} the expected charge at the ith PMT due to multiple scattering. Similarly for muons one has.

$$q_i^{exp}(\mu) = \left(\alpha_\mu \frac{\sin^2 \theta_{x_i}}{r_i \left(\sin \theta_{x_i} + r_i \frac{d\theta_t}{dx} \Big|_{x=x_i} \right)} + q_i^{knock} \right) e^{-\frac{r_i}{\lambda(run)}} f(\theta_i) + q_i^{scatt} \quad (6.13)$$

where α_μ is a normalisation factor, x the muon track length, x_i an estimate of the track length when photons hitting the ith PMT were emitted, θ_t and θ_t^i the Čerenkov angles at position x of the muon track at positions x and x_i respectively, and q_i^{knock} the expected charge deposited in the ith PMT due to electrons that were ionised by the muon as it travelled through water.

This pattern comparison is used alone for MultiRing events and together with a Čerenkov angle comparison for single ring events. For the latter, pattern distributions are defined for electrons and muons similarly as for the charge patterns. With all this information the PID likelihood distribution is built as follows.

$$P_{PID} = \sqrt{-logP(\mu)} - \sqrt{-logP(e)} \quad (6.14)$$

The PID distributions for all fully contained samples are shown in Figs. 6.22, 6.23, 6.24 and 6.25.

Tau leptons from ν_τ may also occur in the detector, although they are extremely difficult to detect due to its short lifetime (2.9×10^{-13} s). Despite that, SK has developed and is improving a neural network based algorithm to try to identify these events, [Abe13].

Neutral pions are also detected in SK by taking advantage of their characteristic decay into $\gamma\gamma$ and their invariant mass. These events are of particular importance

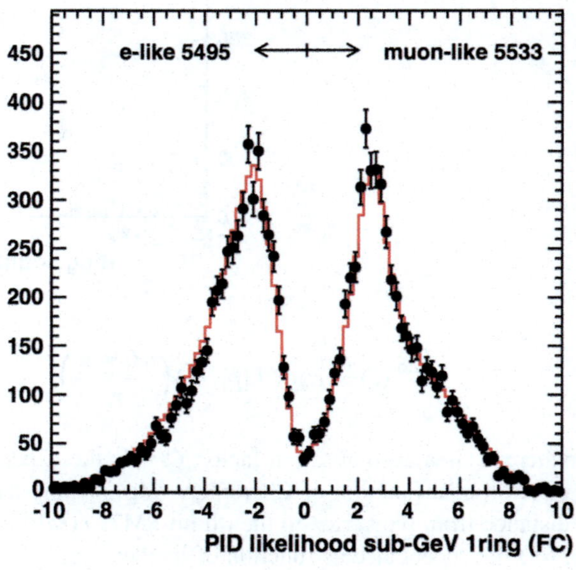

Fig. 6.22 Particle identification likelihood for the SubGeV single ring events for Monte Carlo (red) and SK-IV data (black dots)

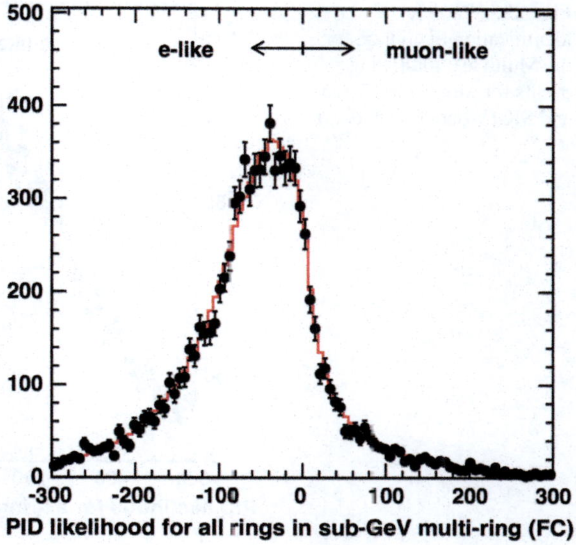

Fig. 6.23 Particle identification likelihood for the SubGeV multi-ring events for Monte Carlo (red) and SK-IV data (black dots)

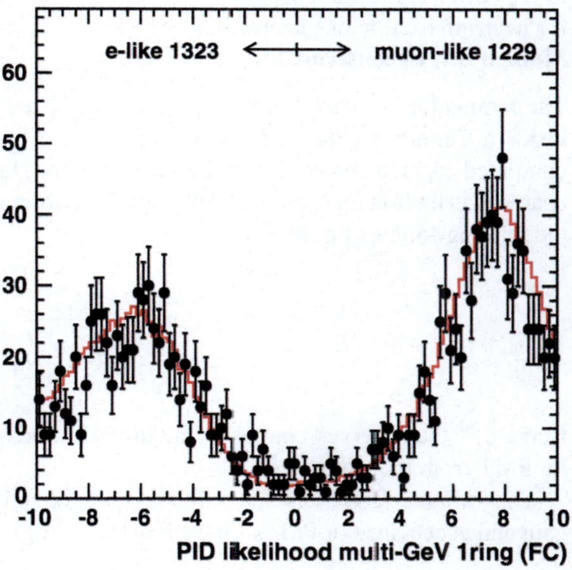

Fig. 6.24 Particle identification likelihood for the MultiGeV single ring events for Monte Carlo (red) and SK-IV data (black dots)

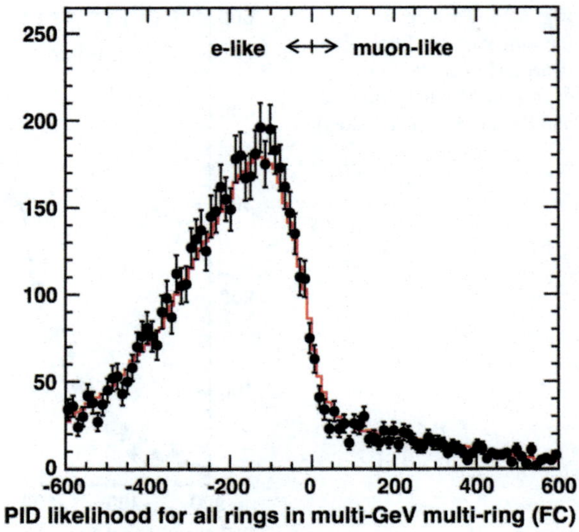

Fig. 6.25 Particle identification likelihood for the MultiGeV multi-ring events for Monte Carlo (red) and SK-IV data (black dots)

as they often come from neutral current single π^0 interactions, a background source for neutrino oscillations analyses.

Momentum Reconstruction

The momentum of each particle (ring) is calculated from the total charged PMTs inside a Čerenkov cone with half opening angle of 70°. Therefore, the energy is computed separately for each of the reconstructed rings, and based on the expected charge distribution for each ring. The estimated observed charge of the ith PMT from the nth ring follows Eq. (6.15).

$$q_{i,n}^{obs} = q_i^{obs} \frac{q_{i,n}^{exp}}{\sum_m q_{i,m}^{exp}} \tag{6.15}$$

Being q_i^{obs} the observed charge by the ith PMT and $q_{i,m}^{exp}$ the expected charge of the ith PMT from the mth ring.

The variable $RTOT_n$ is defined to correct for light attenuation, light reflection, gain and acceptance of PMTs for each of the n rings.

$$RTOT_n = \frac{G_{MC}}{G_{data}} \left(\alpha \sum_{\substack{\theta_{i,n}<70^\circ \\ -50\,\text{ns}<t_i<250\,\text{ns}}} q_i^{obs} e^{\frac{r_i}{\lambda(run)}} \frac{\cos\theta_i}{f(\theta_i)} - \sum_{\theta_{i,n}<70^\circ} \left(S_i + R_i \right) \right) \tag{6.16}$$

where, G_{MC} and G_{data} are the the relative gain PMT factors for Monte Carlo and data, α is a normalisation factor based on the energy scale calibration, t_i is the hit timing of the ith PMT after ToF is subtracted, S_i is the expected charge from scattered photons and R_i the correction for reflected photons with black sheets in the detector.

Decay-Electrons Reconstruction

These are electrons coming from the decay of $\mu^{\pm}$, which, in its turn, may come from the decay of a charged pion. This muons are very likely to come from π^{+} since π^{-} are quickly absorbed by water. This way, decay-electron (decay-e) is also a very useful tool to identify pion production neutrino interactions.

Decay-electrons are searched in an independent time window of 900 ns from that of the primary interaction. In SK-IV, the detection efficiency is 89.1%.

6.6.2 *Neutral Current and Charged Current Separation for Fully Contained Samples*

In this section, the NC-CC separation described in Sect. 5.3 is applied for SuperK-Gd atmospheric fully contained events. This method is developed only for the MultiRing e-like sample because, it is where neutral current and charged current deep inelastic interactions are most present. This two-step method, first tries to separate neutral current events from the rest of the charged current ones. In the second step, muon neutrino events in this sample, which interact mainly through charged current deep inelastic scattering, are separated from the electron neutrino ones. This has been schematically explained in Fig. 5.28.

For the construction of the likelihood distribution, in both separation methods eight variables are used. Amongst these, the neutron multiplicity is the most relevant. The rest of the variables are the number of rings, the MultiRing likelihood, the number of decay-electrons, the distance to the furthest decay-electron, the sum of all rings ID likelihood, the rapidity and the energy fraction of the most energetic ring. All this, has been previously described in Sect. 5.3.

Figure 6.26 shows the distribution for neutron multiplicity and normalised to the present SK-IV running time, 2519.9 days, shown in Table 7.1.

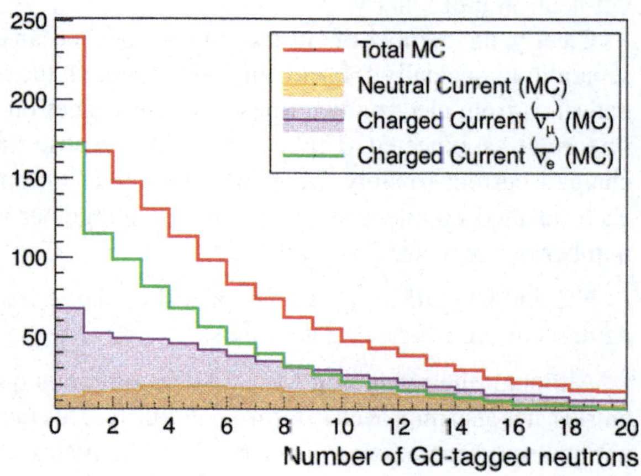

Fig. 6.26 Number of Gd-tagged neutrons for NC (orange), CCDIS ν_{μ} (violet), CC ν_e (green) and total atmospheric Monte Carlo (red) in the MultiRing e-like sample, normalised to 2519.9 days

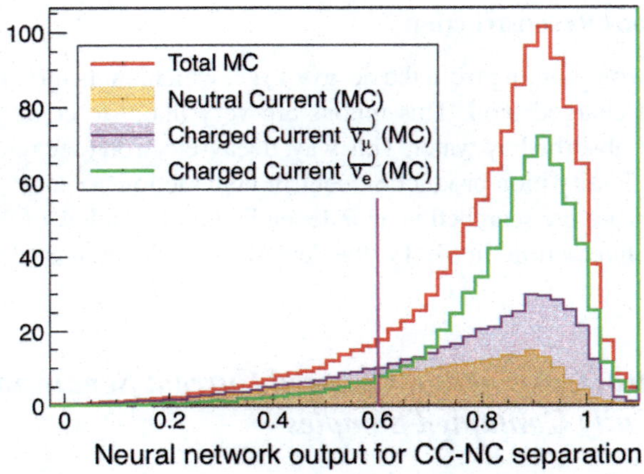

Fig. 6.27 Neural network output for the NC-CC separation ffor NC (orange), CCDIS ν_μ (violet), CC ν_e (green) and total atmospheric Monte Carlo (red) normalised to 2519.9 days (SK-IV running time). The pink vertical line shows the selection point (0.6) for the distribution, which optimises product $\epsilon_{CC} \times p_{CC}$, being the NC-like sample formed by those events at its left and the CC-like by those at its right

NC interactions show a completely different distribution as compared to the charged current ones. It is wider and peaked around five, whereas the rest of interactions reach their peak at zero.

The distributions corresponding to the rest of the variables used in the construction of the both likelihood methods are shown in Figs. 7.7–7.13, and compared with the actual SK-IV data.

The neural network architectures consist in one hidden layer with $N + 5$ neurones, where N is the number of variables involved in the computation. In Figs. 6.27 and 6.28, the neural network outputs for both steps of the selection are shown.

In Table 6.2, the values of the relative importance for each of the variables are presented in order to see the role of each variable in these selections, specially the Gd-neutron multiplicity.

Clearly, the number of Gd-tagged neutrons has an enormous importance in these separations, specially for selecting NC events. In the case of the separation of muon neutrinos from electron neutrinos, the decay-electron variables become much more important as expected. This is, precisely, because the muon neutrinos interacting charged current (mainly CC DIS), produce a prompt μ. The relative contents of each selected sample is shown in Table 6.3, together with the expected unoscillated number of events for 2519.9 days.

NC and CC DIS $\overset{(-)}{\nu}_\mu$-like samples show large fractions of muon neutrinos and neutral current interacting neutrinos.

Although the remaining CC $\overset{(-)}{\nu}_e$-like sample is quite reduced in the number of entries, it is also improved in terms of its purity. This fact will be helpful in its neutrino-antineutrino separation, in Sect. 6.6.3.3. The method explained here, is remarkable

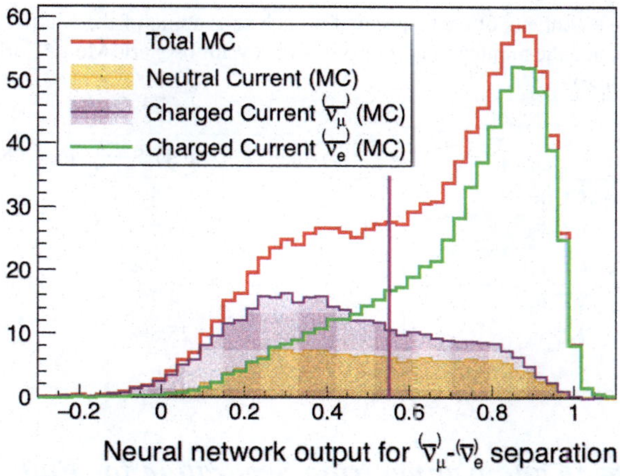

Fig. 6.28 Neural network output for the CCDISν_μ-ν_e separation for NC (orange), CCDIS ν_μ (violet), CC ν_e (green) and total atmospheric Monte Carlo (red) normalised to 2519.9 days (SK-IV running time). The pink vertical line shows the selection point (0.55) for the distribution, which optimises product $\epsilon_{CC-e} \times p_{CC-e}$, being the CCDIS ν_μ-like sample formed by those events at its left and the CC e-like by those at its right

Table 6.2 Relative importance of each variable in the neural network for the NC-CC and CCDISν_μ-ν_e separations

		Relative separation power	
		NC-CC separation (%)	$\overset{(-)}{\nu}_\mu$-$\overset{(-)}{\nu}_e$ separation (%)
Variables	Number of Gd-tagged neutrons	57.5	26.0
	Number of rings	0.8	1.3
	MultiRing likelihood	0.6	0.6
	Number of decay-e	8.7	27.5
	Distance to furthest decay-e	40.5	32.2
	Pseudorapidity	3.0	2.9
	Energy fraction	1.3	0.1
	Sum of all rings PID likelihood	1.5	1.2

due to the complexity and variety of interactions and neutrino flavours in the MultiRing e-like sample. It improves the understanding these events, being able to separate them into three, well-defined, groups of interactions.

Table 6.3 Relative contents of each species for each subsample of the MultiRing e-like sample, after the selection, and the number of entries predicted by atmospheric Monte Carlo for 2519.9 days (SK-IV running time)

		Samples		
		CC $\overset{(-)}{\nu}_e$-like	CCDIS $\overset{(-)}{\nu}_\mu$-like	NC-like
MC true ν	CC ν_e	57.6%	28.4%	27.9%
	CC $\overline{\nu}_e$	21.5%	3.5%	3.9%
	CC ν_μ	8.2%	38.1%	33.5%
	CC $\overline{\nu}_\mu$	0.9%	4.2%	4.2%
	NC ν	11.8%	25.8%	30.6%
MC unosc.		648.8	388.0	617.7

6.6.3 Neutrino and Antineutrino Separation for Fully Contained Samples

In the present section, all the samples affected by the neutrino-antineutrino separation are shown in detail. The distributions and event rates are normalised to the 2519.9 days of SK-IV operation.

The scheme of the samples treated here can be found in Fig. 5.29.

The developments of this section will be included in the oscillation analysis studies in Sect. 6.6.5.

6.6.3.1 SubGeV and MultiGeV e-Like

In this section, the ν-$\overline{\nu}$ separation is implemented for atmospheric single-ring e-like events. These are split into SubGeV ($E_{vis} < 1.33$ GeV) and MultiGeV ($E_{vis} \geq 1.33$ GeV).

Events in these samples are divided into ν_e-like, $\nu_e + \overline{\nu}_e$-like and $\overline{\nu}_e$-like. The separation is based on the neutron and decay-electron multiplicities as described next.

SubGeV e-like samples

Figure 6.29 shows the Gd-neutron multiplicity for this sample.

The neutron multiplicity is typically larger for antineutrinos than for neutrinos, and the number of decay-electrons is usually larger for neutrinos, as shown in Fig. 7.17. The cuts on the neutron and decay-electron multiplicities which define the three regions follow the next expressions:

- ν_e-like: Number of decay-e>0
- $\nu_e + \overline{\nu}_e$-like: Number of decay-e=0 and Number of Gd-tagged neutrons = 0
- $\overline{\nu}_e$-like: Number of decay-e=0 and Number of Gd-tagged neutrons>0

The contents an expected number of unoscillated events for 2519.9 days of exposure are summarised in Table 6.4.

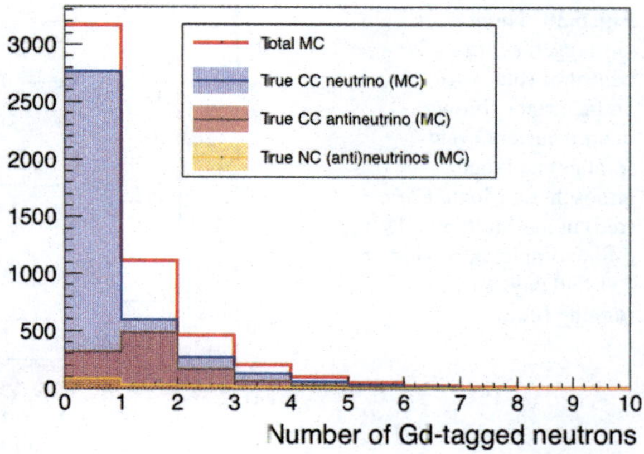

Fig. 6.29 Number of Gd-tagged neutrons for true neutrinos (blue), true antineutrinos (brown), neutral current events (orange) and total atmospheric Monte Carlo (red) in the SubGeV 1Ring e-like sample, normalised to 2519.9 days (SK-IV running time)

Table 6.4 Relative contents of each species for each subsample of the SubGeV e-like sample, after the selection, and the number of entries predicted by atmospheric Monte Carlo normalised to 2519.90 days (SK-IV running time)

		Samples		
		ν_e-like (dcy-e > 0)	$\nu_e + \overline{\nu}_e$-like (dcy-e = 0 and neutrons = 0)	$\overline{\nu}_e$-like (dcy-e = 0 and neutrons>0)
MC true ν	CC ν_e	77.1%	85.8%	51.9%
	CC $\overline{\nu}_e$	1.7%	11.1%	43.9%
	CC ν_μ	11.6%	0.3%	0.3%
	CC $\overline{\nu}_\mu$	3.3%	0.0%	0.0%
	NC ν	6.3%	2.8%	3.9%
MC unosc.		582.9	2543.8	1615.2

MultiGeV e-like samples

Events in this category follow the same logic as in the SubGeV case, being the cuts for the sample selection the same as before:

- ν_e-like: Number of decay-e>0
- $\nu_e + \overline{\nu}_e$-like: Number of decay-e=0 and Number of Gd-tagged neutrons = 0
- $\overline{\nu}_e$-like: Number of decay-e=0 and Number of Gd-tagged neutrons>0

The distribution of the neutron multiplicity is shown in Fig. 6.30 and the distribution for the number of decay-electrons, alongside the SK-IV data, is presented in Fig. 7.19.

The contents and expected rates of the defined samples are shown in Table 6.5.

In both selection procedures, summarised in Tables 6.4 and 6.5, it is clearly seen that the purity of the ν_e-like sample is very high and with low statistics. $\overline{\nu}_e$-like samples contain, approximately, the same amount of neutrinos and antineutrinos, which given the differences in their fluxes, confirms the good performance of the sample definition. Finally, in the $\nu_e + \overline{\nu}_e$-like sample, the neutrino purity is high, but, in this

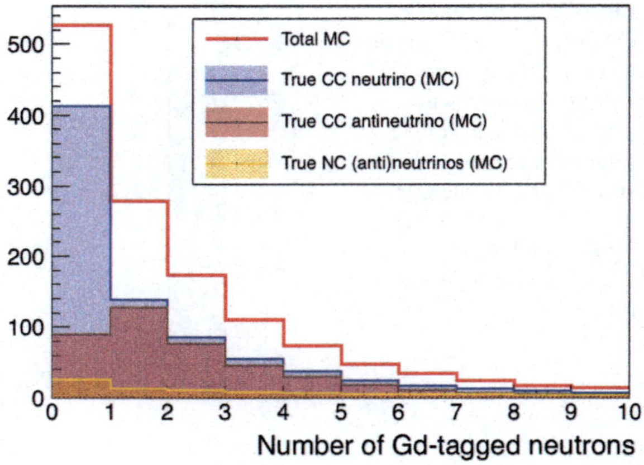

Fig. 6.30 Number of Gd-tagged neutrons for true neutrinos (blue), true antineutrinos (brown), neutral current events (orange) and total atmospheric Monte Carlo (red) in the MultiGeV 1Ring e-like sample, normalised to 2519.90 days (SK-IV running time)

Table 6.5 Relative contents of each species for each subsample of the MultiGeV e-like sample, after the selection, and the number of entries predicted by atmospheric Monte Carlo normalised to 2519.90 days (SK-IV running time)

		Samples		
		ν_e-like (dcy-e > 0)	$\nu_e + \overline{\nu}_e$-like (dcy-e = 0 and neutrons = 0)	$\overline{\nu}_e$-like (dcy-e = 0 and neutrons>0)
MC true ν	CC ν_e	61.4%	73.4%	41.6%
	CC $\overline{\nu}_e$	9.1%	20.1%	49.8%
	CC ν_μ	11.7%	1.0%	1.0%
	CC $\overline{\nu}_\mu$	2.0%	0.1%	0.4%
	NC ν	15.7%	5.4%	7.2%
MC unosc.		359.0	401.1	557.8

case, mainly due to the flux dissimilarities between neutron and antineutrinos. In addition, it is interesting to notice the worsening in the separation with increasing energy. This is because at higher energies the neutron multiplicity is larger, which dims the differences between neutrinos and antineutrinos in terms of neutron production.

6.6.3.2 SubGeV and MultiGeV μ-like

Although it could be tempting to subdivide these events into three categories as in the previous case. It is not done because this sample is not as important for the oscillation analysis and because the corresponding ν_μ-like sample would have a huge systematic error, ~30%, induced by the uncertainty in the decay-e tagging. Therefore, these events are simply separated into two, ν_μ-like and $\overline{\nu}_\mu$-like samples. dims

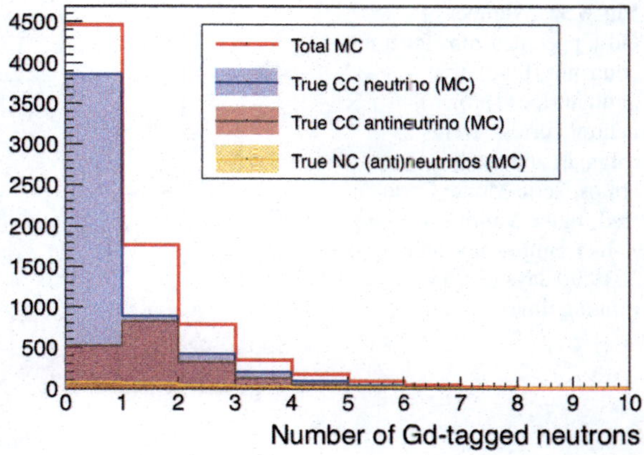

Fig. 6.31 Number of Gd-tagged neutrons for true neutrinos (blue), true antineutrinos (brown), neutral current events (orange) and total atmospheric Monte Carlo (red) in the SubGeV 1Ring μ-like sample, normalised to 2519.90 days (SK-IV running time)

Table 6.6 Relative contents of each species for each subsample of the SubGeV μ-like sample, after the selection, and the number of entries predicted by atmospheric Monte Carlo normalised to 2519.90 days (SK-IV running time)

		Samples	
		ν_μ-like (dcy-e $\neq$ 1 or (dcy-e = 1 and neutrons = 0))	$\overline{\nu}_\mu$-like (dcy-e = 1 and neutrons > 0)
MC true ν	**CC ν_e**	0.8%	0.1%
	CC $\overline{\nu}_e$	0.2%	0.0%
	CC ν_μ	83.8%	47.2%
	CC $\overline{\nu}_\mu$	11.5%	49.5%
	NC ν	3.7%	3.3%
MC unosc.		4743.5	2462.7

SubGeV μ-like samples

The cuts which define the two regions in this sample are the following:

- ν_μ-like: $\begin{cases} \text{Number of decay-e} \neq 1 \\ \text{Number of decay-e} = 1 \text{ and Number of Gd-tagged neutrons} = 0 \end{cases}$
- $\overline{\nu}_\mu$-like: Number of decay-e=1 and Number of Gd-tagged neutrons>0

The distribution for the number of Gd-tagged neutrons is shown in Fig. 6.31. The simulated decay-electron multiplicity for this sample is presented in Fig. 7.21 and compared with the SK-IV data.

Table 6.6 summarises the contents of the selected samples.

MultiGeV μ-like samples

For MultiGeV events, the analogous is done, being the selection criteria for the sample separation as follows.

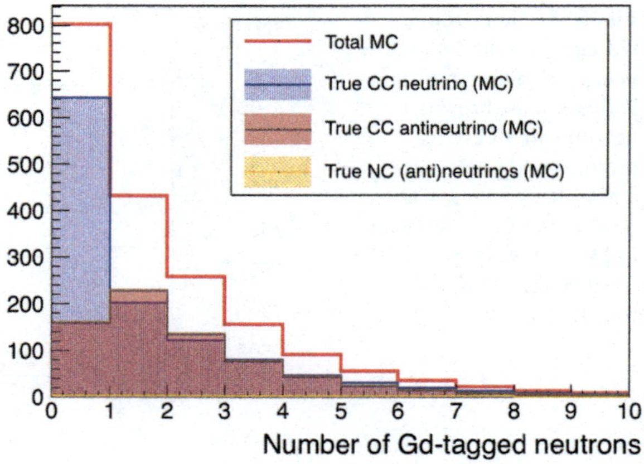

Fig. 6.32 Number of Gd-tagged neutrons for true neutrinos (blue), true antineutrinos (brown), neutral current events (orange) and total atmospheric Monte Carlo (red) in the MultiGeV 1Ring μ-like sample, normalised to 2519.90 days (SK-IV running time)

Table 6.7 Relative contents of each species for each subsample of the MultiGeV μ-like sample, after the selection, and the number of entries predicted by atmospheric Monte Carlo normalised to 2519.90 days (SK-IV running time)

		Samples	
		ν_μ-like (dcy-e $\neq$ 1 or (dcy-e = 1 and neutrons = 0))	$\overline{\nu}_\mu$-like (dcy-e = 1 and neutrons > 0)
MC true ν	CC ν_e	0.3%	0.1%
	CC $\overline{\nu}_e$	0.1%	0.0%
	CC ν_μ	74.9%	41.5%
	CC $\overline{\nu}_\mu$	24.6%	58.3%
	NC ν	0.1%	0.1%
MC unosc.		1090.0	655.5

- ν_μ-like: $\begin{cases} \text{Number of decay-e} \neq 1 \\ \text{Number of decay-e} = 1 \text{ and Number of Gd-tagged neutrons} = 0 \end{cases}$
- $\overline{\nu}_\mu$-like: Number of decay-e=1 and Number of Gd-tagged neutrons>0

The distributions of the Gd-tagged neutrons and the decay-electrons are presented in Figs. 6.32 and 7.23 and normalised to the 2519.90 days of SK-IV data taking.

The contents for each of the resulting samples are calculated in Table 6.7.

Similarly to Sect. 6.6.3.1, in the e-like cases, ν_μ-like samples possess a very high purity, and, in this case, with larger number of entries due to the only two sample-division. The $\overline{\nu}_\mu$-like samples contain, again, similar number f neutrinos and antineutrinos.

The energy dependence of the separation method is also reproduced in these cases. The efficiency of the selection is depleted as the neutrino energy increases.

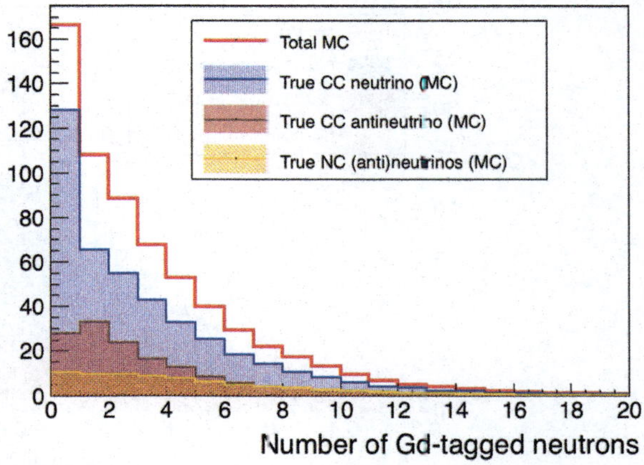

Fig. 6.33 Number of Gd-tagged neutrons for true neutrinos (blue), true antineutrinos (brown), neutral current events (orange) and total atmospheric Monte Carlo (red) in the MultiRing CC ν_e-like sample, normalised to 2519.90 days (SK-IV running time)

6.6.3.3 MultiRing CC ν_e-like

After the NC-CC separation, explained in Sect. 6.6.2, is done, the remaining events in the MultiRing CC $\overset{(-)}{\nu}_e$-like sample are further divided into neutrino and antineutrino-like samples. This separation is performed using a neural network method with one hidden layer and $N + 5$ nodes. The selection point is, then, defined to make the separation.

In this method, the most important variable is the Gd-tagged neutron multiplicity, whose distribution is displayed in Fig. 6.33.

The rest of the variables used in the separation, for the atmospheric Monte Carlo and the SK-IV dataset, are the number of decay-electrons, the distance and time to the furthest decay-electron, the number of rings, the sum of all ring particle ID likelihood, the rapidity and the energy fraction of the most energetic ring. They are shown in Figs. 7.25–7.31.

The distribution of the neural network output is shown in Fig. 6.34.

In order to show the importance of the neutron multiplicity as compared with the rest of the variables used, Table 6.8 shows their relative importance in the separation procedure.

In this separation, the amount of neutrons in the final state plays a crucial role in the separation. In importance, it is followed by the decay-electron variables, but with much less separation power.

Lastly, Table 6.9 shows the contents for all four samples defined for MultiRing e-like events.

The separation is not as good as in the rest of e-like sample due to the remaining neutral current and $\overset{(-)}{\nu}_\mu$ events in the sample. Despite of that, the neutron multiplicity improves the categorisation of neutrinos and antineutrinos as seen in Table 6.8, which will have an important effect in the mass hierarchy sensitivity, analysed in Sect. 6.6.5.2.

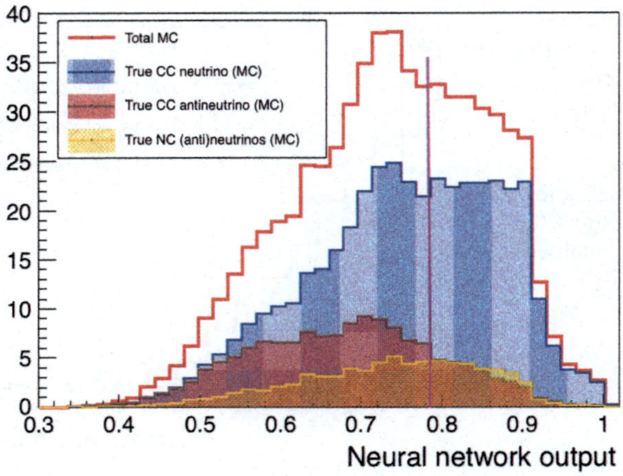

Fig. 6.34 Neural network output for the ν-$\overline{\nu}$ separation in the MultiRing e-like sample, for for true neutrinos (blue), true antineutrinos (brown), neutral current events (orange) and total atmospheric Monte Carlo (red) normalised to 2519.90 days (SK-IV running time). The pink vertical line shows the selection point, 0.8 for the distribution, which separates the events into two categories, from left to right, MultiRing $\overline{\nu}_e$-like and MultiRing ν_e-like

Table 6.8 Relative importance of each variable in the neural network for the $\overline{\nu}$-ν separation in the CCν_e-like sample

		Relative separation power
		ν-$\overline{\nu}$ separation (%)
MC ν information	Number of Gd-tagged neutrons	77.8
	Number of decay-e	9.3
	Distance to furthest decay-e	9.2
	Number of rings	1.8
	Sum of all rings PID likelihood	1.1
	Pseudorapidity	0.3
	Energy fraction of most energetic ring	0.4

Table 6.9 Relative contents of each species for each subsample of the MultiRing e-like sample, after the selection, and the number of entries predicted by atmospheric Monte Carlo normalised to 2519.90 days (SK-IV running time)

		Samples			
		CC ν_e-like	CC $\overline{\nu}_e$-like	CCDIS $\overset{(-)}{\nu}_\mu$-like	NC-like
MC true ν	CC ν_e	63.1%	53.6%	28.4%	27.9%
	CC $\overline{\nu}_e$	10.8%	29.2%	3.5%	3.9%
	CC ν_μ	12.7%	4.9%	38.1%	33.5%
	CC $\overline{\nu}_\mu$	1.4%	0.6%	4.2%	4.2%
	NC ν	12.0%	11.7%	25.8%	30.6%
MC unosc.		270.8	378.0	388.0	617.7

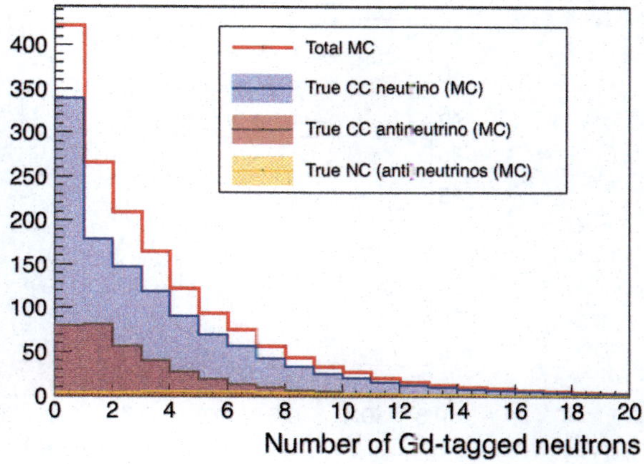

Fig. 6.35 Number of Gd-tagged neutrons for true neutrinos (blue), true antineutrinos (brown), neutral current events (orange) and total atmospheric Monte Carlo (red) in the MultiRing CC e-like sample, normalised to 2519.90 days (SK-IV running time)

6.6.3.4 MultiRing μ-like

As it has been done in the rest of μ-like samples, MultiRing μ-like events will be separated in two samples. However, and in a similar fashion to the MultiRing CC $\overset{(-)}{\nu}_e$-like samples, the selection is done based on a likelihood distribution constructed with a neural network of one hidden layer and $N + 5$ neurones. The neutron multiplicity plays a very important role. The distribution for the number of Gd-tagged neutrons is shown in Fig. 6.35.

The rest of the variables entering the likelihood construction are the the number of decay-electrons, the distance and time to the furthest decay-electron, the number of rings, the sum of all ring particle ID likelihood, the rapidity and the energy fraction of the most energetic ring. All of them are shown in Figs. 7.34–7.40 for atmospheric Monte Carlo and SK-IV dataset.

The neural network output distribution is plotted in Fig. 6.36 and, in Table 6.10, the relative importance in the method for each variable is shown

In this case, the number of decay-electrons have greater importance than neutrons. This is because the in distribution of the latter the differences of neutrinos and antineutrinos become smaller at these high energies. Meanwhile, the decay-electrons maintain their separation power as their origin does not depend so much on the neutrino energy and the secondary interactions.

Table 6.11 shows the contents of each sample and the expected unoscillated event rates for 2519.90 days.

The amount of neutrinos is much bigger than that of antineutrinos, therefore, the $\overline{\nu}_{\mu}$-like sample is largely contaminated with neutrinos. On the other hand, the ν_{μ}-like sample is very pure, but with low statistics.

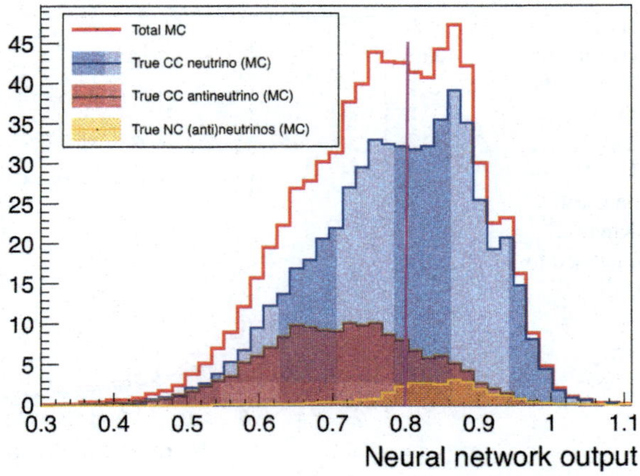

Fig. 6.36 Neural network output for the ν-$\overline{\nu}$ separation in the MultiRing μ-like sample, for for true neutrinos (blue), true antineutrinos (brown), neutral current events (orange) and total atmospheric Monte Carlo (red) normalised to 2519.90 days (SK-IV running time). The pink vertical line shows the selection point, 0.762 for the distribution, which separates the events into two categories, from left to right, MultiRing $\overline{\nu}_\mu$-like and MultiRing ν_μ-like

Table 6.10 Relative importance of each variable in the neural network for the ν-$\overline{\nu}$ separations in the MultiRing μ-like sample

		Relative separation power
		ν-$\overline{\nu}$ separation (%)
Variables	Number of Gd-tagged neutrons	36.7
	Number of decay-e	51.9
	Distance to furthest decay-e	5.0
	Time to furthest decay-e	4.7
	Number of rings	0.6
	Sum of all rings PID likelihood	0.1
	Pseudorapidity	0.3
	Energy fraction of most energetic ring	0.7

Table 6.11 Relative contents of each species for each subsample of the MultiRing μ-like sample, after the selection, and the number of entries predicted by atmospheric Monte Carlo normalised to 2519.90 days (SK-IV running time)

		Samples	
		ν_μ-like	$\overline{\nu}_\mu$-like
MC true ν	CC ν_e	3.7%	1.8%
	CC $\overline{\nu}_e$	0.4%	0.3%
	CC ν_μ	82.6%	71.1%
	CC $\overline{\nu}_\mu$	7.6%	23.3%
	NC ν	5.8%	3.5 %
MC unosc.		167.0	584.4

6.6.4 *Neutron-Corrected Reconstructed Neutrino Energy for Fully Contained Samples*

The energy corrections with the neutron multiplicity, explained in Sect. 5.5, are applied to all the fully contained samples of the atmospheric analysis. The neutron correction function, f, is computed for each of these samples. For the computation of these functions, events with very poorly reconstructed energy are not taken into account in order to avoid the distortion of the distribution. If these are not removed, it would become considerably different due to a tiny fraction of events and worsening the performance of the energy correction functions.

The averaged fraction of invisible energy ($(E_\nu - E_{vis})/E_{vis}$) for all cases, together with the polynomial fit function, are shown in Figs. 6.37 and 6.38.

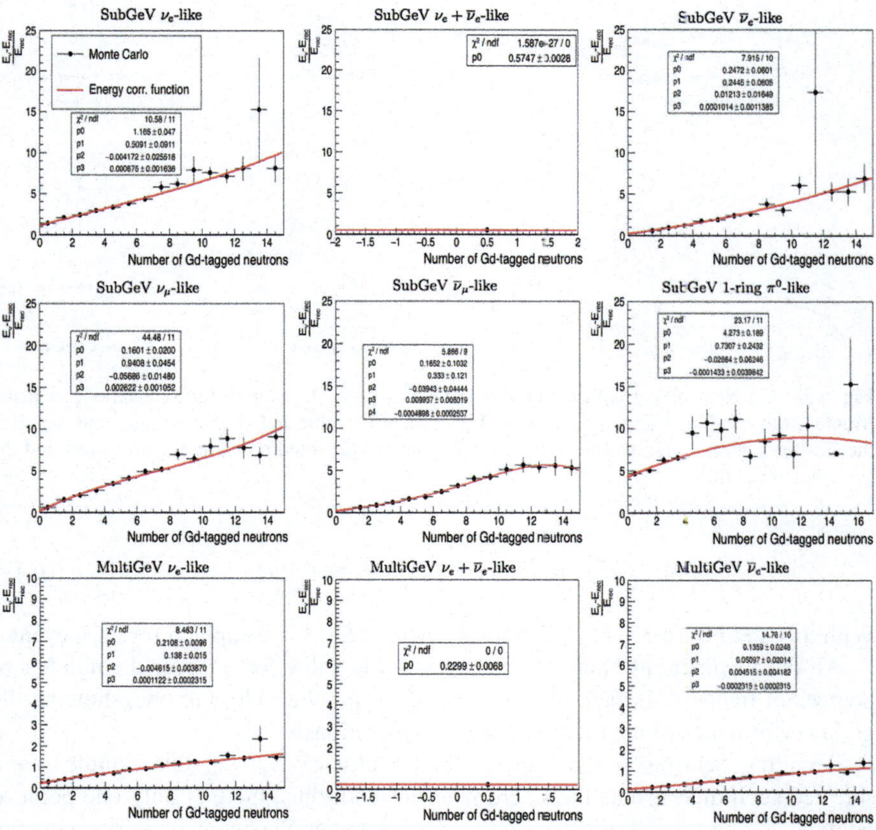

Fig. 6.37 Gd-Neutron correction functions for each of the FC samples for atmospheric neutrino Monte Carlo (1 of 2). Y-axis is the averaged fraction of invisible and visible energies and X-axis is the Gd-neutron multiplicity. The box in each plot shows the fitted polynomial parameters and the χ^2 value of the fit

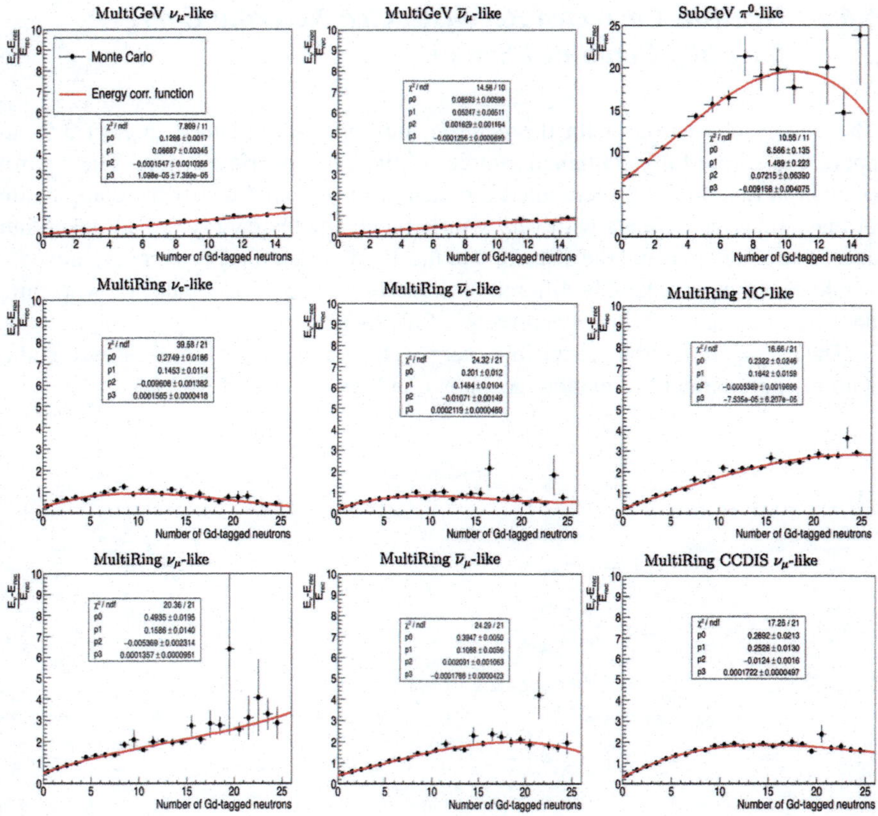

Fig. 6.38 Gd-Neutron correction functions for each of the FC samples for atmospheric neutrino Monte Carlo (2 of 2). Y-axis is the averaged fraction of invisible and visible energies and X-axis is the Gd-neutron multiplicity. The box in each plot shows the fitted polynomial parameters and the χ^2 value of the fit

$$E_{rec}^{Gd} = E_{vis}(1 + f(Gd - neutron)) \tag{6.17}$$

With this and Eq. (6.17) the Gd-neutron energy (E_{rec}^{Gd}) is computed for each event.

All the fit functions show a rather reasonable value for χ^2 over the number of degrees of freedom. In fact, most of the fits show values close to one, showing the good performance of the neutron-correction functions.

Recalling Sect. 6.6.3, the SubGeV and MultiGeV $\nu_e + \overline{\nu}_e$-like samples were selected such their events have zero neutrons and, thus, there is only one point to be fitted. 1-ring π^0 and π^0-like samples show a larger Y-axis intercept as compared with the other samples. This is because this sample is mainly populated by neutral current events, which do not produce any prompt charged lepton, hence, being very hard to reconstruct in a water Čerenkov detector. This is very important because it means that neutrons may have a considerably effect in the energy reconstruction of these samples.

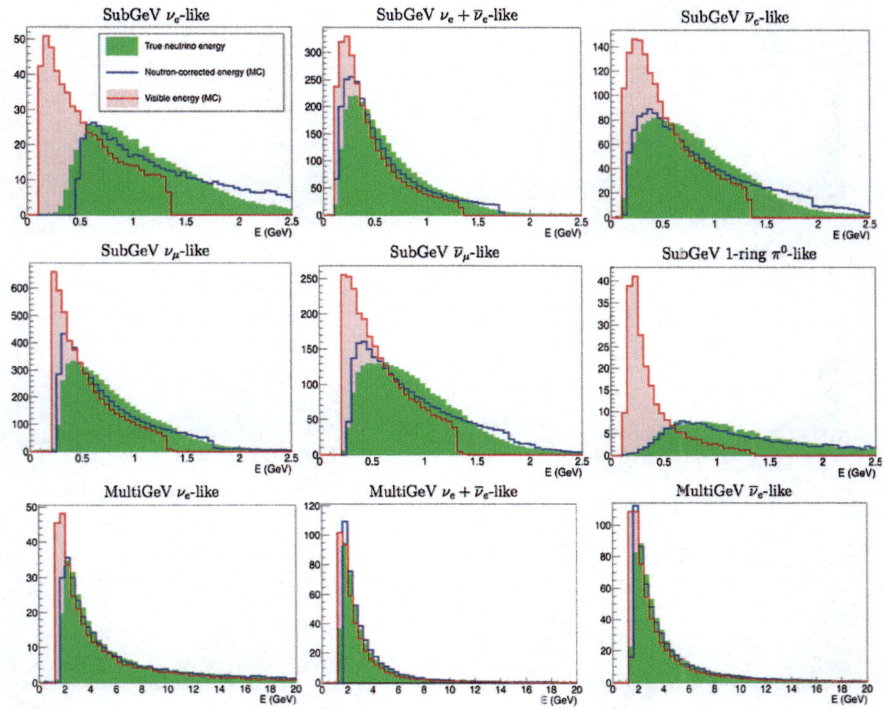

Fig. 6.39 True (green), visible (red) and corrected (blue) energy spectra of the different FC samples for atmospheric neutrino Monte Carlo (1 of 2). The distributions are normalised to 2519.90 days (SK-IV running time)

Another feature from Figs. 6.37 and 6.38, is that for MultiRing NC, CCDIS $\overset{(-)}{\nu}_e$, ν_e and $\overline{\nu}_e$-like samples the correction function levels up for high neutron multiplicities. Figures. 6.39 and 6.40 show the distributions of the true, visible and neutron-corrected energies for the distributions of the different FC samples. The improvement using the neutron correction is apparent and for all samples, being the distributions of the corrected energy much closer to the true neutrino spectrum than the usual visible energy. A feature of the neutron-corrected energy are the steps seen in the SubGeV samples, which are due to the inherited visible energy cut at $E_{vis} = 1.33$ GeV. Of special mention are the cases of the SubGeV 1-ring π^0 and π^0-like samples and the MultiRing samples, where the differences between visible and corrected energies are most noticeable.

The π^0-like samples consist, mostly, on neutral current 1π interactions, meaning that their visible energy is reconstructed just with the photons from the π^0 decay. By using the neutron-corrections a large fraction of the energy invisible to the detector is recovered. This feature could have an important effect in the search for sterile neutrinos in the neutral current disappearance channel, [Adamson11].

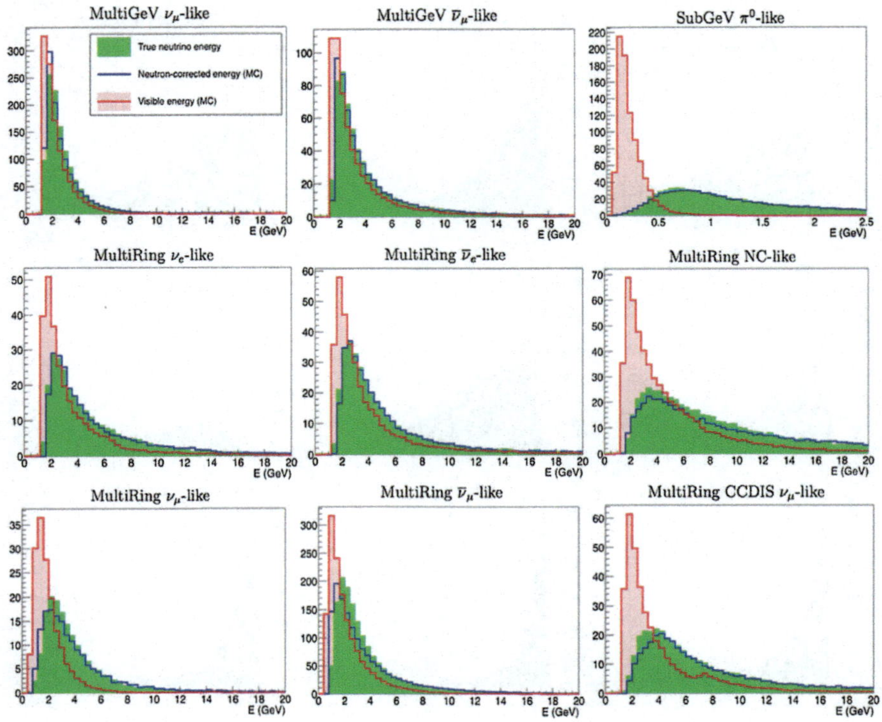

Fig. 6.40 True (green), visible (red) and corrected (blue) energy spectra of the different FC samples for atmospheric neutrino Monte Carlo (2 of 2). The distributions are normalised to 2519.90 days (SK-IV running time)

As for the MultiRing samples, these contain events with larger energy than those with a single ring, implying that their cross-section is not dominated by CCQE interactions and, thus, the energy fraction left in the nucleus enabling the production of numerous secondary particles, is larger. This samples lie on the energy of the Earth's matter resonance, which can better detected with the corrected energy, providing better sensitivity to the neutrino mass hierarchy.

In order to quantify the improvement in the neutrino energy reconstruction, Table 6.12 shows the variables of the corrected energy pulls $((E_\nu - E_r ec)/E_\nu)$ compared to those for the visible energy. These variables characterising the energy pulls are the maximum of the distribution (mode), the mean value, the RMS and the skewness. The mode and mean of the distribution indicate the accuracy with which the energy is reconstructed.

The RMS in conjunction with the skewness of the distribution tell the resolution of the energy reconstruction. Considering both parameters is key for the understanding of the effect of the energy correction. In the case of the visible energy, the energy pulls tend to be very asymmetric with long tails for positive values. On the contrary, energy corrected pull distributions are closer to a Gaussian distribution, being their

Table 6.12 Mode, mean, RMS and skewness of the pull distributions for the visible energy and the Gd-neutron corrected energy

Samples	E_{vis}				E_{rec}^{Gd}			
	Mode	Mean	RMS	Skewness	Mode	Mean	RMS	Skewness
SubGeV ν_e-like	0.548	0.538	0.181	−0.052	0.020	0.005	0.412	−0.037
SubGeV $\nu_e + \overline{\nu}_e$-like	0.198	0.289	0.170	0.541	−0.020	0.084	0.219	0.474
SubGeV $\overline{\nu}_e$-like	0.233	0.346	0.207	0.548	−0.060	−0.070	0.350	−0.029
SubGeV ν_μ-like	0.233	0.319	0.196	0.441	−0.060	0.032	0.286	0.321
SubGeV $\overline{\nu}_\mu$-like	0.233	0.336	0.196	0.530	−0.100	0.002	0.285	0.359
SubGeV 1ring-π^0-like	0.863	0.685	0.211	−0.842	−0.024	−0.919	1.403	−1.224
SubGeV π^0-like	0.898	0.786	0.158	−0.708	0.060	−0.444	0.758	−0.665
MultiGeV ν_e-like	0.178	0.311	0.195	0.686	−0.060	−0.046	0.282	0.048
MultiGeV $\nu_e + \overline{\nu}_e$-like	0.058	0.127	0.131	0.529	0.020	0.087	0.137	0.491
MultiGeV $\overline{\nu}_e$-like	0.058	0.189	0.155	0.848	−0.060	0.015	0.178	0.420
MultiGeV ν_μ-like	0.058	0.161	0.117	0.879	−0.060	0.010	0.122	0.573
MultiGeV $\overline{\nu}_\mu$-like	0.093	0.179	0.118	0.735	−0.060	0.010	0.127	0.548
MultiRing NC-like	0.478	0.455	0.245	−0.093	0.020	−0.229	0.512	−0.408
MultiRing DIS-ν_μ-like	0.408	0.449	0.206	0.201	0.023	−0.058	0.380	−0.211
MultiRing ν_e-like	0.233	0.300	0.185	0.362	−0.013	0.006	0.249	0.075
MultiRing $\overline{\nu}_e$-like	0.163	0.253	0.177	0.513	−0.013	0.013	0.236	0.106
MultiRing ν_μ-like	0.478	0.439	0.178	−0.214	−0.013	−0.037	0.289	−0.084
MultiRing $\overline{\nu}_\mu$-like	0.338	0.345	0.201	0.036	−0.013	−0.046	0.312	0.107

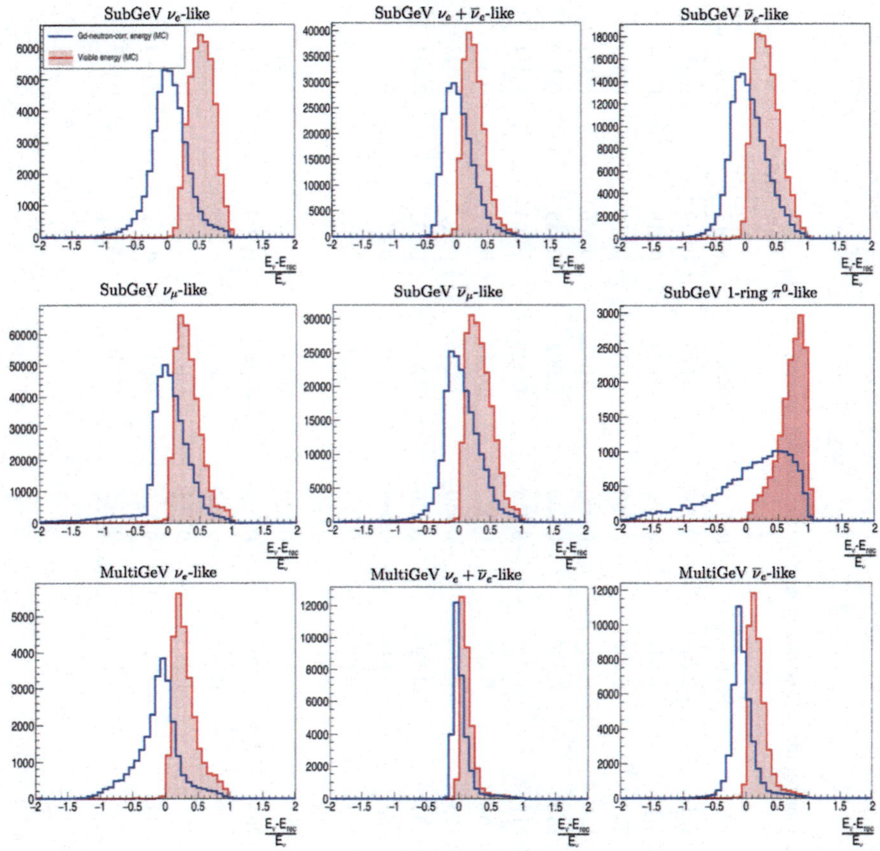

Fig. 6.41 Energy pulls of visible (red) and Gd-neutron-corrected (blue) energies of the different FC samples for atmospheric Monte Carlo (1 of 2)

skewness smaller. This affects the value of the RMS, being, usually, slightly larger for the corrected energy cases, despite of having smaller tails.

In general, the neutron corrected energy improves the energy reconstruction of all samples, which will be beneficial for the oscillation analysis done in Sect. 6.6.5. However, two samples, not very relevant for the standard oscillation analysis, behave in a different fashion to the rest of them, these are the 1ringπ^0 and π^0-like samples. As already noted, the correction function in these samples is much larger than for the rest. Although the Gd-neutron-corrected spectra for these samples matches nicely the true neutrino energy, in Figs. 6.39 and 6.40, the pull distributions are much wider than those for the visible energy, in Figs. 6.41 and 6.42.

In Sect. 6.6.5, the energy correction with neutrons is also included to the oscillation analysis, where its improvements, together with the previous ones, will be quantified in terms of the sensitivity to the atmospheric oscillation parameters.

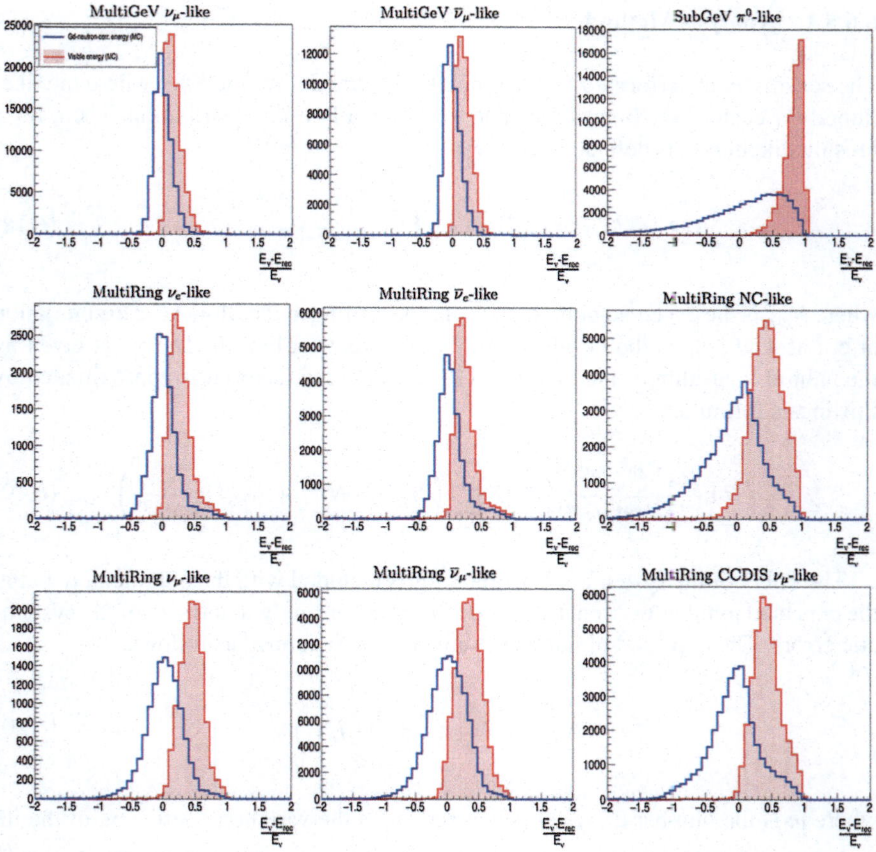

Fig. 6.42 Energy pulls of visible (red) and Gd-neutron-corrected (blue) energies of the different FC samples for atmospheric Monte Carlo (2 of 2)

6.6.5 Oscillation Analysis Sensitivity

All the previous improvements are implemented in the atmospheric oscillation analysis, this will be referred as "Gd-tag" analysis. The current official SK analysis scheme is used for comparison purposes and will be referred as *Standard analysis*. To compute the sensitivity analysis the official tools of SK were modified in order to accommodate the new sample scheme and the corrected energy for fully contained events.

Before the oscillation analysis is performed in Sect. 6.6.5.2, the analysis method and the sample binning are described.

6.6.5.1 Analysis Method

The sensitivity is performed by the binned χ^2 method, [Wendell08], where the likelihood for each of the bins is assumed to follow a Poisson distribution. Then, for n bins the likelihood is defined as follows.

$$L(N^{exp}, N^{obs}) = \prod_{i=1}^{n} \frac{e^{N_i^{exp}} (N_i^{exp})^{N_i^{obs}}}{N_i^{obs}!} \tag{6.18}$$

where N_{exp} is the expected number of events according to the Monte Carlo simulation at bin ith and N_{obs} is the observed number of events at bin ith. The χ^2 is given by the natural logarithm of the likelihood ratio where the factorial is approximated by Stirlings's formula.

$$\chi^2 = -2 \ln \frac{L(N^{exp}, N^{obs})}{L(N^{obs}, N^{obs})} = 2 \sum_{i=1}^{n} \left(N_i^{exp} - N_i^{obs} + N_i^{obs} \ln \frac{N_i^{obs}}{N_i^{exp}} \right) \tag{6.19}$$

The effect of the systematic uncertainties is included with the pull method, being the expected number of events allowed to vary accordingly to the size of the systematic errors. The expected of number events is then redefined as follows.

$$N_{exp} \rightarrow N_{exp} \left(1 + \sum_{j=1}^{m} f_j^i \epsilon_j \right) \tag{6.20}$$

where m is the number of systematic errors, ϵ_j is the variation coefficient of the jth systematic uncertainty and f_i^j is the relative change in the event rate at bin ith due to 1σ variation of the systematic jth.

$$f_i^j = \frac{N_i^{exp}(+1\sigma) - N_i^{exp}(-1\sigma)}{2N_i^{exp}(default)} \tag{6.21}$$

In addition, a term for constraining the range of the ϵ is included in the χ^2 expression.

$$\chi^2 = 2 \sum_{i=1}^{n} \left(N_i^{exp} \left(1 + \sum_{j=1}^{m} f_j^i \epsilon_j \right) - N_i^{obs} + N_i^{obs} \ln \frac{N_i^{obs}}{N_i^{exp} \left(1 + \sum_{j=1}^{m} f_j^i \epsilon_j \right)} + \sum_{j=1}^{m} \left(\frac{\epsilon_j}{\sigma_j} \right) \right) \tag{6.22}$$

The best fit corresponds to the minimisation of the χ^2 with respect to each of the ϵ_j. In order to ensure the good behaviour of the χ^2 distribution, each bin is required to have more than seven expected entries, so the Stirling approximation is close enough to the factorial.

Table 6.13 True oscillation parameters used for the sensitivity studies of the atmospheric oscillation analysis

Oscillation param.	True value	Treatment
Δm^2_{32}	2.5×10^{-3} eV^2/c^2	Fitted
$\sin^2\theta_{23}$	0.575	Fitted
Δm^2_{21}	7.65×10^{-5} eV^2/c^2	Fixed
$\sin^2\theta_{12}$	0.309	Fixed
$\sin^2\theta_{13}$	0.0219	Fixed
δ_{CP}	4.19	Fitted

6.6.5.2 Oscillation Sensitivity Study

The true oscillation parameter used for the sensitivity study, unless otherwise stated, are shown in Table 6.13.

Most of the systematic errors taken are those for SK-IV, in Sect. 7.3.1. Only those errors related to the Gd-neutron tagging tools (NC-CC separation, ν-$\overline{\nu}$ separation and neutron-corrected energy) are computed specifically for the Gd-tag analysis, and can be found in Appendix A.

The binning for the analysis is done in the cosine of the zenith angle and the Gd-neutron corrected reconstructed energy. The total amount of bins for the Gd-tag analysis is 701.

Two sensitivity studies are performed, one for each of the neutrino mass orderings. In Fig. 6.43, the χ^2 distributions for $\sin^2\theta_{23}$ and δ_{CP} are shown assuming true normal hierarchy. In Fig. 6.44 the analogous distributions are displayed for true inverted hierarchy.

The Gd-tag analysis improves the sensitivity to the mass hierarchy and the CP phase significantly. These two parameters are the most sensitive to the neutrino-antineutrino separation and the reconstructed energy. The sensitivity to these parameters is summarised for both analyses in Table 6.14.

The Gd-tag analysis shows a major improvement in the sensitivity study for the mass hierarchy, improving the sensitivity 1.06 units of χ^2, for the normal ordering, and 0.59 units of χ^2 for the inverted one. It also improves the sensitivity to the CP phase, improving largely the fraction of δ_{CP} over 1σ level for both hierarchies.

In order to explicitly see the improvement of the Gd sample selection as compared to SK standard analysis, in Fig. 6.45 the mass hierarchy sensitivity for each of the fully contained samples in shown. This figure shows that, the improvement in the neutrino mass ordering, happens mainly on the MultiGeV and MultiRing samples, and for the Gd-tag analysis, an overall enhancement of the sensitivity is observed. Additionally, the newly introduced SubGeV samples also improve their sensitivity to the neutrino mass hierarchy.

Further, Fig. 6.46 shows the wrong mass hierarchy rejection sensitivity is done for various values of θ_{23}. The sensitivity to the mass hierarchy is seen to be improved very significantly by Gd-tag analysis for all considered values of θ_{23}, becoming much more noticeable as θ_{23} increases.

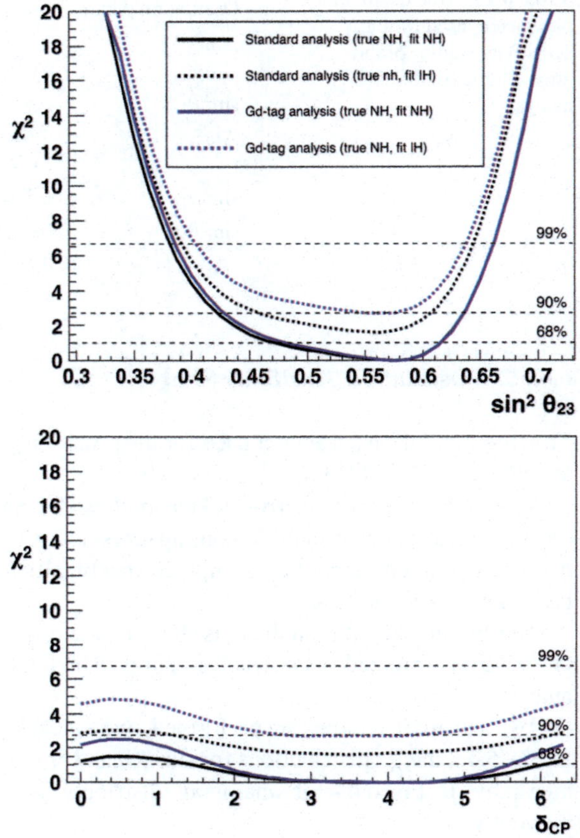

Fig. 6.43 χ^2 distributions for $\sin^2\theta_{23}$ (top) and δ_{CP} (bottom) assuming normal mass hierarchy and for Gd-tag analysis (violet) and current standard SK analysis (black). The horizontal dashed lines represent the 68, 90 and 99% confidence levels, which correspond to χ^2 values of 1, 2.71 and 6.64, respectively. The dataset corresponds to that of SK-IV presented in Table 7.1

6.7 Long-Baseline Neutrinos

The T2K (Tokai-to-Kamioka) experiment, [Abe11b], is a long-baseline neutrino oscillation experiment. The main goals of this experiment are, nowadays, the measurement of the leptonic CP violating phase.

Due to the success of its measurements, in 2016 the T2K collaboration approved the extension of its running period until 20×10^{21} POT are achieved. Meaning that the second run of T2K will be done having SuperK-Gd as far detector and, therefore, the T2K oscillation analyses might benefit from Gd-neutron tagging.

In this section, based on the neutron-tagging abilities of SuperK-Gd, the separation between neutrinos and antineutrinos, presented in Sect. 5.4, is implemented for T2K, Sect. 6.7.4, and applied to the oscillation analysis in Sect. 6.7.6. Furthermore, the neutron multiplicity driven energy corrections are investigated in Sect. 6.7.5, although this feature will not be applied to the oscillation analysis. Instead, a comparison between the neutron-corrected energy and the usual reconstructed energy is done.

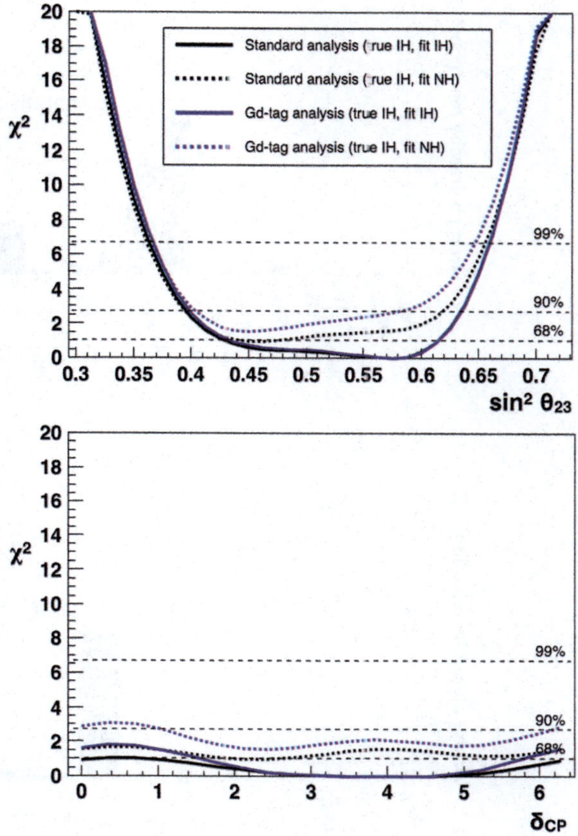

Fig. 6.44 χ^2 distributions for $\sin^2\theta_{23}$ (top) and δ_{CP} (bottom) assuming inverted mass hierarchy and for Gd-tag analysis (violet) and current standard SK analysis (black). The horizontal dashed lines represent the 68, 90 and 99% confidence levels, which correspond to χ^2 values of 1, 2.71 and 6.64, respectively. The dataset corresponds to that of SK-IV presented in Table 7.1

Table 6.14 Mass hierarchy and δ_{CP} sensitivities, assuming as true normal and inverted hierarchies. Comparison between standard SK analysis (black) and Gd-tag analysis (violet)

	Standard (true NH)	Gd-tag (true NH)	Standard (true IH)	Gd-tag (true IH)
$\Delta\chi^2(NH - IH)$	−1.62	−2.68	0.96	1.55
δ_{CP} fraction over 1σ (%)	22.4	39.1	6.0	28.0

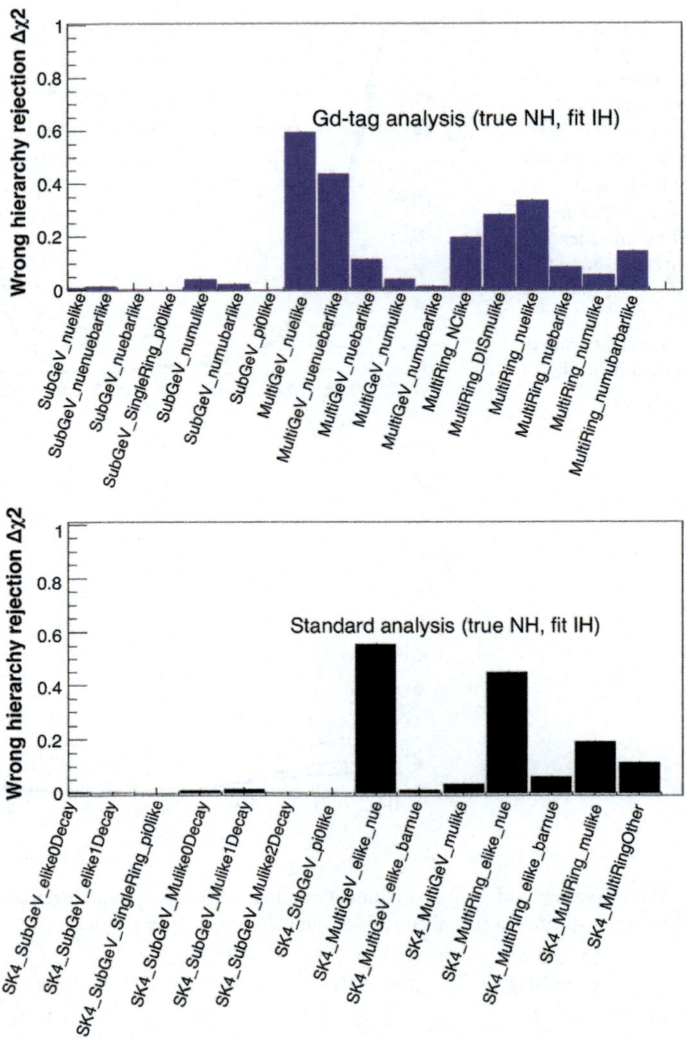

Fig. 6.45 χ^2 values for the inverted mass hierarchy rejection assuming true mass hierarchy is normal. Top plot, in violet, shows the $\Delta\chi^2(IH - NH)$ for each of the defined fully contained samples for SuperK-Gd atmospheric analysis. Bottom plot, in black, shows the $\Delta\chi^2(IH - NH)$ for each of the current fully contained samples in the official SK atmospheric analysis. The exposure assumed is 2519.90 days and the oscillation parameters assumed in Table 6.13

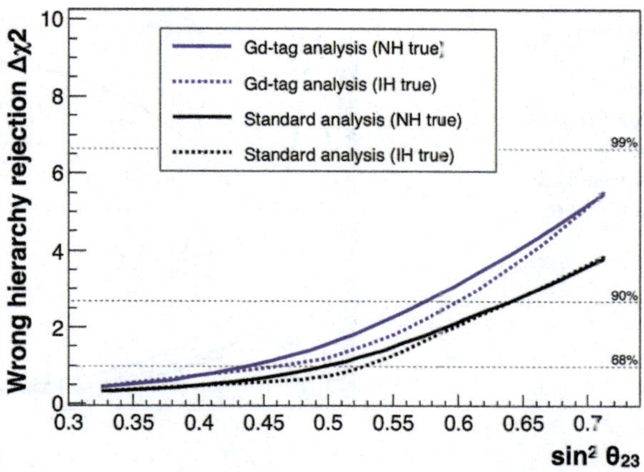

Fig. 6.46 Wrong mass hierarchy sensitivity as function of the θ_{23} oscillation parameter using atmospheric neutrinos for the Gd-neutron tagging (violet) and current SK standard analyses (black). The analysis assumes both possible mass orderings, normal (solid lines) and inverted (dashed), and fixed $\delta_{CP} = 4.19$ and $|\Delta m^2_{32}| = 0.0025$ eV^2. The horizontal dashed lines represent the 68, 90 and 99% confidence levels, which correspond to χ^2 values of 1, 2.71 and 6.64, respectively. The exposure time is considered to be 2519.9 days, the same as currently for the SK-IV dataset and shown in Table 7.1, and for the oscillation parameters listed in Table 6.13

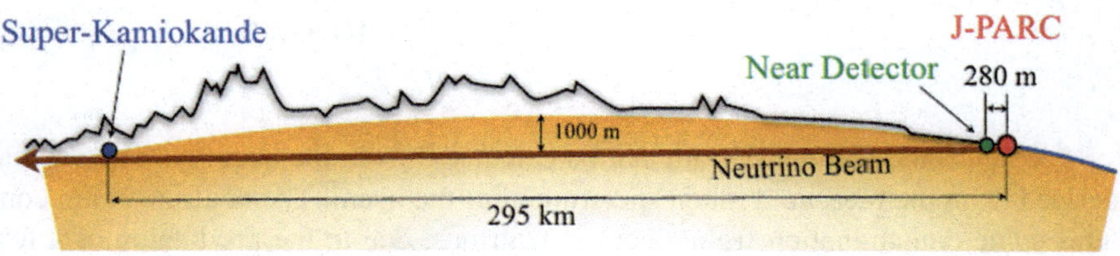

Fig. 6.47 Diagram of T2K neutrino beam produced at J-PARC and measured at ND280 and SK, near and far detectors respectively

6.7.1 The T2K Experiment

T2K neutrinos are produced at the Japan Proton Accelerator Research Complex (J-PARC), 295 km away from its far detector, Super-Kamiokande (Fig. 6.47).

In this facility, protons are accelerated in a synchrotron achieving 30 GeV in the main ring. Then, the proton beam directed towards the neutrino beamline, extracting eight circulating bunches of protons. These protons are made collide with the graphite target, resulting in the production of pions. These are focused with several magnetic horns and let decay along a 96 m tunnel. At the end of the decay volume, a block of graphite and iron is placed for avoiding any particle, except neutrinos and a few muons with $E_\mu > 5$ GeV, go into the neutrino beam. The final result is a nearly pure beam of ν_μ or $\overline{\nu}_\mu$, depending on the neutrino beam mode.

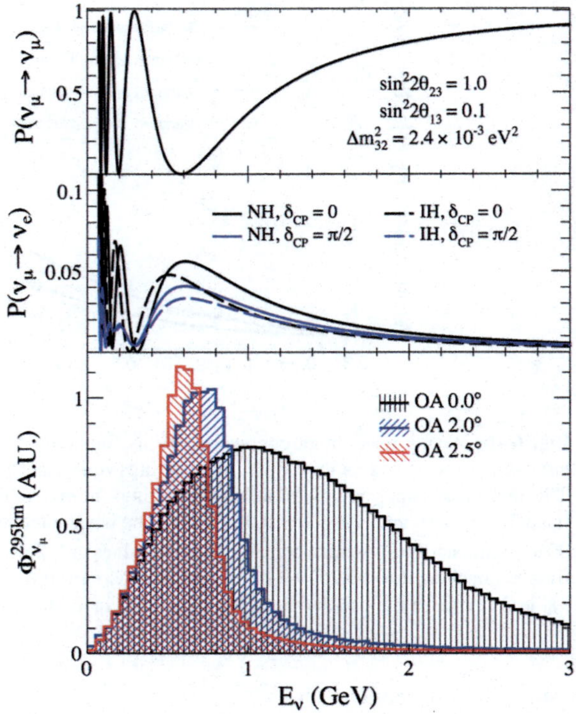

Fig. 6.48 T2K neutrino energy spectrum at Super-Kamiokande for different angles relative to the beam axis. The ν_μ survival probability, and ν_e appearance probability are shown

The neutrino energy spectrum can be calculated from the beam parameters and derived from the measured muon spectrum after the beam. The neutrino beam contains some contamination from electron neutrinos, due to the production of a few kaons, in addition to pions, when protons collide with the graphite target.

The neutrino beam is specially design so the oscillation probability of ν_μ to ν_e reaches its maximum, making easier to measure the neutrino oscillation effects. In order to make the energy spectrum narrower and peaked at maximal oscillation probability, the neutrino beam is displaced 2.5° from the direction to SK, as seen in Fig. 6.48. This facility is equipped with a near detector as well, which carries cross-section, energy spectrum and flux studies of the neutrino beam. This provides great knowledge about the neutrino characteristics, decreasing the associated systematic errors. ND280 is a compendium of various facilities, placed 280 m from J-PARC. It is formed by INGRID, a scintillation light detector made up by sixteen modules of nine iron plates and eleven tracking scintillators each and placed in front of the neutrino beam, and ND280 placed 2.5° off-axis composed by π^0 detector and calorimeter, an electromagnetic calorimeter, a muon detector, fine grained detectors (FGD) and argon time projection chambers (TPC).

6.7.2 *Monte Carlo Simulation and Gd-Neutron Tagging*

The analysis presented in this thesis is done with external simulation tools, not the T2K official ones. The Monte Carlo simulation used is that of atmospheric neutrinos in SK, but weighted to match the T2K spectra, flux and direction for all types of neutrinos. This means that the amount of neutrons detected for each event is previously determined from the atmospheric simulator as in Sect. 6.6.1 on page 109. This simulation was used in Sects. 6.7.4 and 6.7.5, to determine with accuracy the performance and behaviour of the neutrino-antineutrino separation and the neutron-corrected energy.

6.7.3 *T2K Neutrino Event Selection*

The T2K reduction software categorises the beam data into three samples: FC, OD and LE. The former includes the fully contained events. OD category correspond to those events with outer detector activity over the threshold (15 hits). LE events are those with energy below the FC energy threshold.

The reconstruction of T2K events is analogous to that made by SK for atmospheric neutrinos, explained in Sect. 6.6.1.

6.7.4 *Neutrino-Antineutrino Separation*

In the case of T2K events, the neutrino-antineutrino separations are done, only, for fully contained, as for the atmospheric neutrinos.

Due to the precision required by the T2K experiment, the neutrino and antineutrino separation is performed taking into account the different shape of the spectra. For that, the separation is, independently, done for five different energy regions, [0.1, 0.3) GeV, [0.3, 0.5) GeV, [0.5, 0.7) GeV, [0.7, 1.0) GeV and $\geq$1.0 GeV. Both neutrino beam modes have sufficient similar energy spectrum for neutrinos and neutrinos and, therefore, the likelihood distributions for their separation will be the same. As anticipated in Sect. 5.4, for this analysis, a neural network was used to obtain the likelihood distribution using just two variables, the number of neutrons in the final state and the cosine of the charged lepton scattering angle, θ_{scat}. The neural network architecture is the analogous to that used in the previous atmospheric neutrino section, one hidden layer with $N + 5$ neurones.

The number of decay-electrons is already used in T2K for sample separation and, therefore, it is not considered in this separation method.

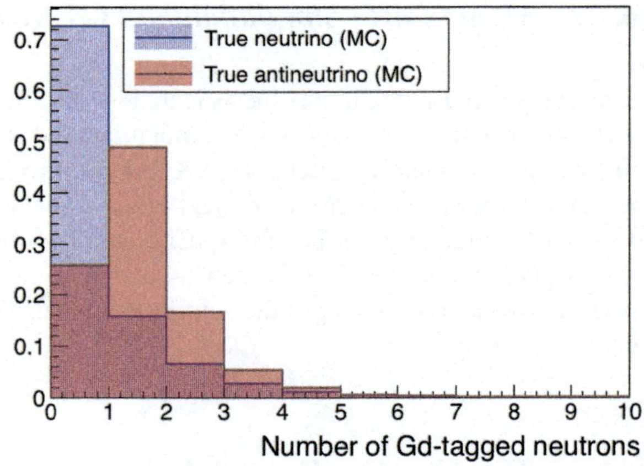

Fig. 6.49 Number of Gd-tagged neutrons for neutrinos (blue) and antineutrinos (red) in the *e*-like sample. Neutrino and antineutrino fluxes are normalised to unit

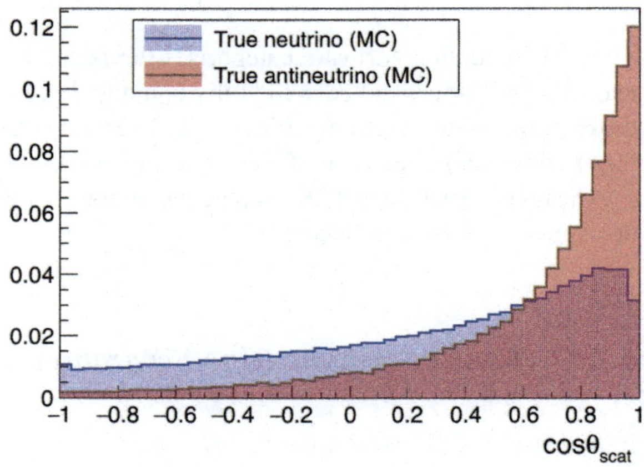

Fig. 6.50 Cosine of the scattering angle for neutrinos (blue) and antineutrinos (red) in the *e*-like sample. Neutrino and antineutrino fluxes are normalised to unit

6.7.4.1 *e*-like Neutrino Sample

The distributions of the variables used in the neural network are shown in Figs. 6.49 and 6.50 for the whole T2K neutrino spectrum.

The neural network outputs, for each of the energy regions, are shown in Figs. 6.51 and 6.52.

Once the selection is done based on the likelihood distributions, events are classified into neutrino or antineutrino. Table 6.15 shows the efficiency of these samples for correctly categorising neutrinos and antineutrinos.

The method is then, very efficient in detecting neutrinos and antineutrinos as such, improving the already high purity of the neutrino and antineutrino beam modes.

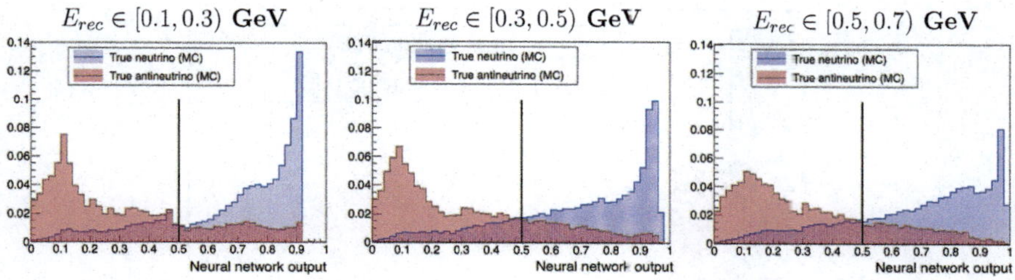

Fig. 6.51 Neural network output distributions for the five energy ranges and for e-like events (1 of 2). Neutrino and antineutrino fluxes are normalised to unit

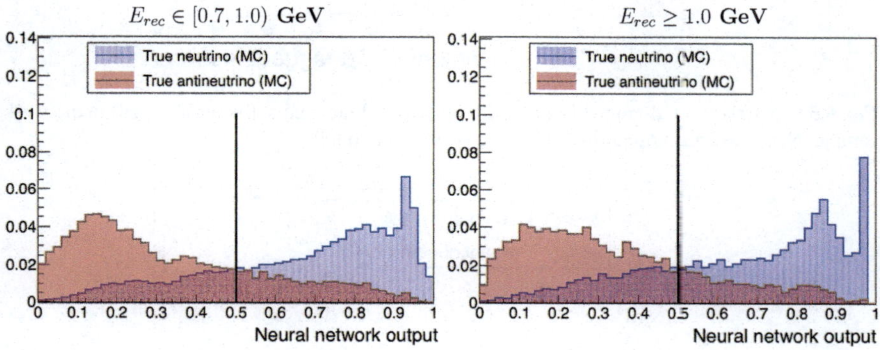

Fig. 6.52 Neural network output distributions for the five energy ranges and for e-like events (2 of 2). Neutrino and antineutrino fluxes are normalised to unit

Table 6.15 Efficiencies of each of the resulting ν_e and $\overline{\nu}_e$-like samples in the five energy ranges

	Efficiency of samples				
	[0.1, 0.3) GeV	[0.3, 0.5) GeV	[0.5, 0.7) GeV	[0.7, 1.0) GeV	≥1.0 GeV
ν_e-like (%)	78.0	78.3	76.8	75.7	72.9
$\overline{\nu}_e$-like (%)	77.1	77.0	77.5	76.5	77.3

6.7.4.2 μ-like Neutrino Sample

For μ-like events the procedure is, basically, the same as before The corresponding distributions for the neutron multiplicity and the cosine of the scattering angle are shown in Figs. 6.53 and 6.54.

The obtained likelihood distributions from the neural network are in Figs. 6.55 and 6.56.

The efficiencies of the method for each energy region is shown in Table 6.16.

For μ-like events the efficiency of the separation is a smaller in general, mainly due to the less separation power of the scattering angle, as compared with the e-like case.

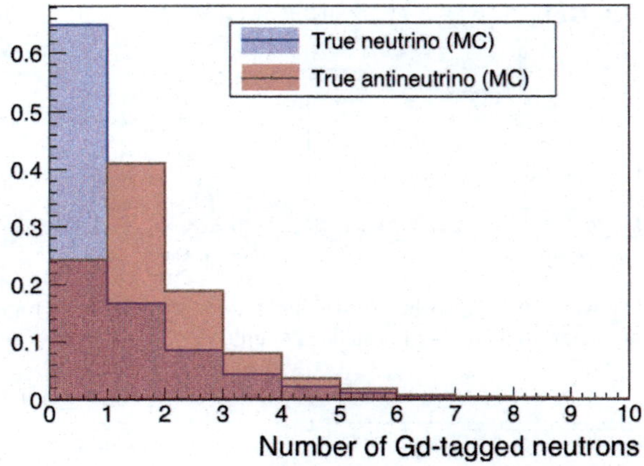

Fig. 6.53 Number of Gd-tagged neutrons for neutrinos (blue) and antineutrinos (red) in the μ-like sample. Neutrino and antineutrino fluxes are normalised to unit

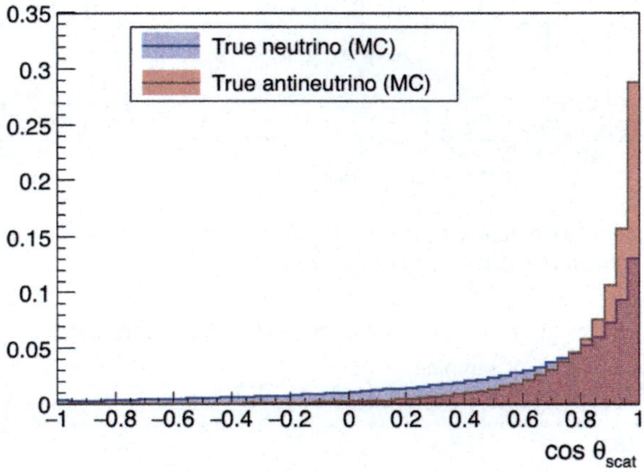

Fig. 6.54 Cosine of the scattering angle for neutrinos (blue) and antineutrinos (red) in the μ-like sample. Neutrino and antineutrino fluxes are normalised to unit

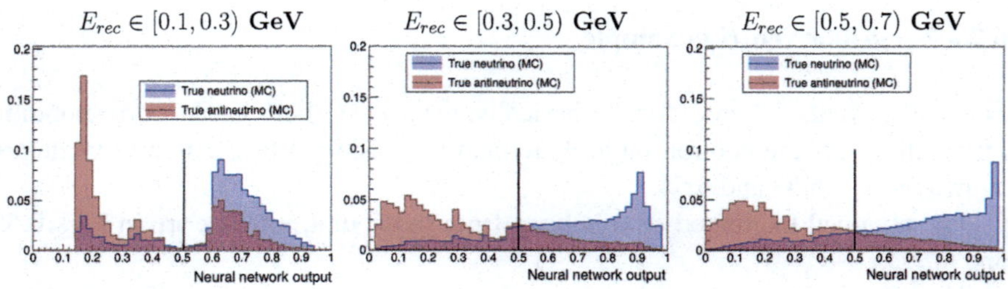

Fig. 6.55 Neural network output distributions for the five energy ranges and for μ-like events (1 of 2). Neutrino and antineutrino fluxes are normalised to unit

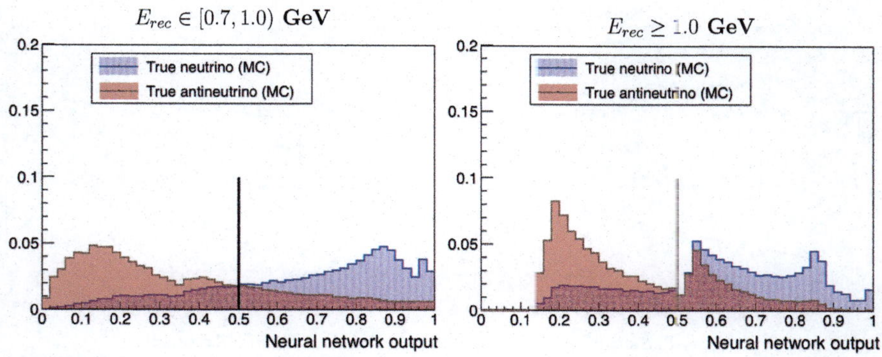

Fig. 6.56 Neural network output distributions for the five energy ranges and for μ-like events (2 of 2). Neutrino and antineutrino fluxes are normalised to unit

Table 6.16 Efficiencies of each of the resulting ν_μ and $\overline{\nu}_\mu$-like samples in the five energy ranges

	Efficiency of samples				
	[0.1, 0.3) GeV	[0.3, 0.5) GeV	[0.5, 0.7) GeV	[0.7, 1.0) GeV	≥1.0 GeV
ν_μ-like (%)	79.7	76.9	76.1	74.8	72.0
$\overline{\nu}_\mu$-like (%)	64.9	73.3	76.5	75.8	68.8

6.7.5 *Neutron-Corrected Reconstructed Neutrino Energy*

The energy correction is also studied just for the fully contained events, where neutron tagging is available.

The neutrino energy is very well reconstructed in T2K because of the accurate knowledge in the direction of the interacting neutrino. Thus, the lepton momentum can be corrected by applying the information about the scattering angle to Eq. (6.23), where it is assumed that the interaction is a quasi-elastic scattering.

$$E_{rec}^{\theta_{scat}} = \frac{(M_n - V)E_\mu - m_\mu^2/2 + (M_p^2 - M_n^2)/2}{M_n - V - E_\mu + |\mathbf{p}_\mu| \cos\theta_{scat}} \tag{6.23}$$

Although, there is very little room for improvement in the energy reconstruction, the neutron corrections to the energy are applied in order to compare their performance with that of Eq. (6.23).

$$E_{rec}^{Gd} = E_{vis}(1 + f(\text{Gd-neutrons})) \tag{6.24}$$

The neutron correction function was computed analogously to the atmospheric case for each of the previously defined samples. This is shown in Fig. 6.57.

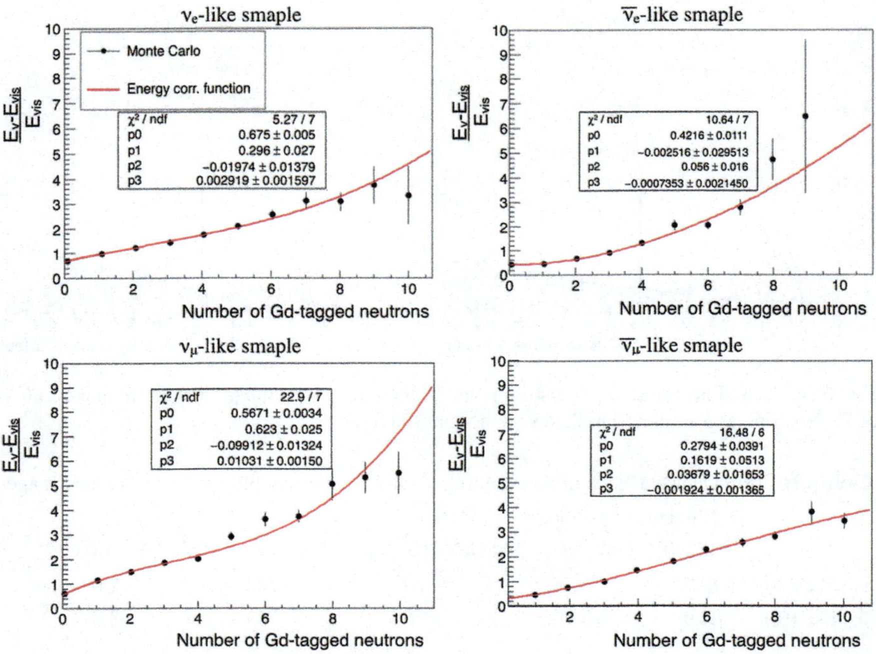

Fig. 6.57 Plot showing the dependence of the quotient of the non-reconstructed energy and the visible energy with the number of Gd-tagged neutrons for *e*-like T2K neutrinos

In Fig. 6.58, the different spectra of the true neutrino energy, the angle-corrected energy and the neutron-corrected are shown. Furthermore, in order to compare their performance, the energy pull distributions for both corrected energies are plotted in Fig. 6.59.

In all four samples, the reconstructed energy from the neutron multiplicity exhibits a similar performance than that obtained from the scattering angle. However, the pull distribution parameters, listed in Table 6.17, show that the neutron-corrected energy tends to have mean values than the angle-corrected one. This is because, although both distributions peak near zero, the neutron-corrected ones have larger tails.

Despite of being slightly worse than the angle energy correction, the neutron corrected energy improves very significantly the energy reconstruction. This study confirms the validity and good performance of the neutron corrections to the neutrino energy, and can be used as motivation for using them in the atmospheric neutrino analyses.

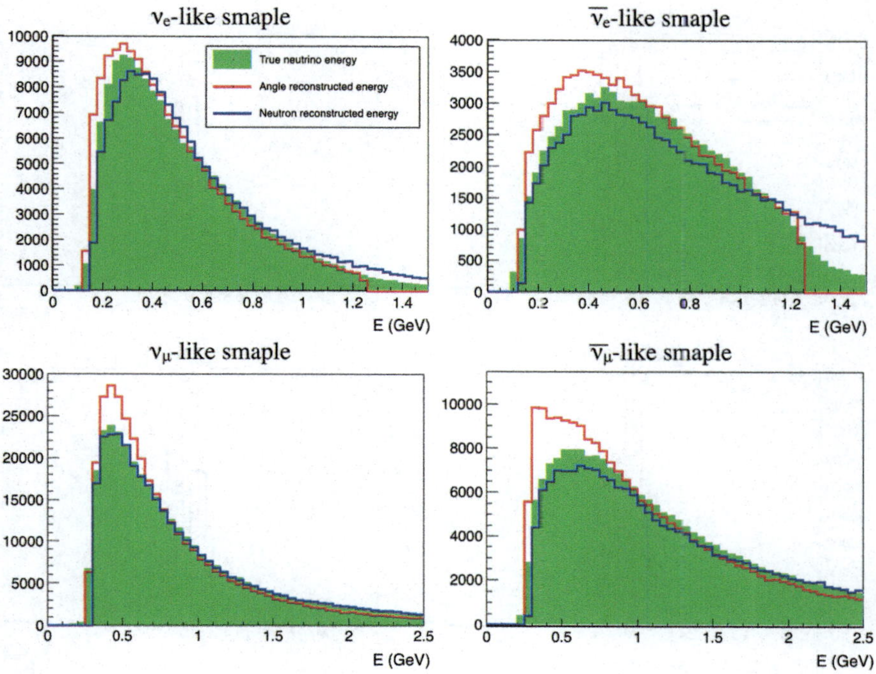

Fig. 6.58 T2K neutrino spectra for true, angle reconstructed and neutron reconstructed energies assuming the beam in neutrino mode. Flux normalised to arbitrary units

6.7.6 Oscillation Analysis

In order to see how the neutrino and antineutrino separation affects the oscillation analysis, sensitivity studies for the CP violation phase were done because it is the parameter most affected by the ν or $\overline{\nu}$ purity of the sample.

The oscillation analysis for T2K is done using the program GLoBES (General Long Baseline Experiment Simulator), [Huber05], with the settings for T2K.

The sensitivity studies presented here compare the standard analysis with that using the the neutrino-antineutrino separation with Gd-neutron tagging.

In this sensitivity study, the assumed mixing parameters are, unless otherwise stated, shown in Table 6.18. In addition, since the neutrino mass hierarchy is yet to be known, several scenarios are explored.

The beam configuration is such that the exposure to the neutrino beam is three times the time that the antineutrino beam is on, summing up in total 3.9×10^{21} POT. The sensitivity studies are performed with no systematic errors. The new uncertainties associated to the ν-$\overline{\nu}$ separation are considered negligible in this analysis, given the high purity of the neutrino beam.

The sensitivity is obtained via a χ^2 method, analogous to that used for the atmospheric neutrino case and described in Sect. 6.6.5.1.

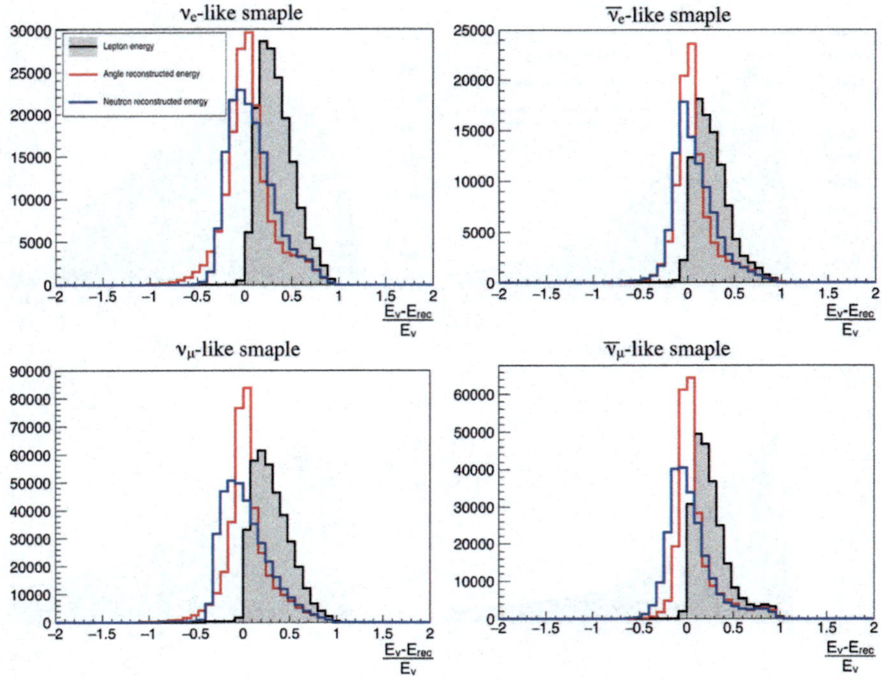

Fig. 6.59 Neutron corrected and angle reconstructed energy pulls for the *e*-like sample with the beam in neutrino mode

Table 6.17 Parameters of the angle and neutron reconstructed neutrino energy pulls for the neutrino mode beam

	$E_{rec}^{\theta_{scat}}$			E_{rec}^{Gd}		
	Mode	Mean	RMS	Mode	Mean	RMS
ν_e-like	0.014	0.031	0.256	−0.002	0.152	0.215
$\overline{\nu}_e$-like	0.027	0.086	0.208	0.009	0.098	0.214
ν_μ-like	0.007	0.035	0.240	0.012	0.093	0.248
$\overline{\nu}_\mu$-like	−0.009	0.097	0.207	0.006	−0.028	0.316

Table 6.18 True oscillation parameters used for the sensitivity studies of the atmospheric oscillation analysis

Oscillation parm.	Value	Treatment
Δm_{32}^2	$2.40 \cdot 10^{-3}$ eV^2/c^2	Fitted
$\sin^2\theta_{23}$	0.50	Fitted
Δm_{21}^2	7.60×10^{-5} eV^2/c^2	Fixed
$\sin^2\theta_{12}$	0.319	Fixed
$\sin^2\theta_{13}$	0.0257	Fitted
δ_{CP}	0	Fitted

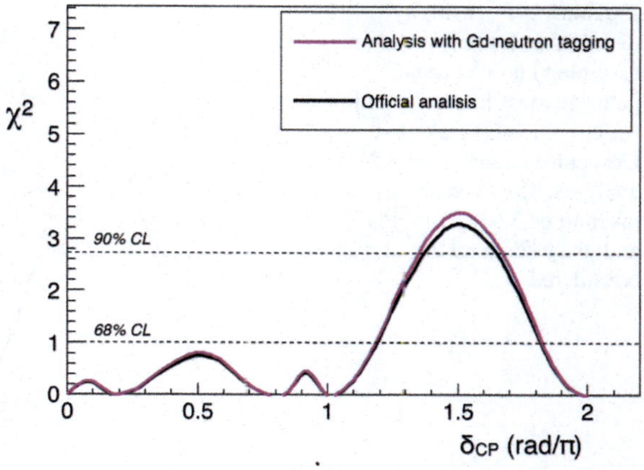

Fig. 6.60 χ^2 sensitivity distribution for δ_{CP} violation assuming unknown normal neutrino mass hierarchy and for the official (black) and Gd-neutron tagging (pink) analyses. The exposure is taken to be 3.9×10^{21} POT and no systematics are considered

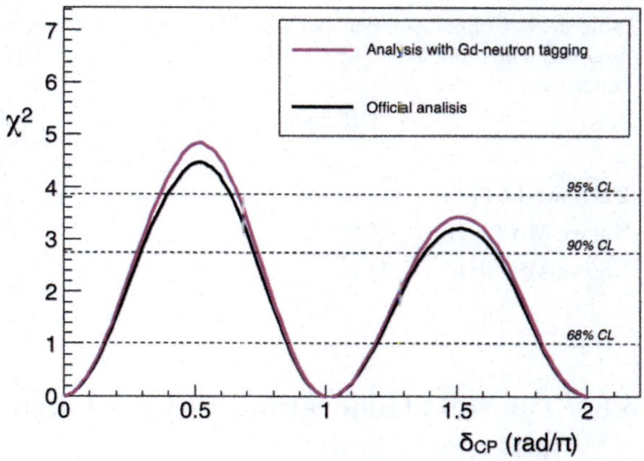

Fig. 6.61 χ^2 sensitivity distribution for δ_{CP} violation assuming known inverted neutrino mass hierarchy and for the official (black) and Gd-neutron tagging (pink) analyses. The exposure is taken to be 3.9×10^{21} POT and no systematics are considered

6.7.6.1 Oscillation Sensitivity Study

Three possible scenarios for the mass hierarchy are assumed: unknown normal mass hierarchy, known normal mass hierarchy and known inverted hierarchy. The sensitivity studies for these are shown in Figs. 6.60, 6.61 and 6.62 respectively. These studies are done for the sensitivity to the CP conservation rejection.

There exists an slight improvement in the sensitivity to δ_{CP} by using the neutrino and antineutrino classification in all three scenarios. This can be better seen in Table 6.19. Here, the fraction of δ_{CP} excluded at 90% confidence level (CL) is compared for both analyses and the three scenarios.

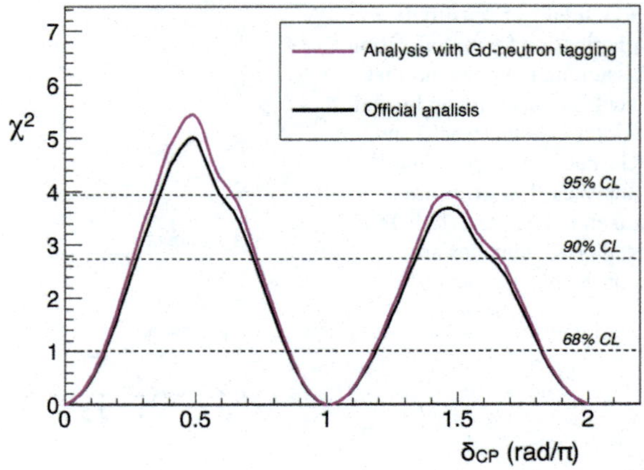

Fig. 6.62 χ^2 sensitivity distribution for δ_{CP} violation assuming known normal neutrino mass hierarchy and for the official (black) and Gd-neutron tagging (pink) analyses. The exposure is taken to be 3.9×10^{21} POT and no systematics are considered

Table 6.19 Comparison between the official and the Gd-neutron analyses of the fraction of δ_{CP} discarded with more than 90% CL and maximum of the χ^2 distribution for the three mass hierarchy scenarios

	Official analysis		Analysis with Gd-tagging	
	δ_{CP} over 90% CL (%)	Max(χ^2)	δ_{CP} over 90% CL (%)	Max(χ^2)
Unknown MH (NH)	13.9	3.31	16.3	3.53
Known MH (NH)	34.8	4.43	38.8	4.84
Known MH (IH)	34.6	5.03	38.2	5.55

6.8 The Next Generation: Physics Potential of Gd-Doping in Hyper-Kamiokande

Hyper-Kamiokande is a next generation underground water Cherenkov, projected as the successor of Super-Kamiokande.

The main goals of Hyper-Kamiokande (HK) are the neutrino oscillation studies from long-baseline (from the upgraded J-PARC beam), atmospheric and solar neutrinos, as well as proton-decay searches. Amongst the neutrino parameters, HK aims to measure the neutrino mass ordering and the leptonic CP violating phase.

In addition, HK will observe an enormous number of neutrino events if a supernova occurs nearby, providing much needed experimental results to researchers seeking to understand the mechanism of the explosion.

In this section, the Hyper-Kamiokande next-generation experiment is reviewed in Sect. 6.8.1. Additionally, the section focuses on the improvements in the sensitivity, due to Gd-neutron tagging, for atmospheric, in Sect. 6.8.2.1, and long-baseline, in Sect. 6.8.3, neutrinos.

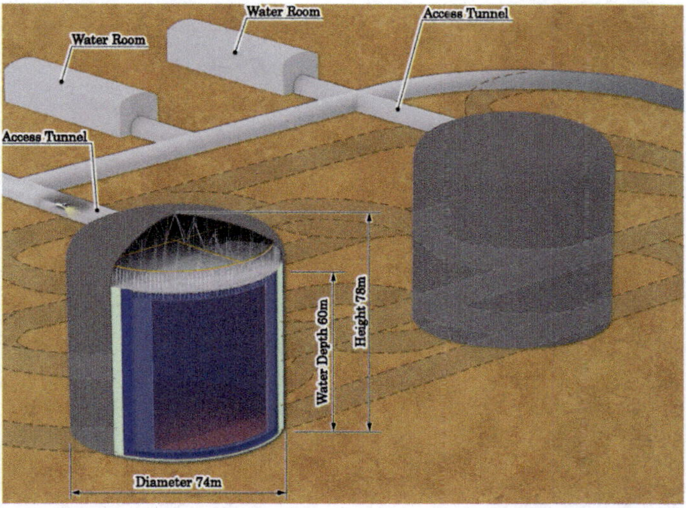

Fig. 6.63 Schematic view for the configuration of two cylindrical tank instrumented with high density (40% photocoverage) PMTs

6.8.1 Detector Description

Three detector configurations are still being considered for HK, in addition to the initial proposal described in [Abe11a, Abe16a]. The latter consists in two tanks with 20% photocoverage, summing up a volume of 0.99 megatons.

On the other hand, the current three detector proposals are: three tanks with 13% photocoverage and a volume of 774 kton, one tank with 40% photocoverage and 258 kton and two tanks with 40% photocoverage and a total volume of 526 kton (more that 10 times larger than SK). The latter is currently the most favoured one, but the construction is planned in two stages. The second tank would be built around 6 years after finishing the first one (Fig. 6.63).

This variety of possibilities open a bunch of possible sensitivity studies. However, in this section, it is assumed that the volume is 258 kton.

The photocoverage is also very important for the implementation of the neutron-tagging improvements. In this case, the neutron-tagging methods assume 20% PMT coverage and the addition of Gd at 0.2% concentration, as in the case of SuperK-Gd. The neutron-tagging efficiency, with this assumption, is estimated to be 70%, according to [Abe14b] (Fig. 6.64).

Despite of this, recent studies within the HK collaboration, indicate that hydrogen neutron tagging will also have 70% efficiency due to the photo-sensor response improvement, seen in Figs. 6.65, 6.66 and 6.67. The studies presented in this section are, then, compatible with Gd-doped HK with 20% photocoverage, with Gd-neutron tagging, and HK with pure water and 40% PMT coverage, with H-neutron tagging.

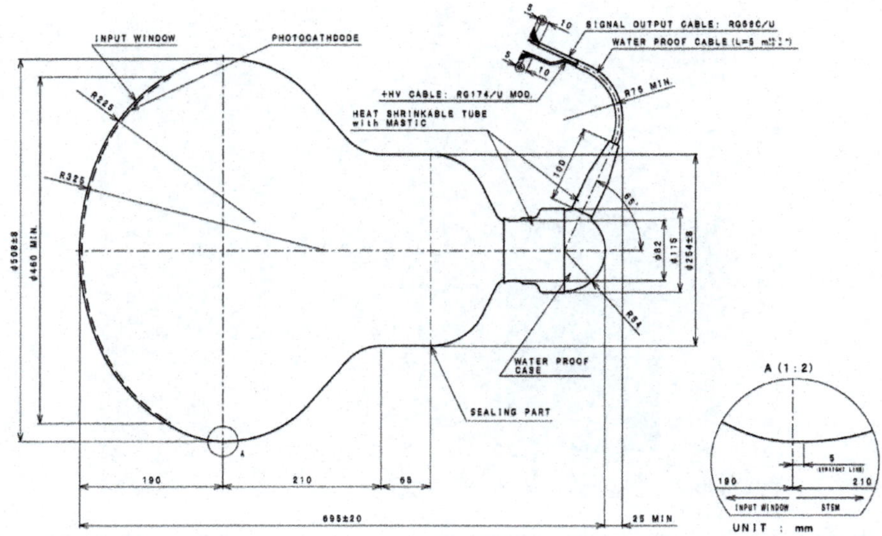

Fig. 6.64 Design of the HQE box-and-line R12860 PMT

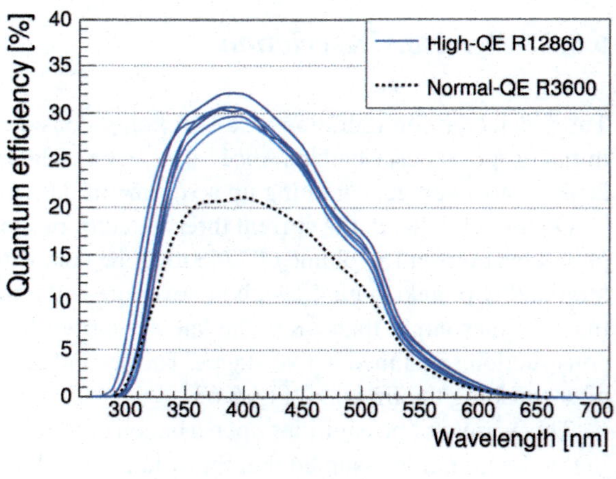

Fig. 6.65 Measured QE for six high-QE R12860 (solid lines) and a normal R3600 (black dashed line)

The ID photosensors are newly developed for HyperK (Hamamatsu R12860-HQE), achieving a higher quantum efficiency (QE) and optimised box-and-line dynode. These PMTs are developed jointly with Hamamatsu and their properties are listed in Table 6.20.

One of the most important characteristics of these new PMTs is the enhancement of the quantum efficiency as compared to that of SK PMTs, but also charge and timing resolutions are significantly improved as shown in Figs. 6.65, 6.66 and 6.67.

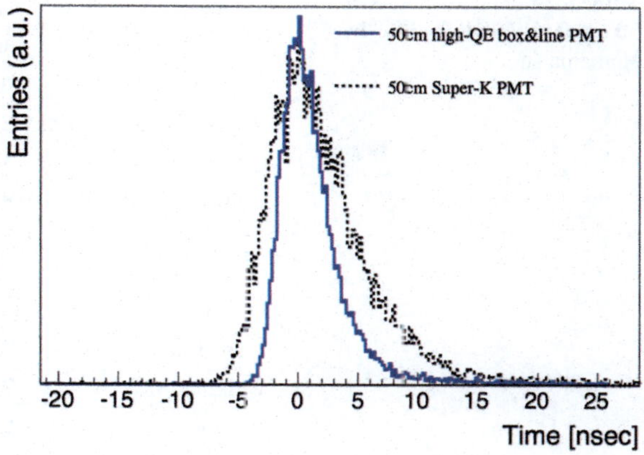

Fig. 6.66 Transit time distribution at 1 p.e. of HQE PMTs (blue) and Super-K PMTs (dotted line)

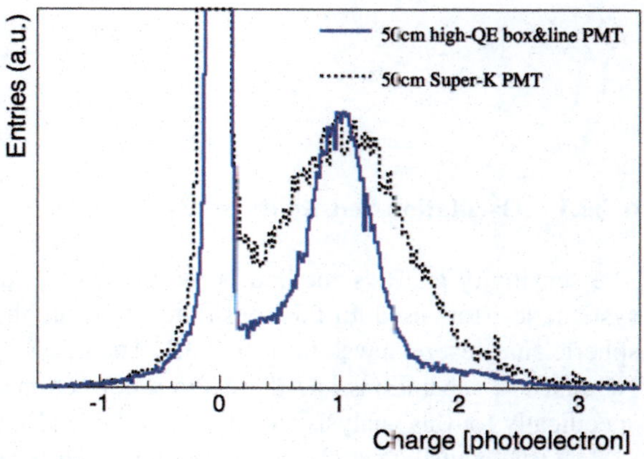

Fig. 6.67 Single photoelectron of HQE PMTs (blue) compared with the Super-K PMT (dotted line)

6.8.2 Atmospheric Neutrinos

The simulation, reconstruction and reduction of atmospheric neutrinos is taken the same as that for Super-Kamiokande, described in Sect. 6.6.1, but scaling the exposure to the HK assumed dimensions. Since the atmospheric oscillation analysis is done for 5 years of exposure, just one tank of 258 kton is considered, the first stage of the HK project.

The improvements based on neutron tagging for atmospheric neutrinos described in Sections 5.3–5.5, are applied for the HK detector assuming 70% neutron tagging efficiency. For the sake of brevity, these are not explicitly shown this time, but the results are fairly similar to those shown in Sects. 6.6.2–6.6.4 for the SuperK-Gd project.

Table 6.20 Specifications of 50 cm R12860-HQE PMT by Hamamatsu

Shape	Hemispherical
Photocathode area	50 cm diameter
Bulb material	Borosilicate glass (~3 mm)
Photocathode material	Bialkali (Sb-K-Cs)
Quantum efficiency	30% at $\lambda = 390$ nm
Collection efficiency	95% at 10^7 gain
Dynodes	10 stage box-and-line
Gain	10^7 at ~2000 V
Dark rate	~8 kHz at ~10^7 gain and 13 °C
Transit time spread	2.7 ns for single PE
Weight	7.5 kg (without cable)
Volume	61050 cm^3
Pressure tolerance	9 kg/cm^2 water proof

6.8.2.1 Oscillation Sensitivity Study

The sensitivity analysis method is the binned χ^2, discussed in Sect. 6.6.5.1. The systematic errors used for this sensitivity study are those from the current SK atmospheric analysis, reviewed in Sect. 7.3.1. The new systematic errors, due to the CC-NC and ν-$\overline{\nu}$ separations and the neutron corrections for the energy, are computed specifically for this analysis and shown in Appendix B.

The true neutrino oscillation parameters used for the sensitivity analysis are the same as in Table 6.13.

The sensitivity study is shown for the $\sin^2\theta_{23}$ and δ_{CP} oscillation parameters, and assuming normal, in Fig. 6.68 and inverted Fig. 6.69 mass orderings.

The 70%-analysis improves very significantly the sensitivity to the neutrino mass ordering and to the CP violating phase. The corresponding significance values for these plots are summarised in Table 6.21.

Furthermore, the wrong mass hierarchy rejection sensitivity is also improved, for all values of θ_{23}, as compared to the standard analysis. This is shown in Fig. 6.70.

6.8.3 Long Baseline Neutrinos

Long baseline neutrinos to HK share most of the features describe for T2K in Sect. 6.7, as the neutrino beam will be located at JPARC and the candidates for the HK site are close to the current location of SK. The baseline, therefore, would be of 295 km as well.

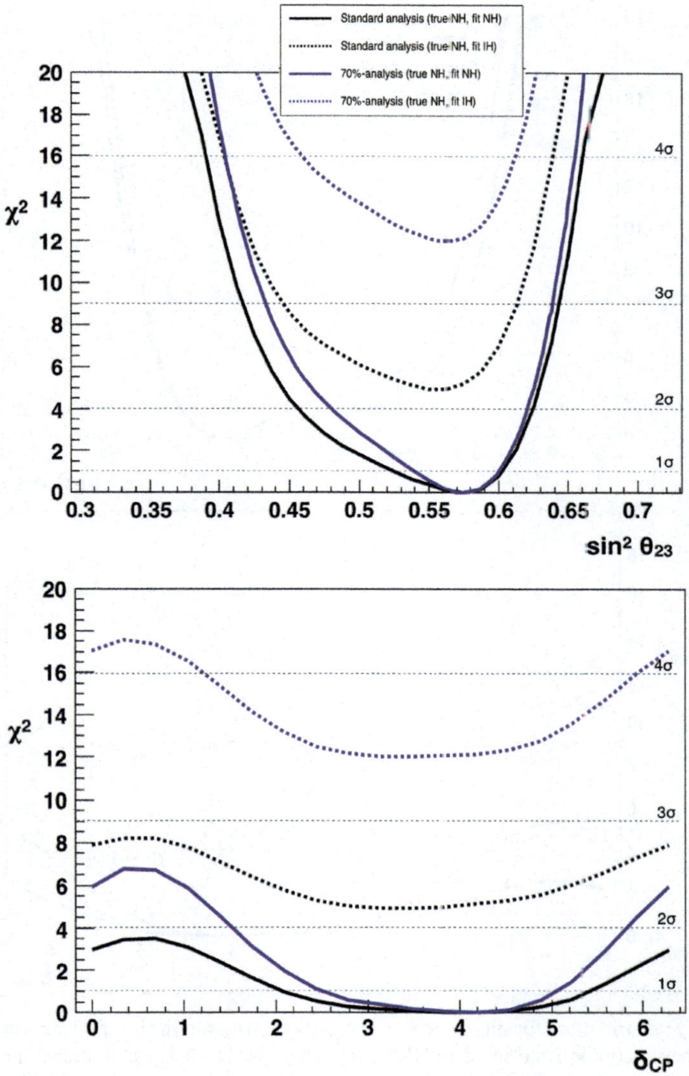

Fig. 6.68 χ^2 distributions for $\sin^2\theta_{23}$ (top) and δ_{CP} (bottom) assuming normal mass hierarchy and for 70%-analysis (violet) and standard HK analysis (black). The horizontal dashed lines represent the 1σ, 2σ, 3σ and 4σ confidence levels. The exposure time is considered to be 5 years of HK with one tank of 258 kton volume

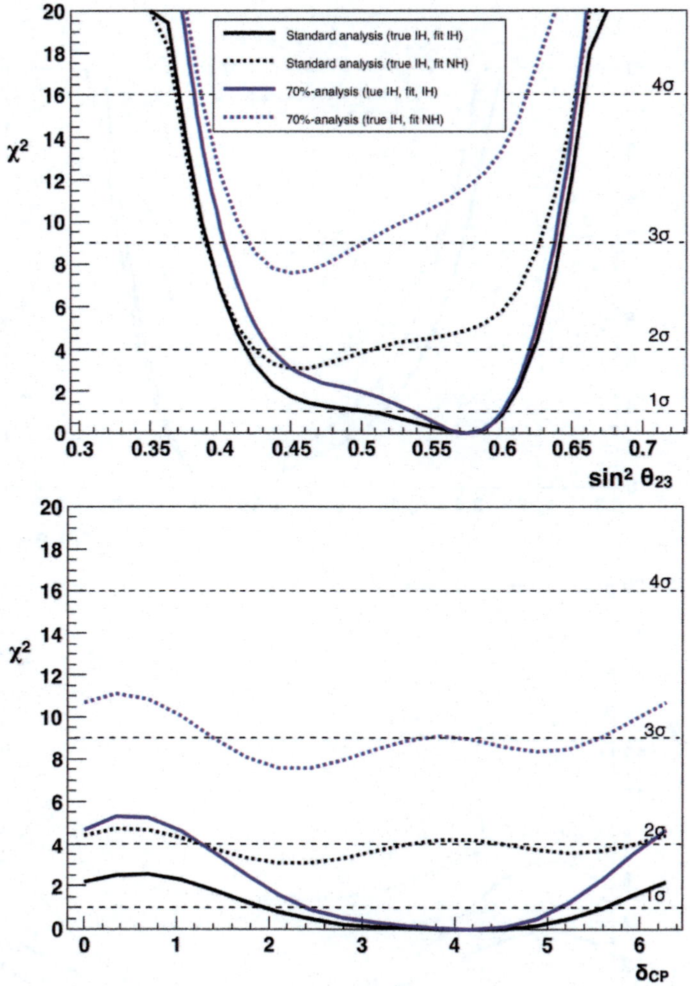

Fig. 6.69 χ^2 distributions for $\sin^2\theta_{23}$ (top) and δ_{CP} (bottom) assuming inverted mass hierarchy and for 70%-analysis (violet) and standard HK (black) analysis. The horizontal dashed lines represent the 1σ, 2σ, 3σ and 4σ confidence levels. The exposure time is considered to be 5 years of HK with one tank of 258 kton volume

Table 6.21 Mass hierarchy and δ_{CP} sensitivities, assuming as true normal and inverted hierarchies. Comparison for the 70%-analysis and standard HK analysis

	Standard (true NH)	70%-tag (true NH)	Standard (true IH)	70%-tag (true IH)
$\Delta\chi^2(NH - IH)$	−4.94	−11.99	3.06	7.55
δ_{CP} fraction over 1σ (%)	32.8	55.6	33.2	49.9

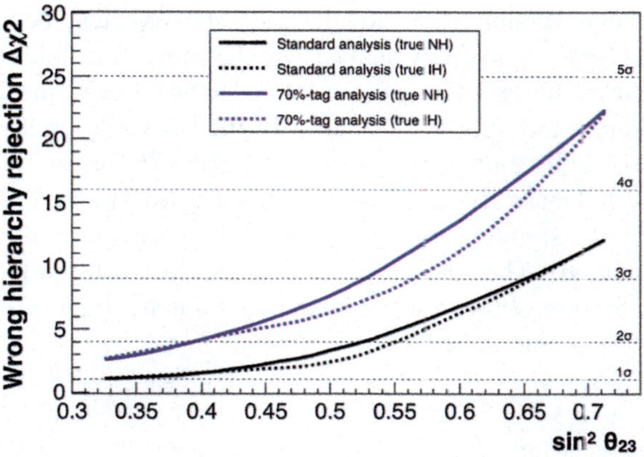

Fig. 6.70 Wrong mass hierarchy sensitivity as function of the θ_{23} oscillation parameter using atmospheric neutrinos for the 70%-neutron tagging (violet) and standard analyses (black). The analysis assumes both possible mass orderings, normal (solid lines) and inverted (dashed), and fixed $\delta_{CP} = 4.19$ and $|\Delta m^2_{32}| = 0.0025$ eV^2. The horizontal dashed lines represent the $1\sigma, 2\sigma, 3\sigma$, 4σ and 5σ confidence levels. The exposure time is considered to be 5 years of HK with one tank of 258 kton volume

The current plan for long baseline neutrinos in HK depends not only on the far detector layout, but also on the future upgrades of the J-PARC facility concerning the neutrino beam. According to Abe et al. [Abe16a], it is planned that, by the time of the Hyper-Kamiokande operation, the J-PARC neutrino beam achieves an intensity of 1.3 MW as result of various upgrades.

In addition, it is expected a considerable reduction of the systematic errors associated to this analysis. This is due to the knowledge to be acquired, in the next years, with the current ND280 near detector of T2K. Furthermore, two new near detectors are being considered, these are water-Čerenkov. The systematic errors are expected to be reduced with the addition of measurements on the water targets and not only on carbon. Namely, two proposals are made, the nuPRISM detector, [Bhadra14], and the TITUS detector, [Andreopoulos16], which is, additionally, a Gd-loaded detector.

As in the case of the atmospheric analysis for HK in Sect. 6.8.2, here the neutron-tagging improvements, developed for long baseline neutrinos in the SuperK-Gd project in Sect. 6.7, are applied to the case of 70% neutron tagging efficiency. These are not explicitly shown, but, instead, are directly implemented into the oscillation sensitivity study.

6.8.3.1 Oscillation Sensitivity Study

The oscillation analysis for long baseline neutrinos at HK, is done using the same tools as done for the T2K analysis, using GLoBES and introducing the improvements due to the H-neutron tagging in HK. The systematic errors considered here are those taken in the last up to date version of the HK design, [Abe16a]. True neutrino oscillation parameters used for the sensitivity analysis are the same as in Table 6.18.

Here, the neutrino mass ordering is considered to be know by the time this facility is running, therefore the sensitivity studies for the CP violation are performed for both hierarchies, but assuming they are known. The CP sensitivity plots assuming normal and inverted mass orderings are shown in Figs. 6.71 and 6.72.

In Table 6.22, the significance for rejecting CP conservation is summarised for both analysis methods. The analysis done with the 70% neutron tagging neutrino-antineutrino separation, shows a slight improvement in the fraction of the δ_{CP} phase rejected at the 5σ confidence level.

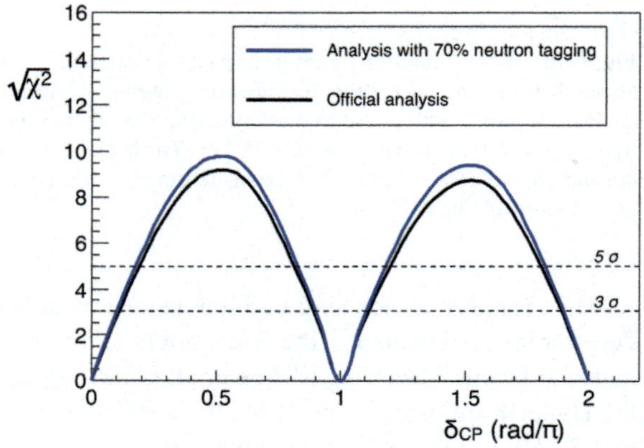

Fig. 6.71 χ^2 distribution for δ_{CP} assuming known inverted neutrino mass hierarchy and for the official (black) and 70% efficiency neutron tagging (blue) analyses

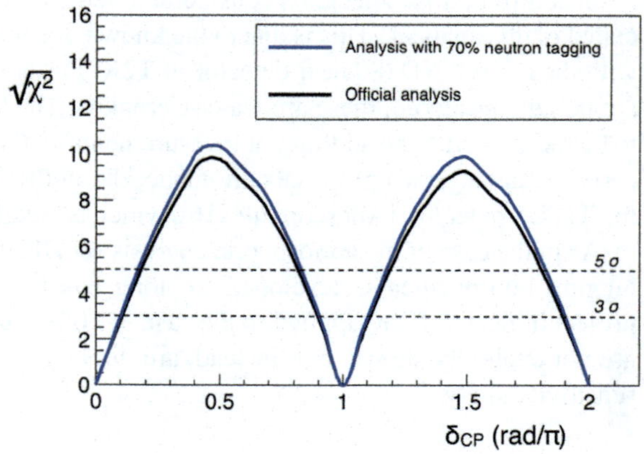

Fig. 6.72 χ^2 distribution for δ_{CP} assuming known normal neutrino mass hierarchy and for the official (black) and 70% efficiency neutron tagging (blue) analyses

Table 6.22 Comparison between the official and the 70%-neutron tagging analyses of the fraction of δ_{CP} discarded with more than 90% CL and maximum of the χ^2 distribution for both mass hierarchy scenarios

	Official analysis		Analysis with H-tagging	
	δ_{CP} over 5σ (%)	Max($\sqrt{\chi^2}$)	δ_{CP} over 5σ (%)	Max($\sqrt{\chi^2}$)
Known MH (NH)	61.3	9.10	64.9	9.71
Known MH (NH)	62.1	9.89	65.5	10.40

References

[Murayama02] H. Murayama et al., Energy spectra of reactor neutrinos at KamLAND. Phys. Rev. D **65**, 013012 (2002). https://doi.org/10.1103/PhysRevD.65.013012

[Li14] S.W. Li et al., First calculation of cosmic-ray muon spallation backgrounds for Mev astrophysical neutrino signals in Super-Kamiokande. Phys. Rev. C **89**, 045801 (2014). https://doi.org/10.1103/PhysRevC.89.045801

[Abe16c] K. Abe et al., Solar neutrino measurements in Super-Kamiokande-IV. Phys. Rev. D **94**, 052010 (2016). https://doi.org/10.1103/PhysRevD.94.052010

[Odrzywolek07] A. Odrzywolek et al., Future neutrino observations of nearby pre-supernova stars before core-collapse. AIP Conf. Proc. **944**, 109 (2007)

[Totani98b] T. Totani et al., Future detection of supernova neutrino burst and explosion mechanism. ApJ **496**, 216 (1998). https://doi.org/10.1086/305364

[Pik12] L.K. Pik, Study of the neutrino mass hierarchy with the atmospheric neutrino data observed in Super-Kamiokande. Ph.D. thesis, University of Tokyo, 2012

[Irvine14] T.J. Irvine, Development of neutron-tagging techniques and application to atmospheric neutrino oscillation analysis in Super-Kamiokande. Ph.D. thesis, University of Tokyo, 2014

[Abe13] K. Abe et al., Evidence for the appearance of atmospheric tau neutrinos in super-kamiokande. Phys. Rev. Lett. **110**(18), 181802 (2013)

[Adamson11] P. Adamson et al., Active to sterile neutrino mixing limits from neutral-current interactions in minos. Phys. Rev. Lett. **107**(1), 011802 (2011)

[Wendell08] R.A. Wendell, Three flavor oscillation analysis of atmospheric neutrinos in Super-Kamiokande. Ph.D. thesis, University of North Carolina, 2008

[Abe11b] K. Abe et al., The T2K experiment. Nucl. Instrum. Meth. **A659**, 106–135 (2011). https://doi.org/10.1016/j.nima.2011.06.067

[Huber05] P. Huber et al., Simulation of long-baseline neutrino oscillation experiments with globes: (general long baseline experiment simulator). Comput. Phys. Commun. **167**(3), 195–202 (2005)

[Abe11a] K. Abe et al., Letter of intent: the hyper-kamiokande experiment—detector design and physics potential (2011). arXiv:1109.3262

[Abe16a] K. Abe et al., Hyper-kamiokande: design report. KEK Preprint, 21 (2016)

[Abe14b] K. Abe et al., Search for proton decay via $p \to \nu k^+$ using 260 kiloton· year data of super-kamiokande. Phys. Rev. D **90**(072005) (2014)

[Bhadra14] S. Bhadra et al., Letter of intent to construct a nuPRISM detector in the J-PARC neutrino beamline (2014)

[Andreopoulos16] C. Andreopoulos et al., TITUS: the Tokai intermediate tank for the unoscillated spectrum (2016)

Chapter 7
Neutron-Tagging with Hydrogen in Super-Kamiokande IV: Global Atmospheric Neutrino Oscillation Analysis with SK

Currently, SK has the capability of tag neutrons through hydrogen-neutron captures. This technique is around four times less efficient than in the proposed SuperK, Gd, i.e. $\sim$20%. However, it is enough for seen the relevance of the improvements, brought by neutron tagging, in the atmospheric oscillation analysis.

Section 7.1 is focused on the review and description of the H-neutron tagging algorithm, only available in SK-IV, originally developed in [Irvine14].

In Sect. 7.2, all the tools developed in the context of the SuperK-Gd atmospheric analysis, in Sects. 6.6.2–6.6.4, are applied to the case of H-neutron tagging, H-tag analysis.

Finally, an oscillation sensitivity study, in Sect. 7.3.2, and the fit to the whole SK atmospheric data using the H-tag analysis is presented in Sect. 7.3.3. Here, the sensitivity comparison to the atmospheric oscillation parameters is done for the whole SK exposure, and comparing the analysis including the neutron-tagging techniques in SK-IV with the current official analysis.

The total exposure time for atmospheric neutrinos in SK is summarised in Table 7.1.

7.1 Hydrogen-Neutron Tagging

Hydrogen is present in the water molecules of SK, behaving almost as free protons, since they are not affected by nuclear forces. Hydrogen has a considerable cross-section, 0.329 ± 0.004 barns, for the capture of thermal neutrons, which, given the amount of H in SK, guarantees that all thermalised neutrons produced inside the tank will be captured by hydrogen. Once hydrogen captures a neutron, it is turned into deuterium, emitting a 2.2 MeV photon from its de-excitation. This is similar to the gadolinium case, but with a signal of significantly lower energy. The mean life-time

P. Fernández Menéndez, *Neutrino Physics in Present and Future Kamioka Water-Čerenkov Detectors with Neutron Tagging*, Springer Theses,
https://doi.org/10.1007/978-3-319-95086-0_7

Table 7.1 Exposure times of the different SK phases for the atmospheric neutrino dataset

SK phase	Sample	Exposure (days)
SK-I	FC and PC	1489.199
	UpMu	1645.910
SK-II	FC and PC	798.599
	UpMu	827.744
SK-III	FC and PC	518.080
	UpMu	635.600
SK-IV	FC and PC	2519.900
	UpMu	2519.900

of the 2.2 MeV photon from the neutron capture is $\sim$200 μs, around six times larger than for Gd (Fig. 7.1).

$$n + p \rightarrow D^* \rightarrow +D + \gamma(2.2\,\text{MeV}) \tag{7.1}$$

Over the last years, the usage and development of neutron tagging on hydrogen is playing an increasing important role on the physics studies of SK. It was firstly developed in [Haibing12] and the optimised in [Irvine14] for atmospheric neutrinos.

Currently, in SK, the H-neutron tagging efficiency 21.7%. This is achieved using a two-staged reconstruction algorithm that is briefly reviewed Sect. 7.1.1.

7.1.1 *H-Neutron Tagging Analysis*

The detection of such weak signal can be done only for SK-IV due to the improvement of the electronics. As in the Gd case, the usage of several characteristics of its signature in the detector, in addition to distance and time to prompt signal of the event, allow to reconstruct some of these neutron captures. The selection method of the H-neutron tagging delayed signals consists in two steps: a rough first selection for detecting delayed candidates and likelihood, done with a neural network, to discriminate as much detector noise as possible from the selected candidates.

7.1.1.1 First Candidate Selection

Based on the reconstructed prompt signal, the PMT hits of the event are corrected by time of flight (ToF) subtraction. Then, a 10 ns sliding window scans from 18 μs after the prompt signal to the end (535 μs) of the event. This sliding window searches a cluster of hits and requires them that the number of hits in that window, N_{10} and in a 200 μs window, N_{200}, to satisfy Eq. (7.2), removing most of the PMT noise and the contamination from cosmic muons.

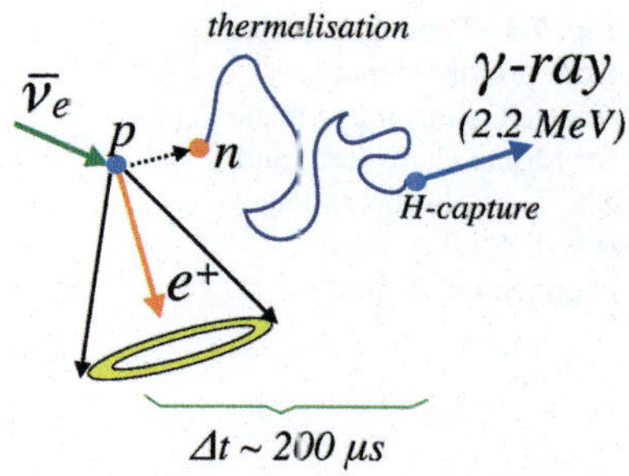

Fig. 7.1 Schematic diagram of Inverse Beta Decay (IBD) with H-neutron capture in SK

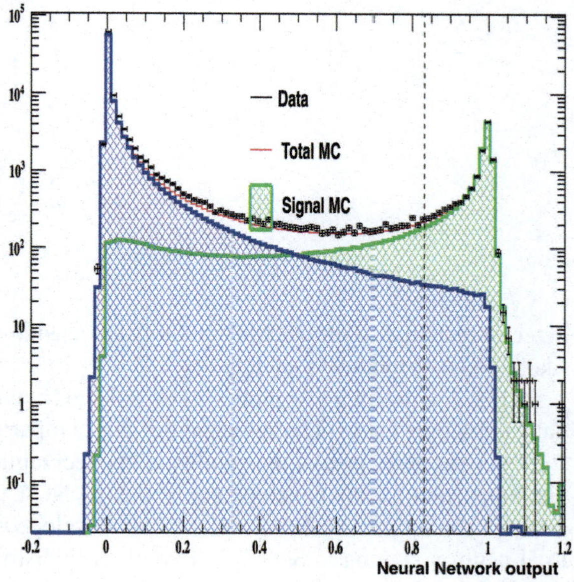

Fig. 7.2 Neural network output of delayed signals for true H-neutron captures (green), background (blue), total Monte Carlo (red) and SK-IV atmospheric data (black dots). Vertical dashed line shows the selection point, events on the right of this line are those selected as H-tagged neutrons

$$7 \leq N_{10} \leq 50, \; N_{200} \leq 200 \tag{7.2}$$

After this selection, and comparing data with Monte Carlo, 97.8% of all neutron captures are selected, of which, 32.7% correspond to actual H-neutron captures and the remaining background is of 4.4 fake candidates per event. The background is estimated as the candidates selected in the T2K dummy trigger data where no neutron-capture is expected, analogous to the Gd-neutron tagging in Sect. 5.2.2.

7.1.1.2 H-Neutron Capture Selection

In this part of the neutron tagging algorithm, a more elaborated selection criterion is constructed to further discern between true neutron captures and miss-tagged fake candidates. For that, sixteen different variables, concerning the physics and the

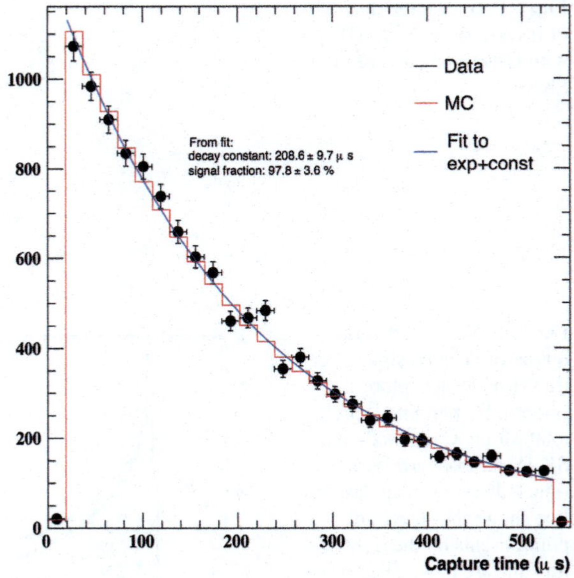

Fig. 7.3 Time distribution from prompt signal to selected neutron candidate for Monte Carlo (red) and SK-IV atmospheric data (black dots) and its exponential fit (blue)

detector response to the 2.2 MeV photon, are used in a neural network to compute a likelihood distribution, shown in Fig. 7.2.

In Fig. 7.3, the time distribution of the selected neutron captures is shown for both, data and Monte Carlo. The agreement of this comparison together with similar plots for other variables proves the validity of the technique. Some of these variables are also important for the development of the the NC-CC and ν-$\overline{\nu}$ separations and the neutron corrections to the visible energy, developed in Sects. 5.3–5.4 and applied to H-neutron tagging in Sect. 7.2. Two of these distributions are the energy dependence of the neutron multiplicity and the averaged number of neutrons for MultiGeV neutrinos, displayed in Figs. 7.4 and 7.5.

7.2 Including Neutron-Tagging Information in SK Atmospheric Neutrino Data Analysis

In this section, the implementation of the neutron-tagging improvements of Sects. 6.6.2–6.6.4 is discussed for atmospheric neutrinos in SK using H-neutron tagging.

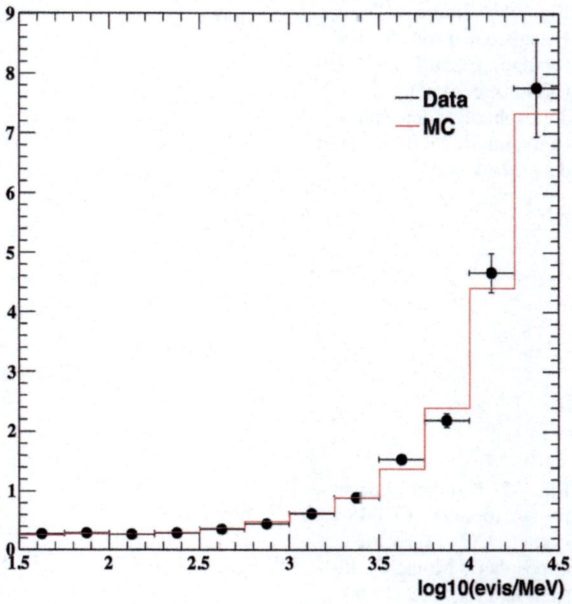

Fig. 7.4 Neutron-tag multiplicity as function of $log_{10}E_{vis}$ for atmospheric Monte Carlo (red) and SK-IV atmospheric data (black dots), for MultiGeV neutrinos ($E_{vis} \geq 1.33$ GeV)

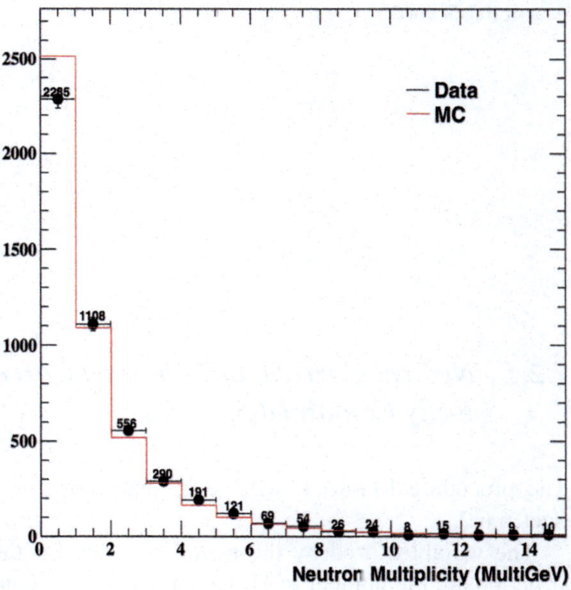

Fig. 7.5 Average number of neutron-tags per event as function of SK days. SK atmospheric data (black dots), fit to the data (blue line) and Monte Carlo (red line), for MultiGeV neutrinos

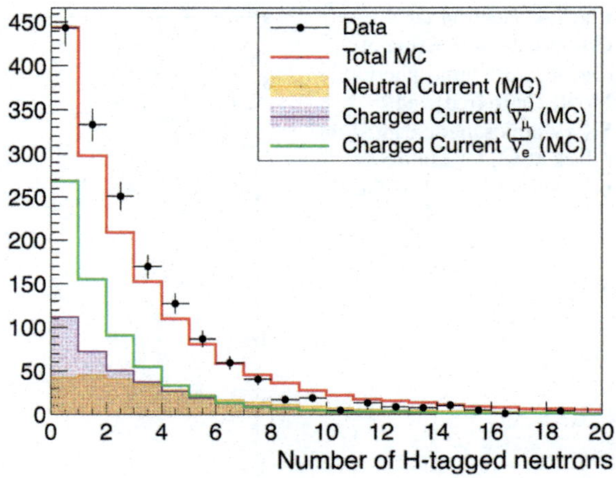

Fig. 7.6 Number of H-tagged neutrons for NC (orange), CCDIS ν_μ (violet), CC ν_e (green), total atmospheric Monte Carlo (red) and SK-IV atmospheric data (black dots)

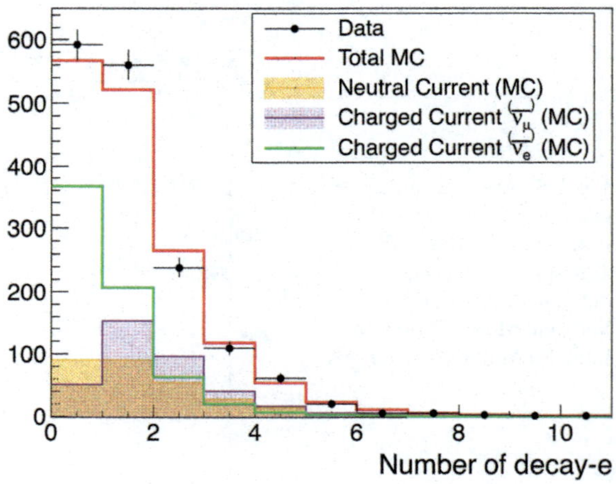

Fig. 7.7 Number of decay-e for NC (orange), CCDIS ν_μ (violet), CC ν_e (green), atmospheric Monte Carlo (red) and for the 2519.90 days of SK-IV atmospheric dataset (black dots)

7.2.1 Neutral Current and Charged Current Separation for Fully Contained Samples

The procedure followed, will be as that done for SuperK-Gd and explained in Sect. 6.6.2.

The variables used by the neural network for the likelihood distribution construction are the number of H-tagged neutrons, the number of rings, the MultiRing likelihood, the number of decay-electrons, the distance to the furthest decay-electron, the sum of all rings ID likelihood, the rapidity and the energy fraction of the most energetic ring. All their corresponding distributions are shown in Figs. 7.6, 7.7, 7.8, 7.9, 7.10, 7.11, 7.12 and 7.13 for the atmospheric Monte Carlo and SK-IV dataset. The description and motivation for the usage of each of these variables is treated in Sect. 5.3.

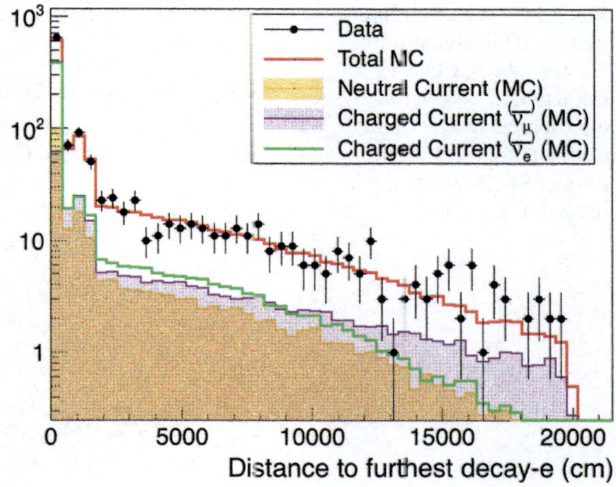

Fig. 7.8 Distance to furthest decay-e for NC (orange), CCDIS ν_μ (violet), CC ν_e (green), atmospheric Monte Carlo (red) and for the 2519.90 days of SK-IV atmospheric dataset (black dots)

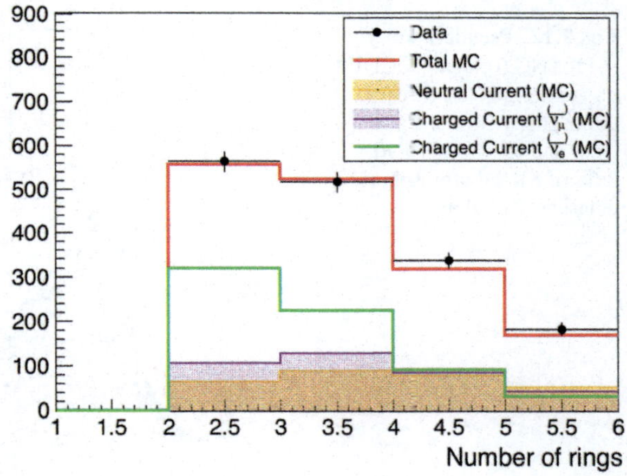

Fig. 7.9 Number of rings for NC (orange), CCDIS ν_μ (violet), CC ν_e (green), atmospheric Monte Carlo (red) and for the 2519.90 days of SK-IV atmospheric dataset (black dots)

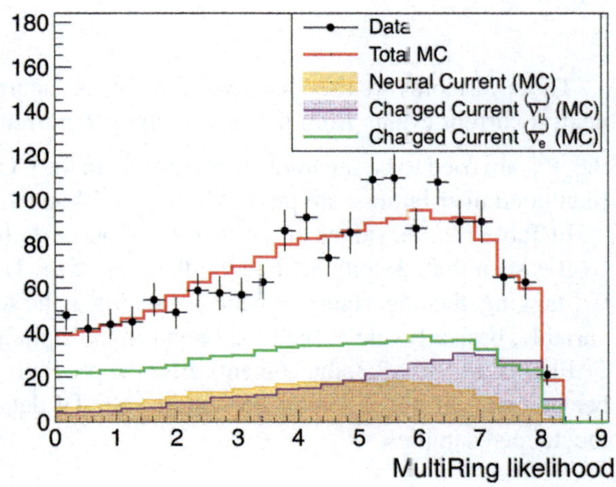

Fig. 7.10 Multi-ring likelihood for NC (orange), CCDIS ν_μ (violet), CC ν_e (green), atmospheric Monte Carlo (red) and for the 2519.90 days of SK-IV atmospheric dataset (black dots)

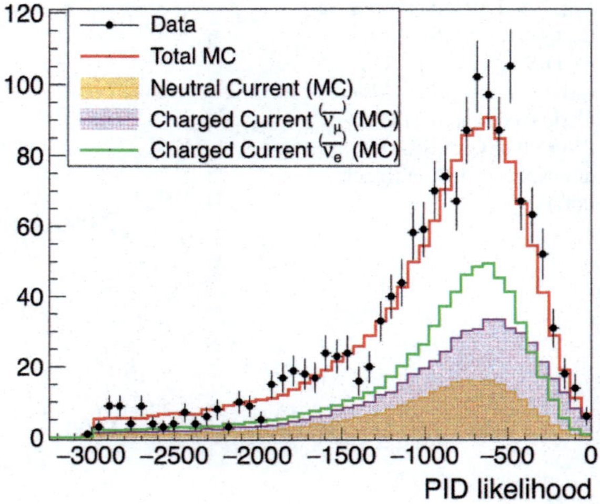

Fig. 7.11 Sum of all ring particle ID likelihoods for NC (orange), CCDIS ν_μ (violet), CC ν_e (green), atmospheric Monte Carlo (red) and for the 2519.90 days of SK-IV atmospheric dataset (black dots)

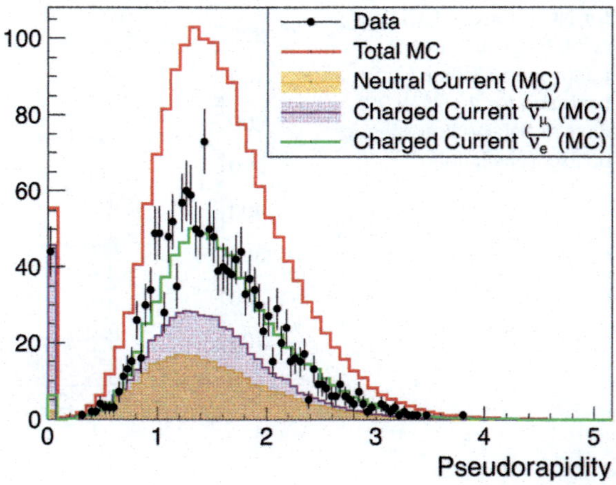

Fig. 7.12 Pseudorapidity (η) for NC (orange), CCDIS ν_μ (violet), CC ν_e (green), atmospheric Monte Carlo (red) and for the 2519.90 days of SK-IV atmospheric dataset (black dots)

These variables are used in a two-step process neural network to discard, first, the neutral current events from the rest of charged current interacting neutrinos. Then, the $\overset{(-)}{\nu}_\mu$ are tried to be separated from the $\overset{(-)}{\nu}_e$ in the CCDIS $\overset{(-)}{\nu}_\mu$-$\overset{(-)}{\nu}_e$ separation. Both likelihood distributions are plotted in Figs. 7.14 and 7.15.

In Table 7.2, the values of the relative importance for each variable are listed.

It is seen that, despite the lower efficiency of the H-neutron tagging compared to Gd-tagging, the importance of neutron tagging in the selections is the most important variable, basically due to the broad neutron multiplicity distributions.

Finally, in Table 7.3, the contents after the previous selections, the expected number of events and its comparison with the SK-IV dataset are calculated for each of the defined samples.

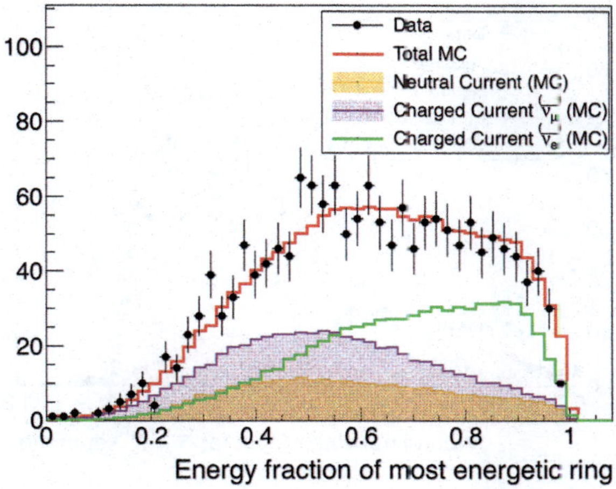

Fig. 7.13 Energy fraction of the most energetic ring for NC (orange), CCDIS ν_μ (violet), CC ν_e (green), Monte Carlo (red) and for the 2519.90 days of SK-IV atmospheric dataset (black dots)

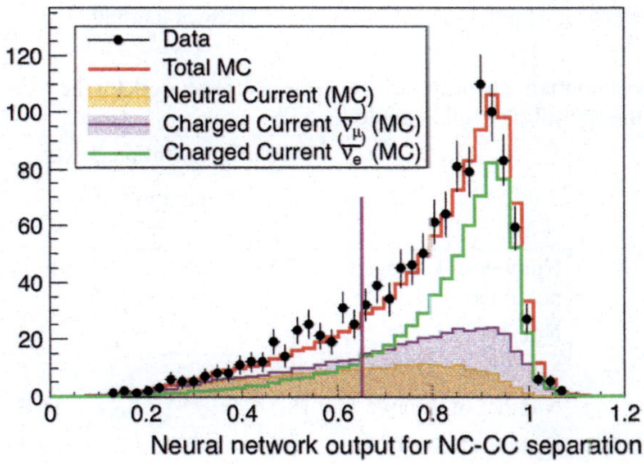

Fig. 7.14 Neural network output for the NC-CC separation for NC (orange), CCDIS ν_μ (violet), CC ν_e (green), atmospheric Monte Carlo (red) and for the 2519.90 days of SK-IV atmospheric dataset (black dots). The pink vertical line shows the selection point (0.55) for the distribution, which optimises product $\epsilon_{CC} \times p_{CC}$, being the NC-like sample formed by those events at its left and the CC-like by those at its right

The sample contents in Table 7.3 are rather alike for these separations, as compared with the Gd case in Table 6.3. This is mainly due to the mentioned large neutron multiplicity of these events, meaning that the effect of the lower neutron tagging efficiency from hydrogen, in the performance of the separations, is rather mild.

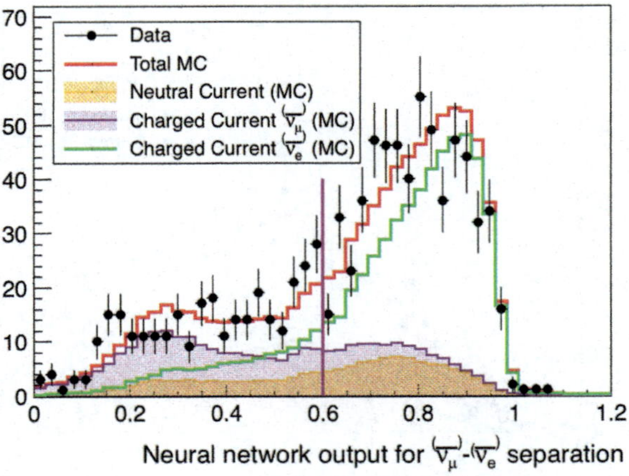

Fig. 7.15 Neural network output for the CCDISν_μ-ν_e separation for NC (orange), CCDIS ν_μ (violet), CC ν_e (green), atmospheric Monte Carlo (red) and for the 2519.90 days of SK-IV atmospheric dataset (black dots) in the CC-like sample. The pink vertical line shows the selection point (0.5) for the distribution, which optimises product $\epsilon_{CC-e} \times p_{CC-e}$, being the CCDIS ν_μ-like sample formed by those events at its left and the CC ν_e-like by those at its right

Table 7.2 Relative importance of each variable in the neural network for the NC-CC and CCDISν_μ-ν_e separations in the MultiRing e-like sample

		Relative separation power	
		NC-CC separation (%)	$\overset{(-)}{\nu}_\mu$-$\overset{(-)}{\nu}_e$ separation (%)
Variables	Number of H-tagged neutrons	57.5	49.5
	Number of rings	1.4	3.2
	MultiRing likelihood	1.7	2.1
	Number of decay-e	18.5	16.8
	Distance to furthest decay-e	5.1	6.3
	Pseudorapidity	1.2	1.4
	Energy fraction	0.9	1.0
	Sum of all rings PID likelihood	12.8	41.0

7.2.2 *Neutrino and Antineutrino Separation for Fully Contained Samples*

The neutrino-antineutrino separations presented in Sect. 5.4 are implemented here for the H-neutron tagging, analogously to the Gd-neutron tagging case in Sect. 6.6.3.

Table 7.3 Relative contents of each species for each subsample of the MultiRing e-like sample, after the selection, and a comparison between the number of entries predicted for the unoscillated atmospheric Monte Carlo and the 2519.90 days of SK-IV data

		Samples		
		CC $\overset{(-)}{\nu}_e$-like	CCDIS $\overset{(-)}{\nu}_\mu$-like	NC-like
MC true ν	CC ν_e	57.9%	36.8%	28.9%
	CC $\overline{\nu}_e$	24.3%	5.8%	4.0%
	CC ν_μ	5.7%	34.5%	31.9%
	CC $\overline{\nu}_\mu$	0.5%	4.0%	3.9%
	NC ν	11.6%	18.9%	32.1%
MC unosc.		503.9	472.7	677.9
SK-IV data		475	528	692

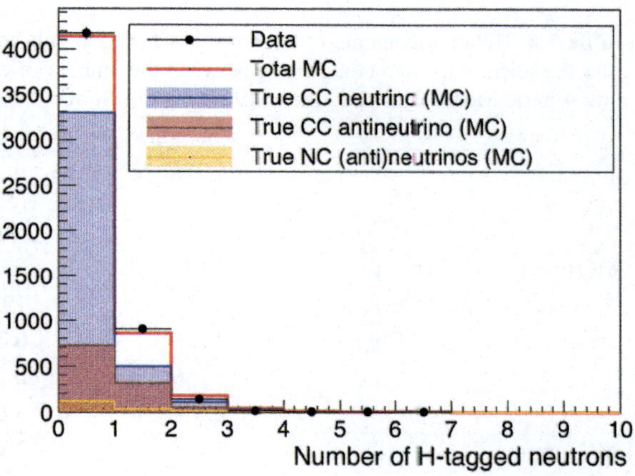

Fig. 7.16 Number of H-tagged neutrons for true neutrinos (blue), true antineutrinos (brown), neutral current events (orange), total atmospheric Monte Carlo (red) and for the 2519.90 days of SK-IV atmospheric dataset (black dots) in the SubGeV e-like sample

7.2.2.1 SubGeV and MultiGeV e-Like

This sample is split into three event categories based on cuts on the neutron and decay electron multiplicities. The cuts which define the three regions in this sample are the following:

- ν_e-like: Number of decay-e > 0
- $\nu_e + \overline{\nu}_e$-like: Number of decay-e $= 0$ and Number of H-tagged neutrons $= 0$
- $\overline{\nu}_e$-like: Number of decay-e $= 0$ and Number of H-tagged neutrons > 0

Next, the separation is implemented for the SubGeV and MultiGeV e-like samples.

SubGeV e-like samples

The distributions for the H-tagged neutron multiplicity and the number of decay-electrons are shown in Figs. 7.16 and 7.17 respectively.

The three samples have the contents and event rates listed in Table 7.4.

Fig. 7.17 Number of μ-decay electrons for true neutrinos (blue), true antineutrinos (brown), neutral current events (orange), total atmospheric Monte Carlo (red) and for the 2519.90 days of SK-IV atmospheric dataset (black dots) in the SubGeV e-like sample

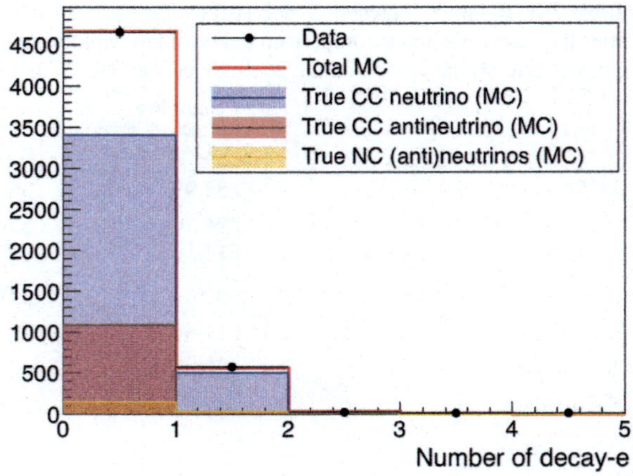

Table 7.4 Relative contents of each species for each subsample of the SubGeV e-like sample, after the selection, and a comparison between the number of entries predicted for the unoscillated atmospheric Monte Carlo and the 2519.90 days of SK-IV data

		Samples		
		ν_e-like (dcy-e > 0)	$\nu_e + \overline{\nu}_e$-like (dcy-e = 0 and neutrons = 0)	$\overline{\nu}_e$-like (dcy-e = 0 & neutrons > 0)
MC true ν	CC ν_e	77.1%	77.6%	54.9%
	CC $\overline{\nu}_e$	1.7%	19.2%	40.6%
	CC ν_μ	11.6%	0.3%	0.2%
	CC $\overline{\nu}_\mu$	3.3%	0.0%	0.1%
	NC ν	6.3%	3.0%	4.2%
MC unosc.		582.9	3649.6	938.8
SK-IV data		629	3961	949

Table 7.4 shows a similar trend in the sample contents than that for the Gd case in Table 6.4, although with considerable worsening of the selection efficiencies and sample purities due to the lower efficiency of the neutron tagging.

MultiGeV e-like samples

The event selection in this sample is very similar to the SubGeV e-like one. The neutron and decay-electron distributions are shown in Figs. 7.18 and 7.19.

The table contents and expected unoscillated event rates for the SK-IV period and for the SK-IV data are shown in Table 7.5.

The contents of the samples follow the same trend as in the SubGeV e-like case and compare in a similar fashion to those for Gd-neutron tagging in Table 6.5.

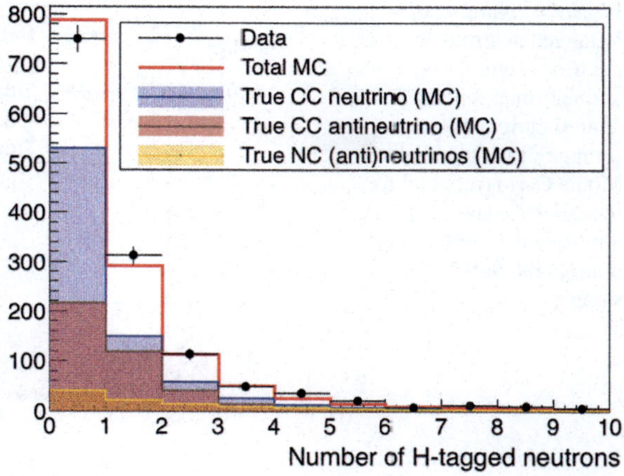

Fig. 7.18 Number of H-tagged neutrons for true neutrinos (blue), true antineutrinos (brown), neutral current events (orange), total atmospheric Monte Carlo (red) and for the 2519.90 days of SK-IV atmospheric dataset (black dots) in the MultiGeV e-like sample

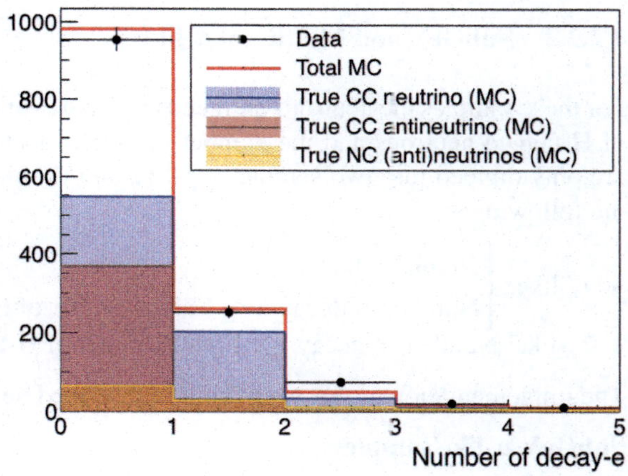

Fig. 7.19 Number of μ-decay electrons for true neutrinos (blue), true antineutrinos (brown), neutral current events (orange), total atmospheric Monte Carlo (red) and for the 2519.90 days of SK-IV atmospheric dataset (black dots) in the MultiGeV e-like sample

Table 7.5 Relative contents of each species for each subsample of the MultiGeV e-like sample, after the selection, and a comparison between the number of entries predicted for the unoscillated atmospheric Monte Carlo and the 2519.90 days of SK-IV data

		Samples		
		ν_e-like (dcy-e > 0)	$\nu_e + \overline{\nu}_e$-like (dcy-e = 0 & neutrons = 0)	$\overline{\nu}_e$-like (dcy-e = 0 & neutrons > 0)
MC true ν	CC ν_e	61.4%	61.4%	42.8%
	CC $\overline{\nu}_e$	9.1%	32.0%	47.3%
	CC ν_μ	11.7%	0.9%	1.1%
	CC $\overline{\nu}_\mu$	2.0%	0.2%	0.3%
	NC ν	15.7%	5.4%	8.4%
MC unosc.		359.0	671.1	361.9
SK-IV data		385	637	375

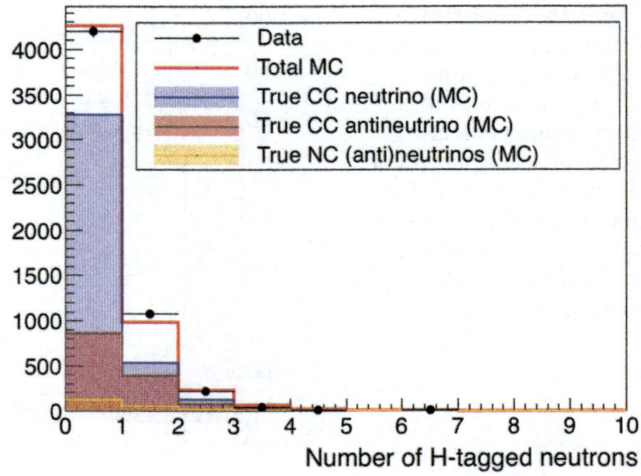

Fig. 7.20 Number of H-tagged neutrons for true neutrinos (blue), true antineutrinos (brown), neutral current events (orange), total atmospheric Monte Carlo (red) and for the 2519.90 days of SK-IV atmospheric dataset (black dots) in the SubGeV μ-like sample

7.2.2.2 SubGeV and MultiGeV μ-like

For these samples the neutrino-antineutrino separation is also done using the number of H-tagged neutrons and the number of decay-electrons. However, this time they are only divided into two samples, ν_μ-like and $\overline{\nu}_\mu$-like. The cuts defining them are the following:

- ν_μ-like: $\begin{cases} \text{Number of decay-e} \neq 1 \\ \text{Number of decay-e} = 1 \text{ and Number of H-tagged neutrons} = 0 \end{cases}$
- $\overline{\nu}_\mu$-like: Number of decay-e =1 and Number of H-tagged neutrons > 0

The implementation of this separation is described below.

SubGeV μ-like samples

Figures 7.20 and 7.21 show the distributions for the H-tagged neutrons and the number of decay-electrons.

The contents of the resulting samples are summarised in Table 7.6.

Again, the selection worse performance due to the lesser neutron tagging efficiency of hydrogen compared to gadolinium. This can be seen in Table 6.6.

MultiGeV μ-like samples

For this sample, the neutron and decay-electron multiplicities distributions are shown in Figs. 7.22 and 7.23.

The contents and expected rates of the defined samples are shown in Table 7.7.

7.2.2.3 MultiRing CC ν_e-like

Once again, here, the analogous technique as in Sect. 6.6.3.3, is used for the neutrino-antineutron separation of the CC $\overset{(-)}{\nu}_e$. A likelihood distribution for the separation is

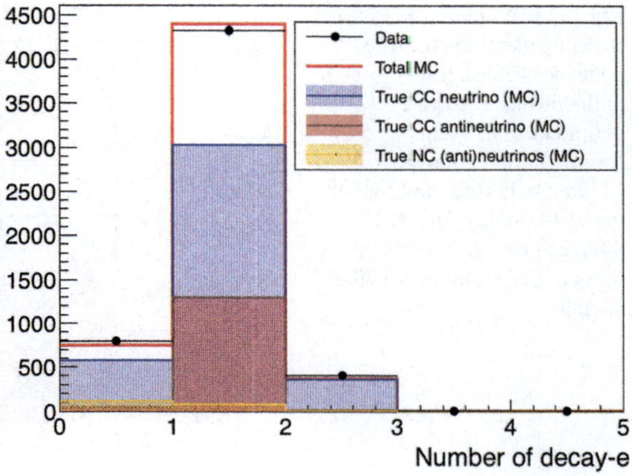

Fig. 7.21 Number of μ-decay electrons for true neutrinos (blue), true antineutrinos (brown), neutral current events (orange), total atmospheric Monte Carlo (red) and for the 2519.90 days of SK-IV atmospheric dataset (black dots) in the SubGeV μ-like sample

Table 7.6 Relative contents of each species for each subsample of the SubGeV μ-like sample, after the selection, and a comparison between the number of entries predicted by atmospheric Monte Carlo and SK-IV data

		Samples	
		ν_μ-like (dcy-e $\neq$ 1 or (dcy-e = 1 & neutrons = 0))	$\overline{\nu}_\mu$-like (dcy-e = 1 & neutrons > 0)
MC true ν	CC ν_e	0.6%	0.1%
	CC $\overline{\nu}_e$	0.2%	0.0%
	CC ν_μ	76.1%	49.8%
	CC $\overline{\nu}_\mu$	19.4%	46.7%
	NC ν	3.6%	3.4%
MC unosc.		6336.0	1427.6
SK-IV data		4799	1006

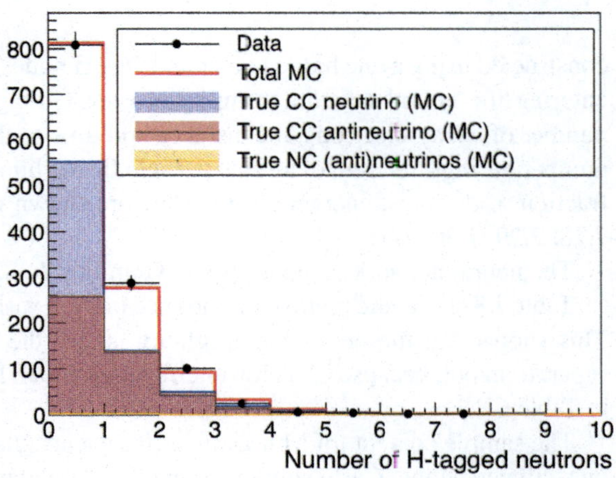

Fig. 7.22 Number of H-tagged neutron for neutrinos (blue), antineutrinos (brown), neutral current events (orange), total atmospheric Monte Carlo (red) and for the 2519.90 days of SK-IV atmospheric dataset (black dots) in the MultiGeV μ-like sample

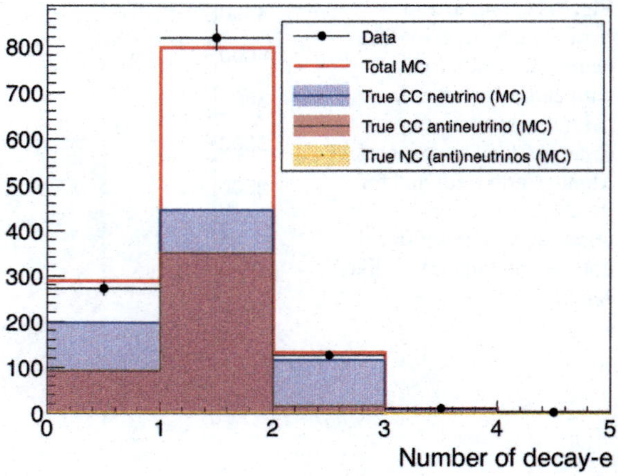

Fig. 7.23 Number of μ-decay electrons for true neutrinos (blue), true antineutrinos (brown), neutral current events (orange), total atmospheric Monte Carlo (red) and for the 2519.90 days of SK-IV atmospheric dataset (black dots) in the MultiGeV μ-like sample

Table 7.7 Relative contents of each species for each subsample of the MultiGeV e-like sample, after the selection, and a comparison between the number of entries predicted for the unoscillated atmospheric Monte Carlo and the 2519.90 days of SK-IV data

		Samples	
		ν_μ-like (dcy-e $\neq$ 1 or (dcy-e = 1 & neutrons = 0))	$\overline{\nu}_\mu$-like (dcy-e = 1 & neutrons > 0)
MC true ν	CC ν_e	0.2%	0.1%
	CC $\overline{\nu}_e$	0.1%	0.0%
	CC ν_μ	67.4%	44.3%
	CC $\overline{\nu}_\mu$	32.2%	55.6%
	NC ν	0.1%	0.1%
MC unosc.		1470.6	410.0
SK-IV data		996	295

constructed using a one hidden layer and $N + 5$ nodes neural network. The variables entering the neural network computation are Gd-tagged neutron multiplicity, the number of decay-electrons, the distance and time to the furthest decay-electron, the number of rings, the sum of all ring particle ID likelihood, the rapidity and the energy fraction of the most energetic ring. They are shown in Figs. 7.24, 7.25, 7.26, 7.27, 7.28, 7.29, 7.30, 7.31.

The neural network response is shown in Fig. 7.32.

Table 7.8 shows the relative importance of each variable in the separation method. This shows that the neutron multiplicity is still the most relevant variable in the separation, but, because of its lower efficiency, much less than in the Gd case shown in Table 6.8.

The sample content for MultiRing e-like events, their expected event rates for the unoscillated Monte Carlo simulation and SK-IV dataset are given in Table 7.9.

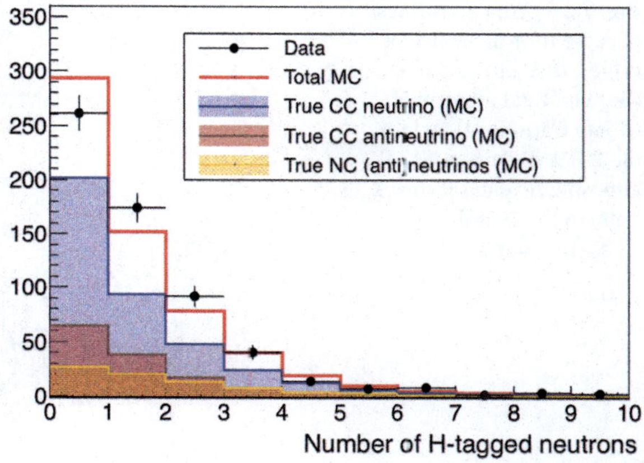

Fig. 7.24 Number of H-tagged neutrons for true neutrinos (blue), true antineutrinos (brown), neutral current events (orange), total atmospheric Monte Carlo (red) and for the 2519.90 days of SK-IV atmospheric dataset (black dots) in the MultiRing CCν_e-like sample

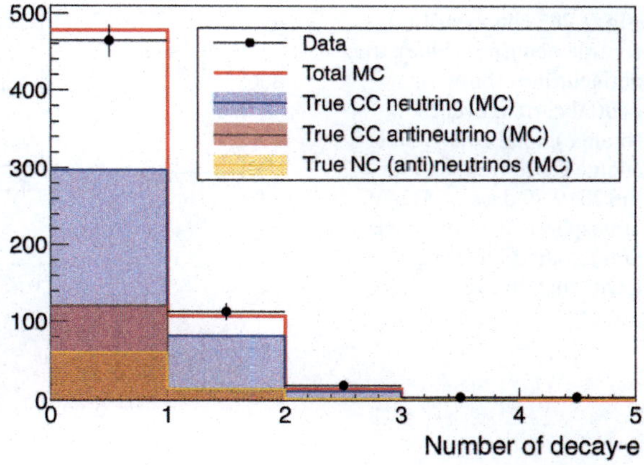

Fig. 7.25 Number of μ-decay electrons for true neutrinos (blue), true antineutrinos (brown), neutral current events (orange), total atmospheric Monte Carlo (red) and for the 2519.90 days of SK-IV atmospheric dataset (black dots) in the MultiRing CCν_e-like sample

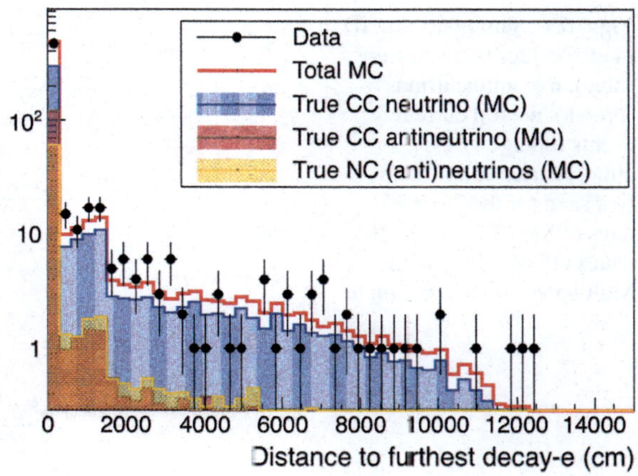

Fig. 7.26 Distance to furthest decay-e for neutrinos (blue), antineutrinos (brown), neutral current events (orange), total atmospheric Monte Carlo (red) and for the 2519.90 days of SK-IV atmospheric dataset (black dots) in the MultiRing CCν_e-like sample

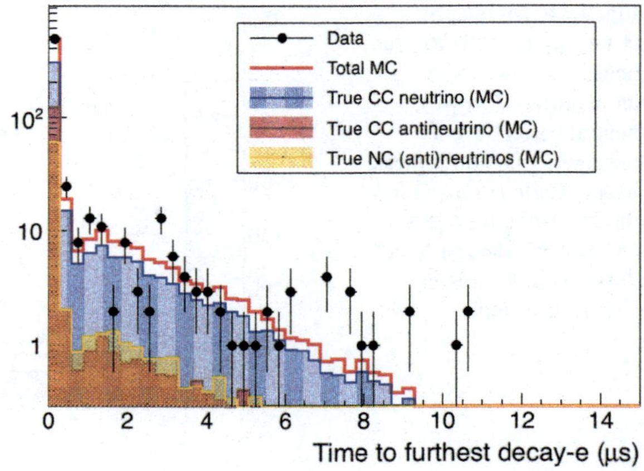

Fig. 7.27 Time to furthest decay-e for true neutrinos (blue), true antineutrinos (brown), total atmospheric Monte Carlo (red) and for the 2519.90 days of SK-IV atmospheric dataset (black dots) in the MultiRing $CC\nu_e$-like sample

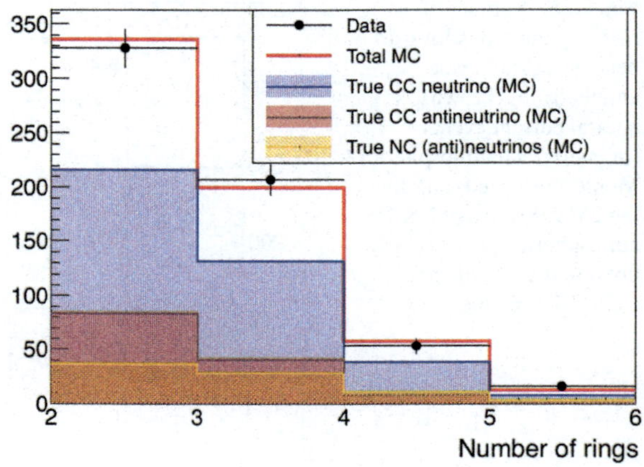

Fig. 7.28 Number of rings for true neutrinos (blue), true antineutrinos (brown), neutral current events (orange), total atmospheric Monte Carlo (red) and for the 2519.90 days of SK-IV atmospheric dataset (black dots) in the MultiRing $CC\nu_e$-like sample

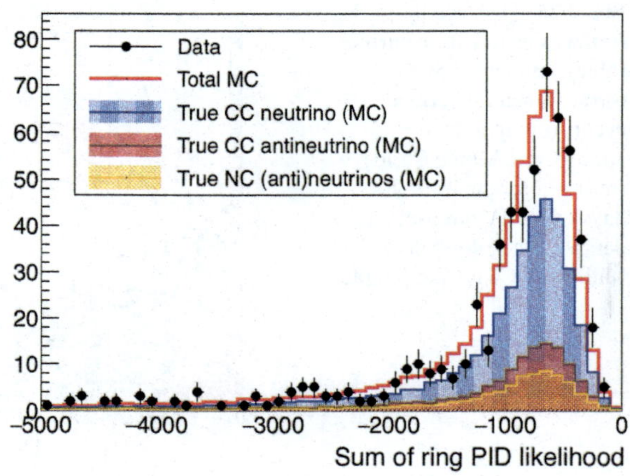

Fig. 7.29 Sum of all ring ID likelihood for true neutrinos (blue), true antineutrinos (brown), neutral current events (orange), total atmospheric Monte Carlo (red) and for the 2519.90 days of SK-IV atmospheric dataset (black dots) in the MultiRing $CC\nu_e$-like sample

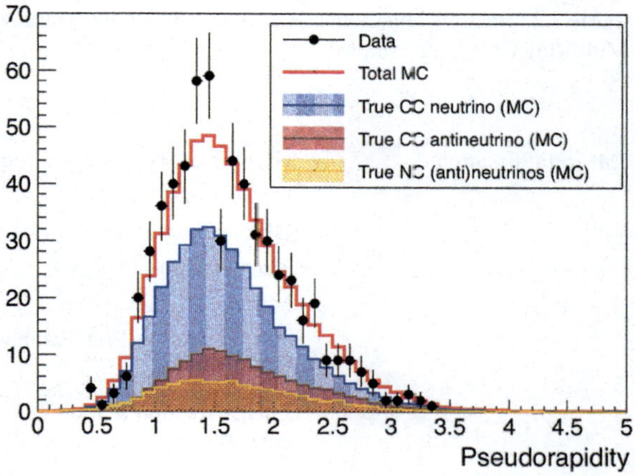

Fig. 7.30 Pseudorapidity for true neutrinos (blue), true antineutrinos (brown), neutral current events (orange), total atmospheric Monte Carlo (red) and SK-IV atmospheric data (black dots) in the MultiRing CCν_e-like sample

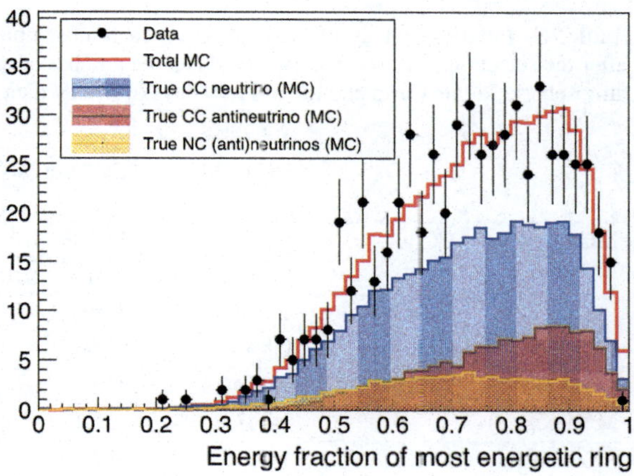

Fig. 7.31 Energy fraction of the most energetic ring for true neutrinos (blue), true antineutrinos (brown), neutral current events (orange), total atmospheric Monte Carlo (red) and for the 2519.90 days of SK-IV atmospheric dataset (black dots) in the MultiRing CCν_e-like sample

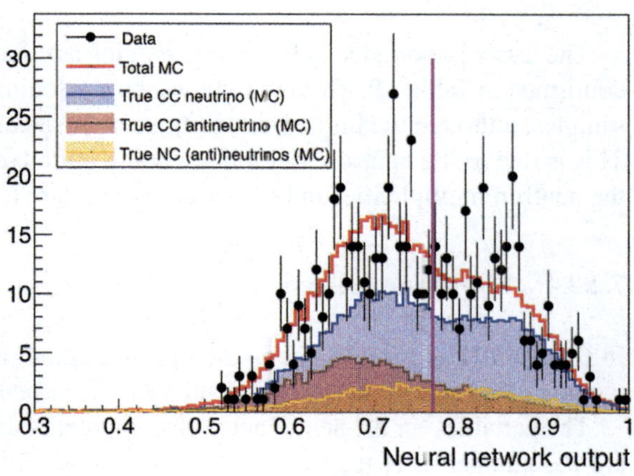

Fig. 7.32 Neural network output for the ν-$\overline{\nu}$ separation in the MultiRing CC ν_e-like sample, for for true neutrinos (blue), true antineutrinos (brown), neutral current events (orange), total atmospheric Monte Carlo (red) and SK-IV atmospheric data (black dots). The pink vertical line shows the selection point, 0.8 for the distribution, which separates the events into two categories MultiRing $\overline{\nu}_e$-like (left) and MultiRing ν_e-like (right)

Table 7.8 Relative importance of each variable in the neural network for the $\overline{\nu}$-ν separation in the MultiRing CCν_e-like sample

		Relative separation power
		ν-$\overline{\nu}$ separation (%)
MC ν information	Number of H-tagged neutrons	34.1
	Number of decay-e	22.8
	Distance to furthest decay-e	9.5
	Time to furthest decay-e	14.5
	Number of rings	3.6
	Sum of all rings PID likelihood	11.6
	Pseudorapidity	3.3
	Energy fraction of most energetic ring	0.7

Table 7.9 Relative contents of each species for each subsample of the MultiRing CCe-like sample, after the selection, and a comparison between the number of entries predicted for the unoscillated atmospheric Monte Carlo and the 2519.90 days of SK-IV data

		Samples			
		CC ν_e-like	CC $\overline{\nu}_e$-like	CCDIS $\overset{(-)}{\nu}_\mu$-like	NC-like
MC true ν	CC ν_e	64.0%	55.3%	36.8%	28.9%
	CC $\overline{\nu}_e$	12.8%	29.1%	5.8%	4.0%
	CC ν_μ	9.9%	4.0%	34.5%	31.9%
	CC $\overline{\nu}_\mu$	0.8%	0.4%	4.0%	3.9%
	NC ν	12.4%	11.3%	18.9%	32.1%
MC unosc.		147.7	356.1	472.7	677.9
SK-IV data		153	322	528	692

The lesser importance of H-neutron tagging as compared to Gd-neutron tagging is confirmed in Table 7.9. This table shows the worsening in purity of the ν_e and $\overline{\nu}_e$-like samples, although the impoverishment is not as sizeable as in the previous samples. This is due to the neutron multiplicity being quite large and, thus, the behaviour of the neutron multiplicities in both cases are similar above two.

7.2.2.4 MultiRing μ-like

In the MultiRing sample, the homologous method, already applied to Gd-neutron tagging in Sect. 6.6.3.4, is implemented for the H-neutron tagging case.

The variables for the neural network computation of the ν-$\overline{\nu}$ separation likelihood are the number of H-tagged neutrons, the number of decay-electrons, the distance and time to the furthest decay-electron, the number of rings, the sum of all ring particle ID likelihood, the rapidity and the energy fraction of the most energetic ring. They are shown in Figs. 7.33, 7.34, 7.35, 7.36, 7.37, 7.38, 7.39, 7.40.

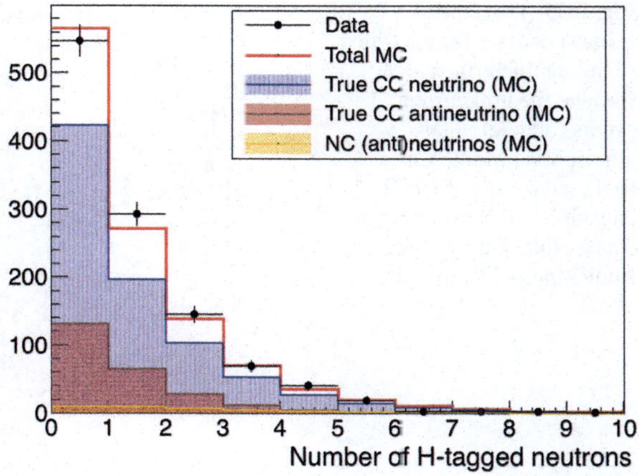

Fig. 7.33 Number of H-tagged neutrons for true neutrinos (blue), true antineutrinos (brown), total atmospheric Monte Carlo (red) and for the 2519.90 days of SK-IV atmospheric dataset (black dots) in the MultiRing μ-like sample

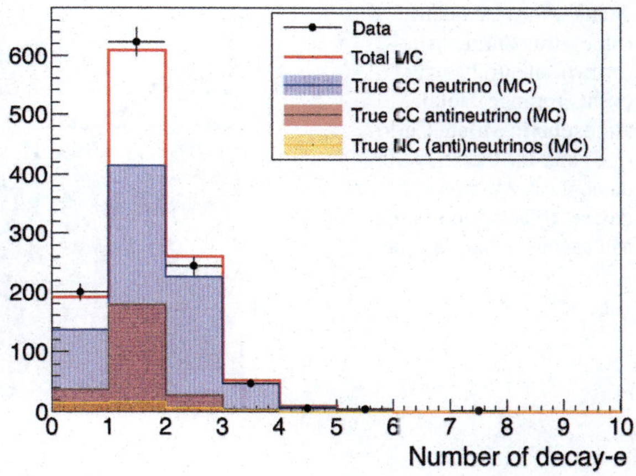

Fig. 7.34 Number of μ-decay electrons for true neutrinos (blue), true antineutrinos (brown), neutral current events (orange), total atmospheric Monte Carlo (red) and for the 2519.90 days of SK-IV atmospheric dataset (black dots) in the MultiRing μ-like sample

The obtained likelihood distribution is shown in Fig. 7.41.

Table 7.10 shows the relative importance of each variable considered in the separation. Here, the same behaviour is seen as in the Gd case, in Table 6.10, the decay-electron variables have more relevance than the neutron multiplicity. In this case, the H-neutron tagging has less relevance in the separation as compared to the Gd-neutron tagging.

The event rates and contents for unoscillated MC and SK-IV data are summarised in Table 7.11. Again, the purity of the samples and efficiency of the separation are slightly diminished as compared to the Gd case, in Table 6.11. However, this worsening is very mild due to the lesser importance, than in the previous separations, of the neutron multiplicity in favour of the number of decay-electrons.

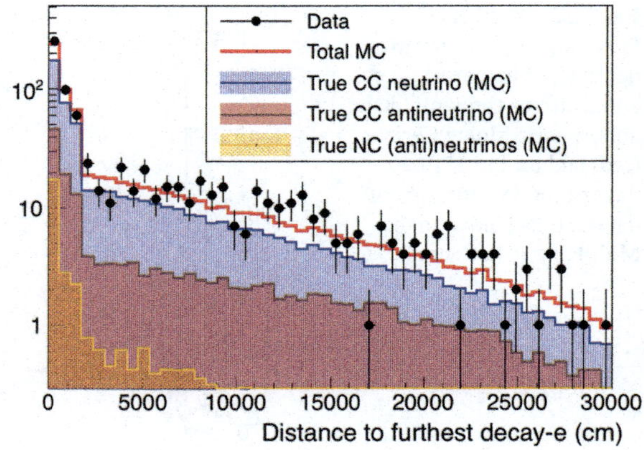

Fig. 7.35 Distance to furthest decay-e for neutrinos (blue), antineutrinos (brown), neutral current events (orange), total atmospheric Monte Carlo (red) and for the 2519.90 days of SK-IV atmospheric dataset (black dots) in the MultiRing μ-like sample

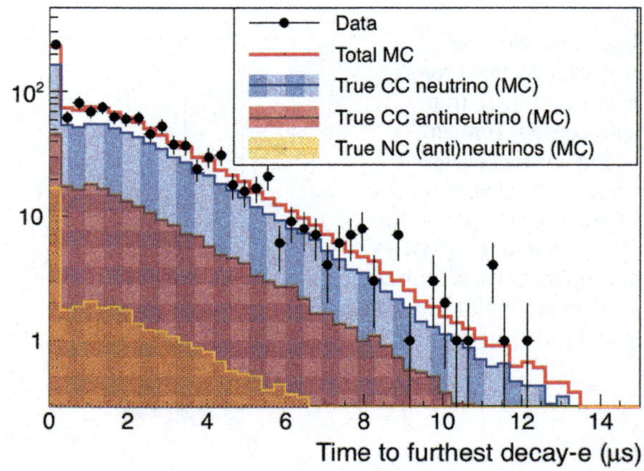

Fig. 7.36 Time to furthest decay-e for true neutrinos (blue), true antineutrinos (brown), neutral current events (orange), total atmospheric Monte Carlo (red) and for the 2519.90 days of SK-IV atmospheric dataset (black dots) in the MultiRing μ-like sample

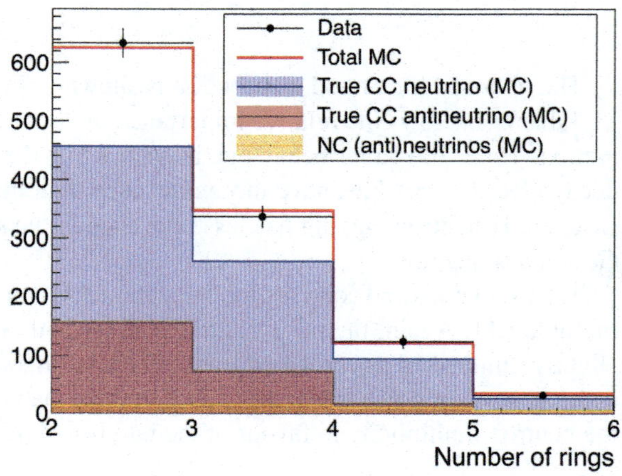

Fig. 7.37 Number of rings for true neutrinos (blue), true antineutrinos (brown), neutral current events (orange), total atmospheric Monte Carlo (red) and for the 2519.90 days of SK-IV atmospheric dataset (black dots) in the MultiRing μ-like sample

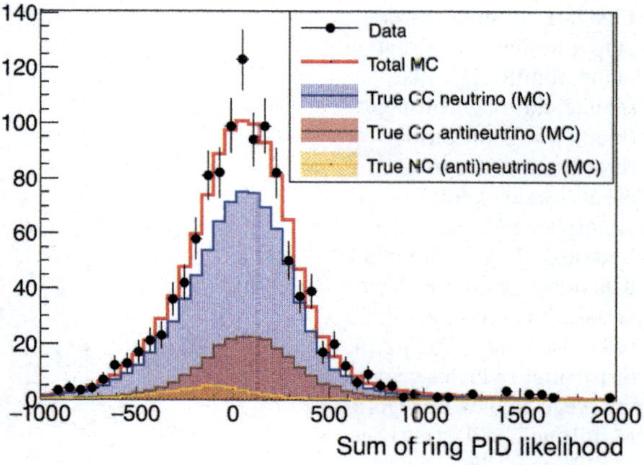

Fig. 7.38 Sum of all ring ID likelihood for true neutrinos (blue), true antineutrinos (brown), neutral current events (orange), total atmospheric Monte Carlo (red) and for the 2519.90 days of SK-IV atmospheric dataset (black dots) in the MultiRing μ-like sample

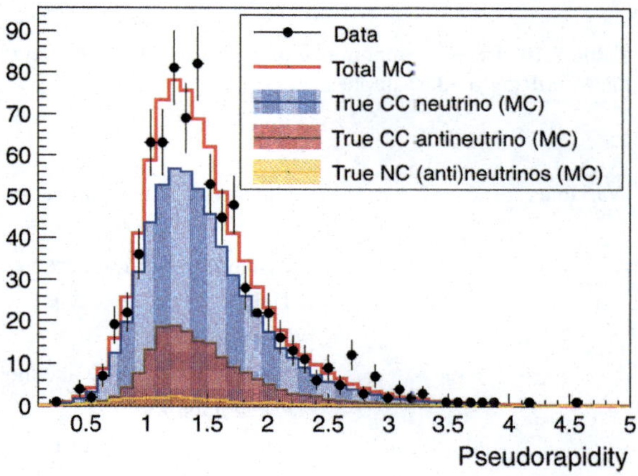

Fig. 7.39 Pseudorapidity for true neutrinos (blue), true antineutrinos (brown), neutral current events (orange), total atmospheric Monte Carlo (red) and for the 2519.90 days of SK-IV atmospheric dataset (black dots) in the MultiRing μ-like sample

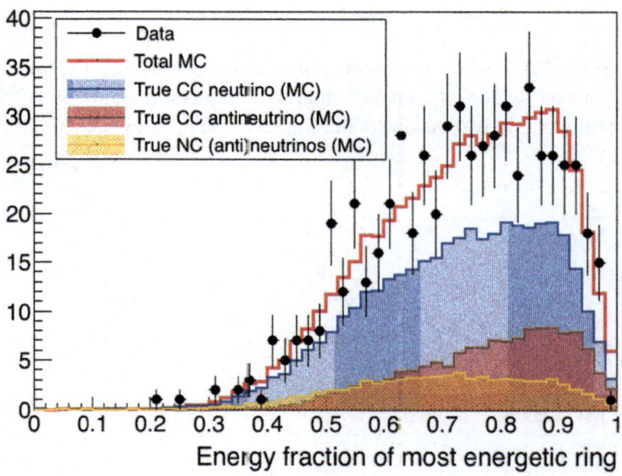

Fig. 7.40 Energy fraction of the most energetic ring for true neutrinos (blue), true antineutrinos (brown), neutral current events (orange), total atmospheric Monte Carlo (red) and for the 2519.90 days of SK-IV atmospheric dataset (black dots) in the MultiRing μ-like sample

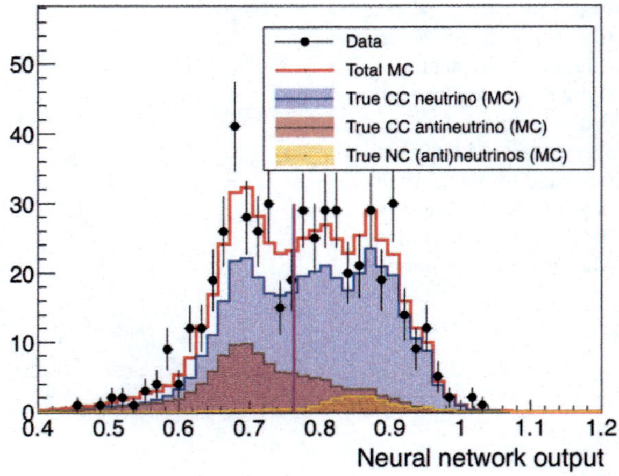

Fig. 7.41 Neural network output for the ν-$\overline{\nu}$ separation in the MultiRing μ-like sample, for true neutrinos (blue), true antineutrinos (brown), neutral current events (orange), total atmospheric Monte Carlo (red) and SK-IV atmospheric data (black dots). The pink vertical line shows the selection point, 0.762 for the distribution, which separates the events into two categories MultiRing $\overline{\nu}_\mu$-like (left) and MultiRing ν_μ-like (right)

Table 7.10 Relative importance of each variable in the neural network for the $\overline{\nu}$-ν separation in the MultiRing μ-like sample

		Relative separation power
		ν-$\overline{\nu}$ separation (%)
Variables	Number of H-tagged neutrons	25.2
	Number of decay-e	43.0
	Distance to furthest decay-e	7.6
	Time to furthest decay-e	14.0
	Number of rings	5.0
	Sum of all rings PID likelihood	3.5
	Pseudorapidity	1.0
	Energy fraction of most energetic ring	0.7

Table 7.11 Relative contents of each species for each subsample of the MultiRing μ-like sample, after the selection, and a comparison between the number of entries predicted for the unoscillated atmospheric Monte Carlo and the 2519.90 days of SK-IV data

		Samples	
		ν_μ-like	$\overline{\nu}_\mu$-like
MC true ν	CC ν_e	3.7%	1.4%
	CC $\overline{\nu}_e$	0.6%	0.2%
	CC ν_μ	77.2%	70.7%
	CC $\overline{\nu}_\mu$	12.7%	24.7%
	NC ν	5.8%	3.0 %
MC unosc.		250.6	1370.0
SK-IV data		300	886

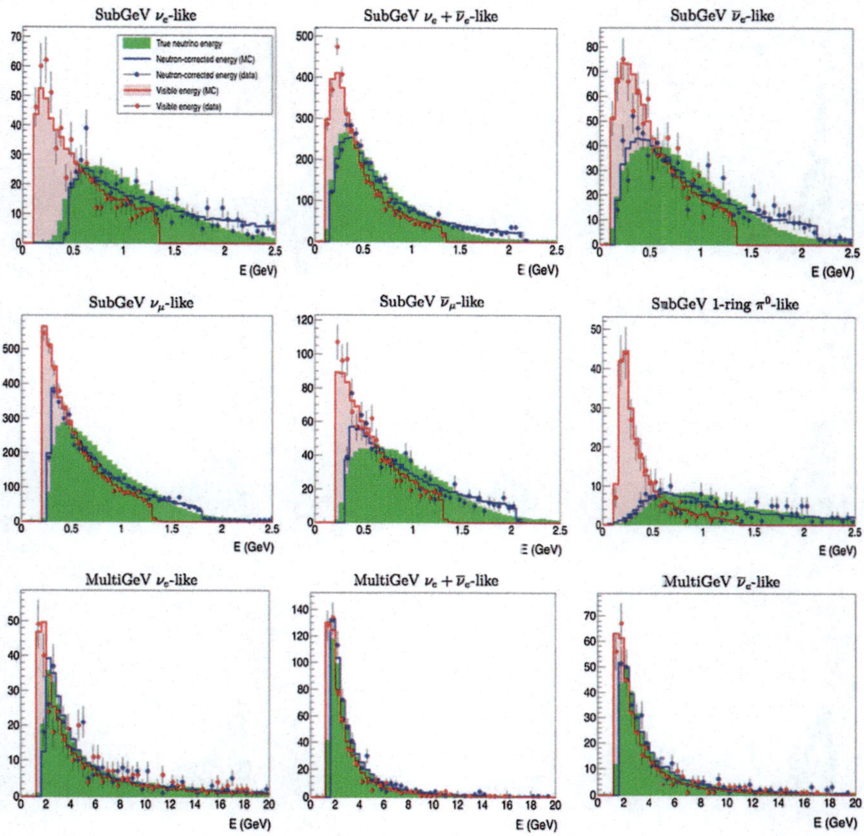

Fig. 7.42 True (green), visible (red) and corrected (blue) energy spectra of the different FC samples for atmospheric neutrino Monte Carlo (solid lines) and the SK-IV data (dots) (1 of 2)

7.2.3 *Neutron-Corrected Reconstructed Neutrino Energy for Fully Contained Samples*

The energy corrections with the neutron multiplicity, explained in Sect. 5.5, are applied to all the, previously defined, fully contained samples of the atmospheric analysis for the case of H-neutron tagging.

In this case, the expression for the H-neutron corrected energy is that of Eq. (7.3).

$$E^{H}_{rec} = E_{vis}(1 + f(H - neutron)) \tag{7.3}$$

The correction functions as well as the performance of the energy correction, in terms of the energy pull distributions, are shown in Appendix C.

Figures 7.42 and 7.43 show the distributions of the true, visible and neutron-corrected energies for the distributions of the different FC samples.

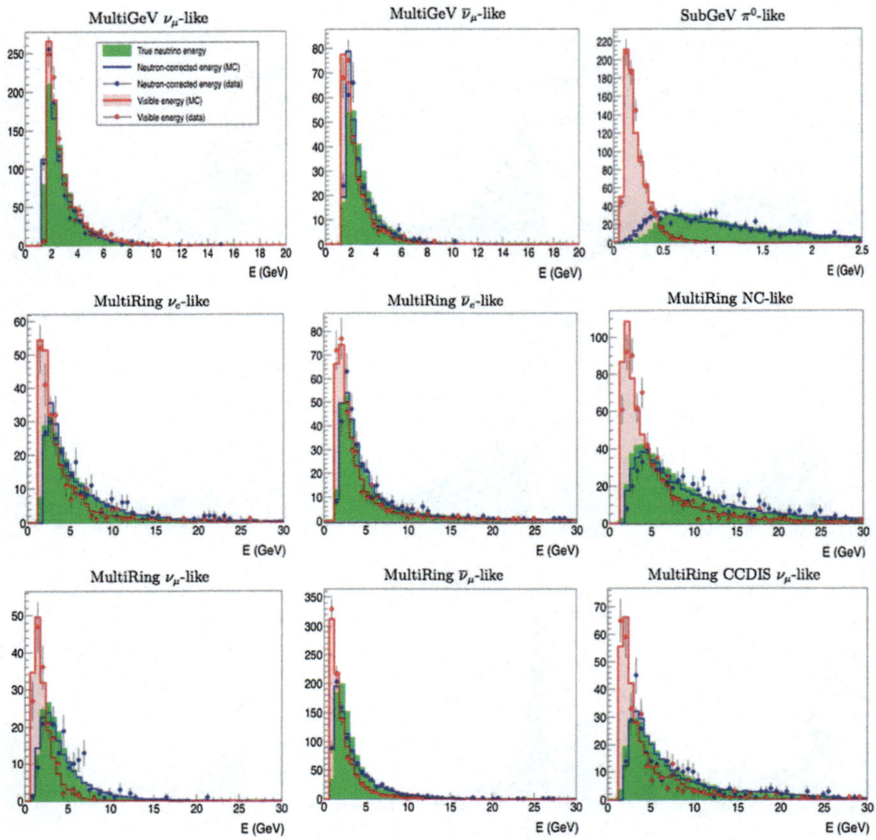

Fig. 7.43 True (green), visible (red) and corrected (blue) energy spectra of the different FC samples for atmospheric neutrino Monte Carlo (solid lines) and the SK-IV data (dots) (2 of 2)

Comparing Figs. 7.42 and 7.43 with those for Gd in Figs. 6.39 and 6.40, it is seen that the neutron correction functions for the H and Gd-tagging cases are similar. This can be further seen in Table 7.12.

The lower neutron tagging efficiency is translated into steeper neutron correction functions (Figs. C.1 and C.2) than for Gd-neutrons (Figs. 6.37 and 6.38). Therefore, the information about the invisible energy is similar for the H-neutron tagging and the Gd-neutron tagging cases, although stored at different typical neutron multiplicities.

Table 7.12 Mode, mean, RMS and skewness of the pull distributions for the visible energy and the H-neutron corrected energy

Samples	E_{vis}				E_{rec}^{H}			
	Mode	Mean	RMS	Skewness	Mode	Mean	RMS	Skewness
SubGeV ν_e-like	0.548	0.538	0.181	−0.052	−0.013	−0.000	0.324	0.038
SubGeV $\nu_e+\overline{\nu}_e$-like	0.223	0.298	0.179	0.422	−0.048	0.087	0.232	0.578
SubGeV $\overline{\nu}_e$-like	0.313	0.361	0.208	0.231	−0.118	0.010	0.325	0.393
SubGeV ν_μ-like	0.223	0.319	0.196	0.492	−0.083	0.024	0.306	0.348
SubGeV $\overline{\nu}_\mu$-like	0.268	0.351	0.199	0.420	−0.118	−0.008	0.312	0.353
SubGeV 1ring-π^0-like	0.863	0.685	0.211	−0.842	0.513	−0.003	0.631	−0.817
SubGeV π^0-like	0.898	0.786	0.158	−0.708	0.618	0.008	0.615	−0.990
MultiGeV ν_e-like	0.178	0.311	0.195	0.686	−0.083	0.090	0.327	−0.023
MultiGeV $\nu_e+\overline{\nu}_e$-like	0.043	0.139	0.135	0.719	−0.048	0.036	0.151	0.553
MultiGeV $\overline{\nu}_e$-like	0.088	0.206	0.163	0.725	−0.048	0.055	0.195	0.526
MultiGeV ν_μ-like	0.088	0.160	0.116	0.630	−0.048	0.033	0.127	0.630
MultiGeV $\overline{\nu}_\mu$-like	0.133	0.194	0.122	0.502	−0.118	0.009	0.144	0.883
MultiRing NC-like	0.493	0.462	0.242	−0.127	0.058	−0.044	0.448	−0.226
MultiRing DIS-ν_μ-like	0.358	0.408	0.203	0.248	0.023	0.059	0.314	0.115
MultiRing ν_e-like	0.223	0.323	0.194	0.518	−0.013	0.041	0.266	0.201
MultiRing $\overline{\nu}_e$-like	0.133	0.254	0.184	0.660	0.023	0.042	0.239	0.082
MultiRing ν_μ-like	0.493	0.442	0.181	−0.277	−0.013	−0.012	0.313	0.002
MultiRing $\overline{\nu}_\mu$-like	0.313	0.338	0.200	0.129	0.058	0.035	0.297	−0.077

7.3 Oscillation Analysis

All the previous developments, due to H-neutron tagging, are applied to SK-IV for proving their benefits to the atmospheric neutrino oscillation analysis. The oscillation analysis is done for all SK periods, being the SK-I, SK-II and SK-III treated as in the current official SK analysis, described in [Pik12]. For the implementation of the new sample definition and the neutron-corrected energy, the SK analysis tools are modified to accommodate them. The sensitivity of this analysis is computed, as explained previously in Sect. 6.6.5.1, by the binned χ^2 method.

First, in Sect. 7.3.1 all the systematic errors involved in this analysis, including those associated to the new sample selection and energy correction for the SK-IV are described. Then, an oscillation study is performed, in Sect. 7.3.2, for all SK periods assuming their corresponding exposure times, shown in Table 7.1. The current official analysis is applied to SK-I, SK-II and SK-III, and the new sample selection and energy corrections of Sect. 7.2, only for the SK-IV phase, where H-neutron tagging is available.

In Sect. 7.3.3, a global oscillation analysis of the actual atmospheric neutrino data including the neutron tagging tools developed for SK-IV, is performed.

7.3.1 Systematic Errors

In this analysis, 77 systematic error sources are considered. Some of them are common to all SK periods, such as those associated to the neutrino flux (except that associated with the solar activity as it is time-dependent) and interactions. The uncertainty in the oscillation parameters is also considered as a systematic error source and shared by all SK phases as well. Other sources of systematics depend on the performance of the detectors and therefore, on the SK phase, like the errors induced by the reduction and reconstruction of events. Finally, three new systematic errors are defined, two of them account for the uncertainty in the new sample selection methods, and the other for the uncertainty in neutron-corrected energy. These systematic errors affect only the fully contained data of SK-IV.

7.3.1.1 Neutrino Flux Errors

The systematic uncertainties related to the neutrino fluxes are shared by all SK phases and summarised in Table 7.13.

Absolute Normalisation

This is the uncertainty of the absolute neutrino flux and calculated in [Sanuki07], taking into account the atmospheric pion and kaon production in the hadronic interaction model, the hadronic interaction cross sections and the atmospheric density profile. The systematic error is divided into two energy regions, below and above 1 GeV, as seen in Figs. 7.44 and 7.45.

Table 7.13 Systematic uncertainties for the atmospheric oscillation analysis related to the neutrino flux

Neutrino flux errors		σ (%)
Absolute normalisation	$E_\nu < 1$ GeV	25.0
	$E_\nu > 1$ GeV	15.0
$\overline{\nu}/\nu$	$E_\nu < 1$ GeV	2.0
	1 GeV $< E_\nu <$ 10 GeV	3.0
	$E_\nu >$ 10 GeV	5.0
$\overline{\nu}_e/\nu_e$	$E_\nu < 1$ GeV	5.0
	1 GeV $< E_\nu <$ 10 GeV	5.0
	$E_\nu >$ 10 GeV	8.0
$\overline{\nu}_\mu/\nu_\mu$	$E_\nu < 1$ GeV	2.0
	1 GeV $< E_\nu <$ 10 GeV	6.0
	$E_\nu >$ 10 GeV	15.0
Up/down ratio		1.0
Horizontal/vertical ratio		1.0
κ/π ratio		10.0
Neutrino path length		10.0
Relative normalisation	FC MultiGeV	5.0
	PC, UpMu stop	5.0
Matter effect		6.8
Solar activity	SK-I	20.0
	SK-II	50.0
	SK-III	20.0
	SK-IV	10.0

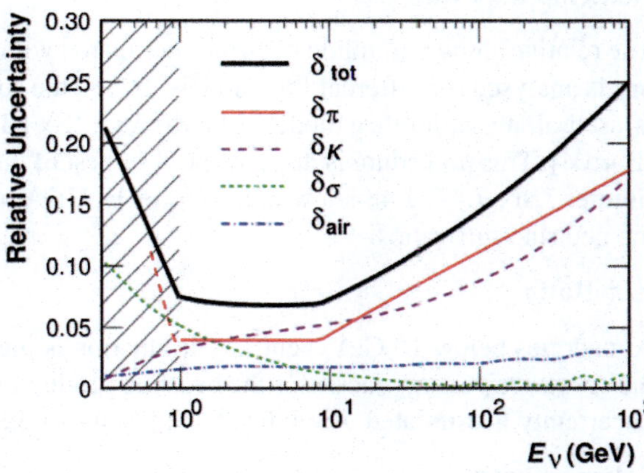

Fig. 7.44 Relative uncertainty of each factor entering the computation of the neutrino flux. δ_π, δ_κ, δ_σ, and δ_{air} are the uncertainties on pion production, kaon production, hadronic interaction cross sections and atmospheric density profile

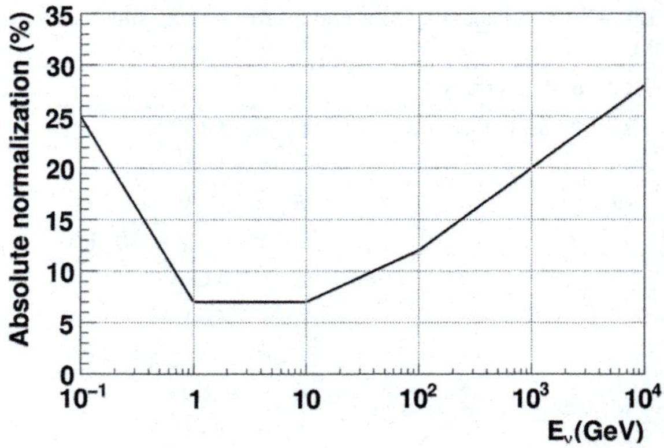

Fig. 7.45 Calculated uncertainty of the absolute normalisation of the neutrino flux as a function of the neutrino energy

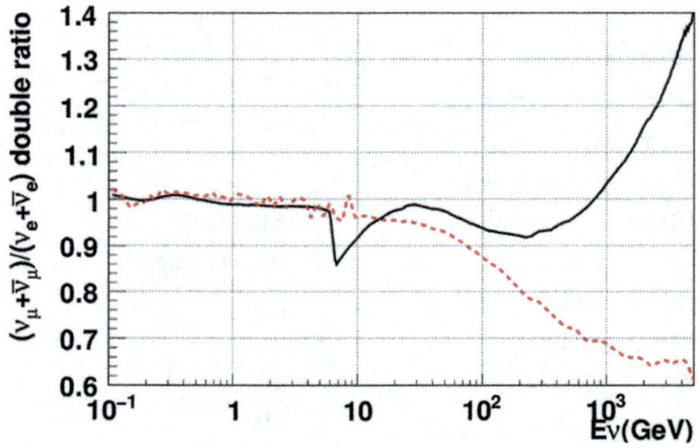

Fig. 7.46 The $(\nu_\mu + \overline{\nu}_\mu)/(\nu_e + \overline{\nu}_e)$ ratio between the FLUKA and Honda models (black solid line) and between Bartol and Honda models (red dashed line)

Neutrino Flux Ratios

The relative amount of muon neutrinos compared with the amount of electron neutrinos is analysed for different flux models. In SK, the so called Honda flux [Honda11] is used along with other models like the FLUKA [Battistoni03] and Bartol fluxes [Barr04]. This procedure is done also for the rest of the flux ratios, $\overline{\nu}_\mu/\nu_\mu$ and $\overline{\nu}_e/\nu_e$. Figures 7.46, 7.47, 7.48 show the ratio for FLUKA and Honda models for each of the neutrino flux ratios.

κ/π Ratio

At energies below 10 GeV, neutrino production is mainly due to pion decay and as the cosmic ray energy increases the neutrino production is dominated by kaons. This uncertainty is estimated using data from [Ambrosini98].

Solar Activity

The 11 year solar cycle slightly changes the Earth's magnetic field, affecting the cosmic ray flux. The uncertainty of the solar cycle is taken to be ± 1 year and it is translated into a neutrino flux systematic error.

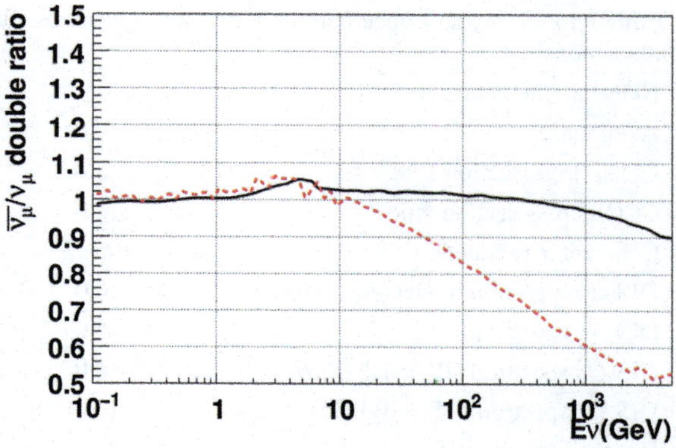

Fig. 7.47 The $\overline{\nu}_\mu/\nu_\mu$ ratio between the FLUKA and Honda models (black solid line) and between Bartol and Honda models (red dashed line)

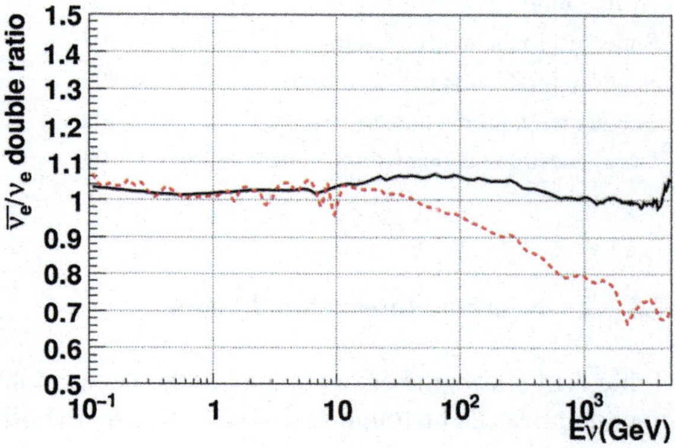

Fig. 7.48 The $\overline{\nu}_e/\nu_e$ ratio between the FLUKA and Honda models (black solid line) and between Bartol and Honda models (red dashed line)

Neutrino Production

This systematic error deals with the height at which the neutrinos are produced. This affects the length travelled by neutrinos and therefore, the oscillation probability. This error is negligible for upward-going neutrinos compared to the matter effect error. The atmosphere density uncertainty is the main reason for this error and estimated by comparing the US-standard 76 and the MSISE90 experiments [Hedin91].

Matter Effect

The oscillation probability of neutrinos passing though the Earth depend on its electron density. The exact composition of the Earth's layers is of great importance. The different composition and its uncertainty is propagated to the oscillation probability uncertainty and used as a systematic error.

Table 7.14 Systematic uncertainties for the atmospheric oscillation analysis related to the neutrino interactions

Neutrino interaction errors	σ (%)
Axial mass	10.0
CC ν_τ cross-section	25.0
CCQE cross-section ratio	10.0
Coherent π production	100.0
DIS cross-section model (E_ν < 10 GeV)	10.0
DIS cross-section	5.0
DIS Q^2 spectrum (W < 1.3 GeV/c^2)	10.0
DIS Q^2 spectrum (W > 1.3 GeV/c^2)	10.0
Hadronic simulation	10.0
NC/CC ratio	20.0
Single π production ($\overline{\nu}/\nu$ ratio)	10.0
Single π production ($\pi^0/\pi^\pm$ ratio)	40.0
Single meson production cross-section	20.0
Decay-electron tagging (from π interaction)	10.0

7.3.1.2 Neutrino Interaction Errors

Table 7.14 shows the errors related to the neutrino interactions. Some of the largest uncertainties can be found in this group due to the little data available from neutrino cross-section experiments.

CCQE Cross-Section

For modelling the CCQE cross-section in SK, the Fermi-gas model in [Smith72] is used. For the estimation of the systematic uncertainty, the difference between this model and [Nieves04] is evaluated.

Axial Mass

The axial mass used in the SK Monte Carlo simulation is $M_A = 1.21$ GeV, and for accounting for the uncertainty on this value various experimental results are considered, [Gran06, Nieves12].

Single Meson Production

The estimate of this error is done by comparing the $1\pi^\pm$ production experimental measurements. For the case of $1\pi^0$ production and due to the lack of enough measurements, the experimental results are also compared with the model in [Hernandez07].

Coherent Pion Production

This uncertainty is set to 50% due to the difference between the Rein&Sehgal model and experimental data [Nakajima11] for NC and CC ν_e interactions. There is no experimental measurement for CC ν_μ coherent pion production, so the uncertainty is taken to be 100%.

Deep Inelastic Scattering

The deep inelastic scattering cross-section expectations agree with experiments within 5%. This uncertainty becomes larger at low energies, making necessary an additional error is estimated by comparison to the CKMT model [Capella01].

Other Cross-Sections

The Hagiwara model is considered for the ν_τ CC cross-section [Hagiwara03].

7.3.1.3 Event Reduction Errors

Table 7.15 shows the errors related to the event reduction, their values and the value of ϵ for the best fit point.

Fiducial Volume

This systematic uncertainty takes into account the migration of events because of the fiducial volume cut.

FC, PC, UpMu Reductions

The reduction uncertainties are estimated by comparing the distributions of each cut variable for the observed data and Monte Carlo and the remaining contamination background.

The separation between FC and PC and between UpMu stopping and through-going muons is done using the number of OD hits. For this, data and Monte Carlo distributions are compared to estimate the uncertainty associated to these migrations.

UpMu Showering and Non-showering Backgrounds

UPMU non-showering and showering events are separated by using the difference between the observed charge and expected charge of non-showering muons and the uncertainty is estimated by comparing this distribution for data and Monte Carlo.

7.3.1.4 Event Reconstrucion Errors

Table 7.16 shows the errors related to the event reconstruction.

The ring separation is based on a likelihood distribution and its uncertainty is estimated by comparing data and Monte Carlo distributions. In a similar fashion, the particle identification systematic errors is extracted from the comparisons, between data and Monte Carlo, of the likelihood distributions used for separating between e-like and μ-like events. The energy calibration error is estimated by using the calibration data from.

Table 7.15 Systematic uncertainties for the atmospheric oscillation analysis related to the event reduction for all SK periods

Reduction errors	SK-I	SK-II	SK-III	SK-IV
	σ	σ	σ	σ
FC reduction	0.2	0.2	0.8	0.3
Fiducial volume	2.0	2.0	2.0	2.0
Non-ν background (μ-like)	1.0	1.0	1.0	1.0
Non-ν background (SubGeV e-like)	1.0	1.0	1.0	1.0
Non-ν background (MultiGeV 1ring e-like)	13.2	38.1	26.7	17.6
Non-ν background (MultiGeV MultiRing e-like)	12.1	11.1	11.4	11.6
FC-PC separation	0.6	0.5	0.9	0.9
PC reduction	2.4	4.8	0.5	1.0
PC stop-through separation (barrel)	7.0	9.4	28.7	8.5
PC stop-through separation (bottom)	22.7	12.9	12.1	6.8
PC stop-through separation (top)	46.1	19.4	86.6	40.3
UpMu stop-through separation	0.4	0.6	0.4	0.6
UpMu reduction	1.0	1.0	1.0	1.0
UpMu stop background	16.0	21.0	20.0	17.0
UpMu stop energy cut	0.9	1.3	2.0	1.7
UpMu through path cut	1.5	2.3	2.8	1.5
UpMu non-showering background	18.0	14.0	24.0	17.0
UpMu showering–non-showering separation	3.4	4.4	2.4	3.0
UpMu showering background	18.0	14.0	24.0	24.0

Table 7.16 Systematic uncertainties for the atmospheric oscillation analysis related to the event reconstruction for all SK periods

Reconstruction errors	SK-I	SK-II	SK-III	SK-IV
	σ	σ	σ	σ
SingleRing PID	1.0	1.0	1.0	1.0
MultiRing PID	10.0	10.0	10.0	10.0
Ring separation	10.0	10.0	10.0	10.0
SingleRing π^0 selection	10.0	10.0	10.0	10.0
TwoRing π^0 selection	5.6	4.4	5.9	5.6
Decay-electron tagging efficiency	10.0	10.0	10.0	(10.0)
Neutron tagging efficiency	–	–	–	(10.0)
Energy calibration	1.1	1.7	2.7	2.3
Energy calibration up/down ratio	0.6	0.6	1.3	0.3

The error associated to the decay-electron tagging for SK-IV is also of 10%, but it will be treated differently in the Sect. 7.3.1.5 because it takes part in the sample separation methods, having different impact on each of the newly defined SK-IV samples. In a similar manner, there is only neutron tagging at SK-IV and is of 10%, but its impact will be also analysed for each sample in Sect. 7.3.1.5.

7.3.1.5 Fully Contained Event Selection Errors

This systematic errors are newly calculated specifically for the new sample selection of fully contained events in SK-IV. These uncertainties come from the event migration between samples after the uncertainty in the variables used for the sample selection is propagated.

In the case of samples selected by applying cuts on the number of neutrons and decay-electrons, the systematic uncertainty is induced by the propagation in the tagging efficiency of neutrons and decay-electrons, which are both of 10% [Irvine14].

$$\sigma^{ntag} = \frac{N(ntag, +10\%) - N(ntag, -10\%)}{2N(ntag)} \tag{7.4}$$

$$\sigma^{decay-e} = \frac{N(decay - e, +10\%) - N(decay - e, -10\%)}{2N(decay - e)} \tag{7.5}$$

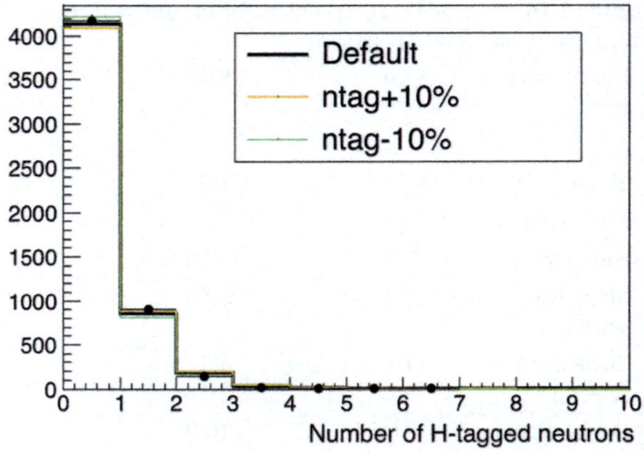

Fig. 7.49 Number of H-tagged neutrons for the default Monte Carlo, plus and minus 10% uncertainty in the tagging efficiency and SK-IV atmospheric data (black dots) for the SubGeV e-like sample

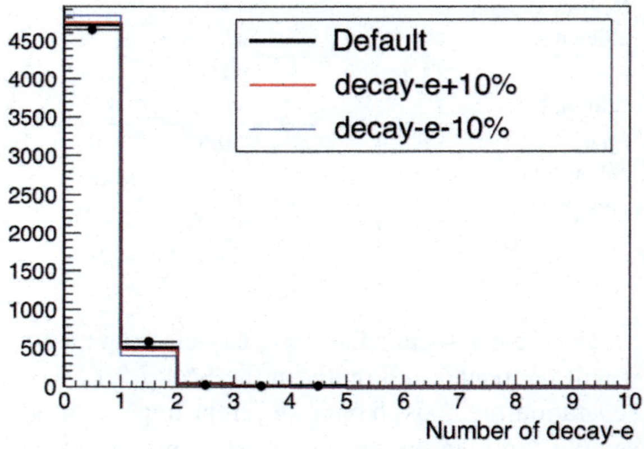

Fig. 7.50 Number of decay-electrons for the default Monte Carlo, plus and minus 10% uncertainty in the tagging efficiency and SK-IV atmospheric data (black dots) for the SubGeV e-like sample

On the other hand, for those samples where multivariate methods are applied for the construction of a likelihood distribution, and, in addition to the errors induced by neutrons and decay-electrons (principal variables for the sample selection), the likelihood distributions are compared for data and Monte Carlo. This last estimate uses two parameters to shift (β) and smear (α) the Monte Carlo distribution to build a χ^2 map. The systematic error is estimated as the relative variation of the number of events in the samples between the point with minimum χ^2 value and the point at 1σ (i.e. $\Delta\chi^2 = 2.3$).

$$\sigma^{shift} = \frac{N(bestfit) - N(1\sigma)}{N(bestfit)} \tag{7.6}$$

SubGeV e-like

Figures 7.49 and 7.50 below show the nominal and $\pm 10\%$ distributions for the neutron and decay-electron multiplicities.

Table 7.17 Systematic uncertainties associated to the H-neutron and decay-electron tagging in the selection of the three SubGeV e-like samples

	σ^{ntag} (%)	$\sigma^{decay-e}$ (%)
SubGeV ν_e-like	0.0	8.2
SubGeV $\nu_e + \overline{\nu}_e$-like	−1.5	−0.8
SubGeV $\overline{\nu}_e$-like	5.8	−1.3

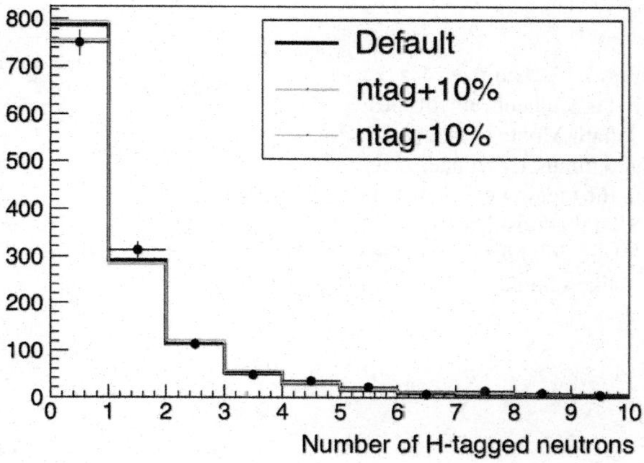

Fig. 7.51 Number of H-tagged neutrons for the default Monte Carlo, plus and minus 10% uncertainty in the tagging efficiency and SK-IV atmospheric data (black dots) for the MultiGeV e-like sample

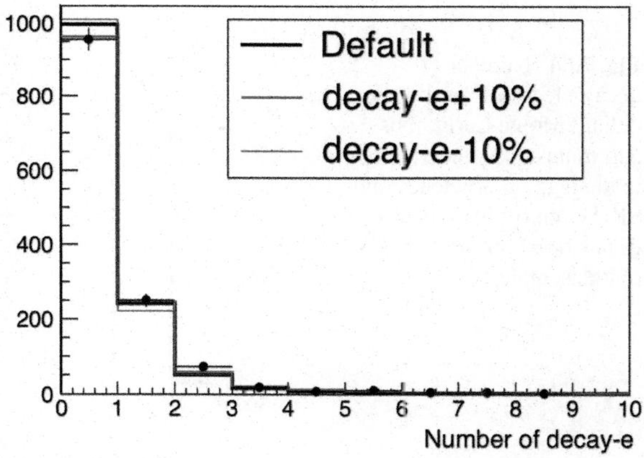

Fig. 7.52 Number of decay-electrons for the default Monte Carlo, plus and minus 10% uncertainty in the tagging efficiency and SK-IV atmospheric data (black dots) for the MultiGeV e-like sample

Table 7.17 shows the estimated systematic errors associated to the SubGeV e-like sample selection. The SubGeV ν_e sample does not require any cut on the number of neutrons and, therefore, the associated systematic is null.

MultiGeV e-like

The results for these samples are very similar to those from the SubGeV e-like case. The ±10% variation and the nominal distributions for the number of H-tagged neutrons and for the number of decay-electrons are shown in Figs. 7.51 and 7.52.

Table 7.18 shows the estimated systematic errors associated to this sample selection.

Table 7.18 Systematic uncertainties associated to the H-neutron and decay-electron tagging in the selection of the three MultiGeV e-like samples

	σ^{ntag} (%)	$\sigma^{decay-e}$ (%)
MultiGeV ν_e-like	0.0	7.3
MultiGeV $\nu_e\overline{\nu}_e$-like	−2.1	−1.8
MultiGeV $\overline{\nu}_e$-like	4.0	−3.6

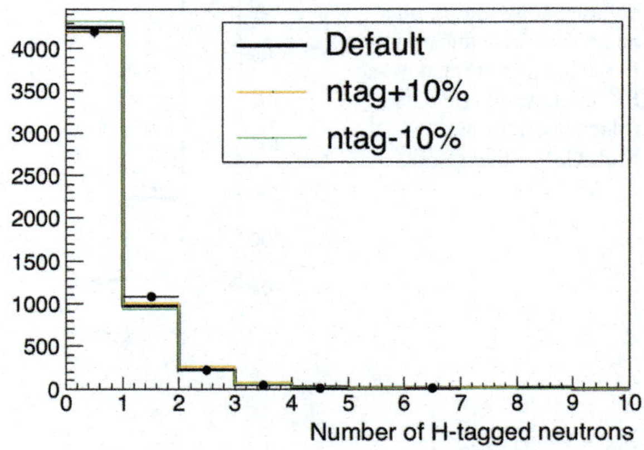

Fig. 7.53 Number of H-tagged neutrons for the default Monte Carlo, plus and minus 10% uncertainty in the tagging efficiency and SK-IV atmospheric data (black dots) for the SubGeV μ-like sample

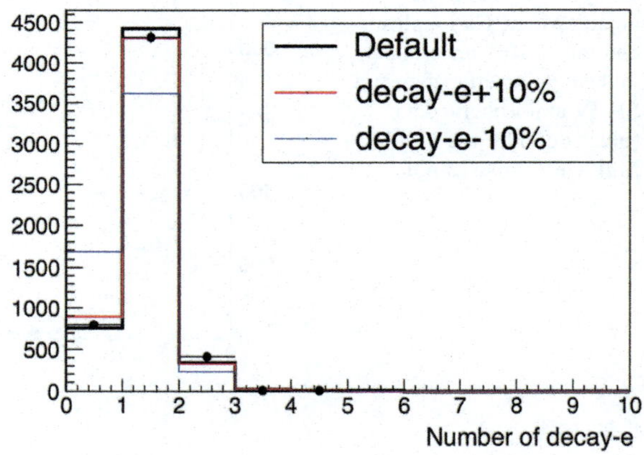

Fig. 7.54 Number of decay-electrons for the default Monte Carlo, plus and minus 10% uncertainty in the tagging efficiency and SK-IV atmospheric data (black dots) for the SubGeV μ-like sample

SubGeV μ-like

Figures 7.53 and 7.54 show the nominal and $\pm$10% distributions for the neutron and decay-electron multiplicities for SubGeV μ-like events.

The estimated systematic errors for both defined samples are summarised in Table 7.19.

Table 7.19 Systematic uncertainties associated to the H-neutron and decay-electron tagging in the selection of the two SubGeV μ-like samples

	σ^{ntag} (%)	$\sigma^{decay-e}$ (%)
SubGeV ν_μ-like	1.2	−1.7
SubGeV $\overline{\nu}_\mu$-like	5.2	7.5

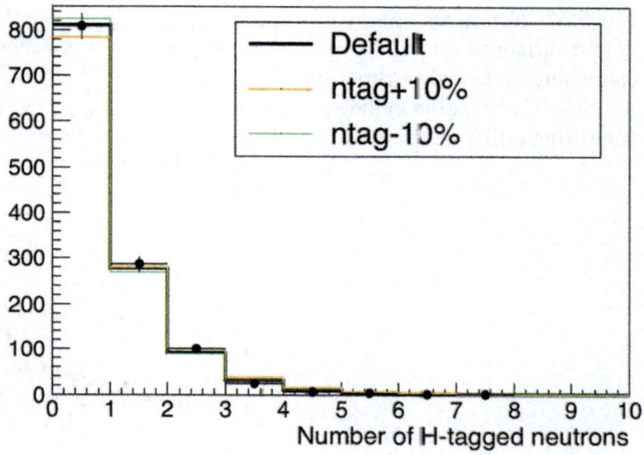

Fig. 7.55 Number of H-tagged neutrons for the default Monte Carlo, plus and minus 10% uncertainty in the tagging efficiency and SK-IV atmospheric data (black dots) for the MultiGeV μ-like sample

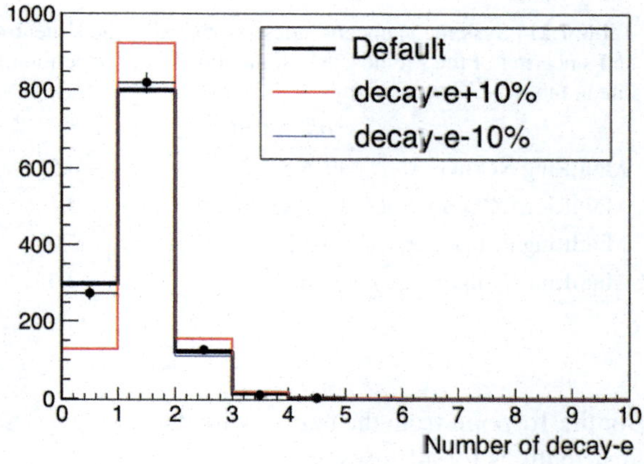

Fig. 7.56 Number of decay-electrons for the default Monte Carlo, plus and minus 10% uncertainty in the tagging efficiency and SK-IV atmospheric data (black dots) for the MultiGeV μ-like sample

MultiGeV μ-like

The results for these samples are very similar to those from the SubGeV μ-like case. The neutron and decay-electron multiplicities are shown in Figs. 7.55 and 7.56.

The estimated systematic errors are presented in Table 7.20.

MultiRing e-like

In this group of samples three neural networks are used for the separation.

The first one is the NC-CC separation. Its corresponding likelihood distribution is shown in Fig. 7.57 for the uncertainty in the neutron and decay-electron tagging, and

Table 7.20 Systematic uncertainties associated to the H-neutron and decay-electron tagging in the selection of the two MultiGeV μ-like samples

	σ^{ntag} (%)	$\sigma^{decay-e}$ (%)
MultiGeV ν_μ-like	−1.4	−1.7
MultiGeV $\overline{\nu}_\mu$-like	5.0	6.6

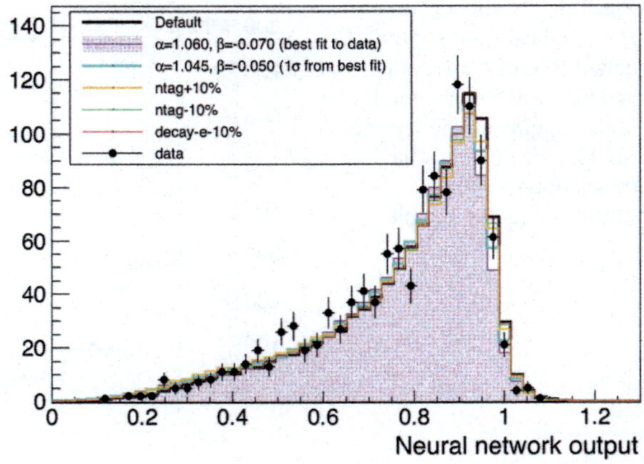

Fig. 7.57 Neural network output variations for estimating systematic errors for NC-CC separation in the MultiRing e-like sample

Table 7.21 Systematic uncertainties associated to the H-neutron, decay-electron tagging and shift and smearing of the Monte Carlo simulation in the selection of the four MultiRing e-like samples due to the NC-CC separation

	σ^{ntag} (%)	$\sigma^{decay-e}$ (%)	σ^{shift} (%)
MultiRing NC-like	5.5	−0.5	−3.9
MultiRing CCDIS ν_μ-like	−2.7	0.4	2.3
MultiRing ν_e-like	−1.3	−0.0	1.1
MultiRing $\overline{\nu}_e$-like	−0.4	−0.0	0.4

for the 1σ point from the best fit with data in the χ^2-map. In Fig. 7.57, the likelihood distributions for all cases are presented.

With all the previous, one can extract the systematic errors from each method. Table 7.21 shows that the neutron and decay-electron tagging errors are larger than those from obtained from the χ^2-map and, hence, considered as the systematic errors associated to this separation.

For the second step, the CC-DIS $\overset{(-)}{\nu}_\mu$-CC $\overset{(-)}{\nu}_e$ separation, the same procedure applies. The neural network output for all the cases considered is shown in Fig. 7.58.

In this case, there are only three of the four samples affected by this uncertainty and, again, the estimate with the larger value for the systematic error is that due to the neutrons and decay-electrons, as shown in Table 7.22.

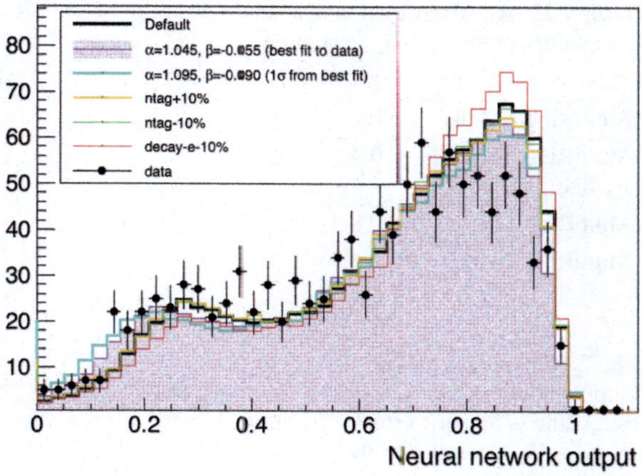

Fig. 7.58 Neural network output variations for estimating systematic errors for $\nu_\mu - \nu_e$ separation in the MultiRing e-like sample

Table 7.22 Relative variation of contents for the three MultiRing e-like samples due to the induced systematics of the $\nu_\mu - \nu_e$ separation

	σ^{ntag} (%)	$\sigma^{decay-e}$ (%)	σ^{shift} (%)
MultiRing NC-like	0	0	0
MultiRing CCDIS ν_μ-like	1.2	3.4	0.9
MultiRing ν_e-like	−1.0	−3.8	−1.0
MultiRing $\overline{\nu}_e$-like	−0.3	−0.1	−0.1

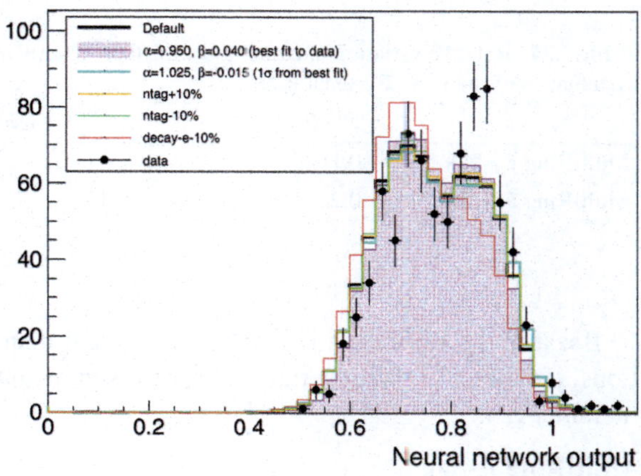

Fig. 7.59 Neural network output variations for estimating systematic errors for $\nu_e - \overline{\nu}_e$ separation in the MultiRing e-like sample

The final step splits the remaining events into neutrino-like and antineutrino-like. Its likelihood, taking into account the neutron and decay-electron tagging uncertainties and that 1σ away from the best fit point of Monte Carlo to the data, is plotted in Fig. 7.59.

Table 7.23 Relative variation of contents for the three MultiRing e-like samples due to the induced systematics of the $\nu_e - \overline{\nu}_e$ separation

	σ^{ntag} (%)	$\sigma^{decay-e}$ (%)	σ^{shift} (%)
MultiRing NC-like	0	0	0
MultiRing CCDIS ν_μ-like	0	0	0
MultiRing ν_e-like	0.8	3.5	1.4
MultiRing $\overline{\nu}_e$-like	−0.5	−2.2	−1.1

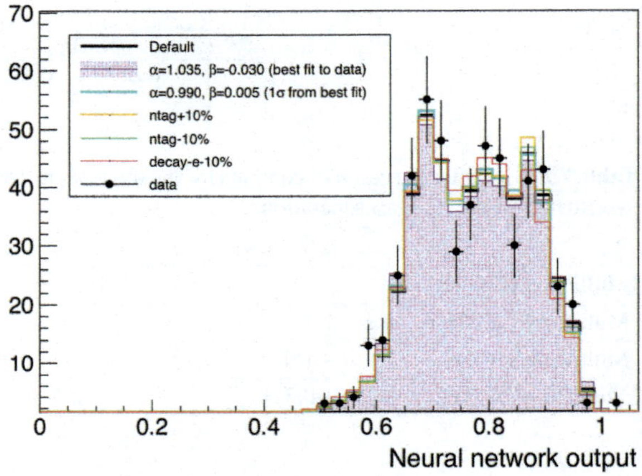

Fig. 7.60 Neural network output variations for estimating systematic errors for $\nu_\mu - \overline{\nu}_\mu$ separation in the MultiRing μ-like sample

Table 7.24 Relative variation of contents for the three MultiRing e-like samples due to the induced systematics of the $\nu_\mu - \overline{\nu}_\mu$ separation

	σ^{ntag} (%)	$\sigma^{decay-e}$ (%)	σ^{shift} (%)
MultiRing ν_μ-like	−0.1	3.0	−0.2
MultiRing $\overline{\nu}_\mu$-like	0.2	−3.4	0.3

Because the separation is much less efficient than in the previous sample selections, the effect of the uncertainties is also significantly smaller. This can be found in Table 7.23.

MultiRing μ-like

In this case, a neural network is also considered, but this time with only one step. The procedure is the analogous as in the paragraph before and the separation likelihood distribution is plotted in Fig. 7.60.

The systematic uncertainties associated to this samples selection are shown in Table 7.24.

Summary

For MultiRing samples, the estimated errors induced by the neutron tagging and the decay-electron tagging efficiency uncertainties are larger than that compute by the χ^2-map method. Therefore, the formers are taken, for all cases, as the two systematic sources for the sample selection.

Table 7.25 summarises the previous study for the estimation of the systematic errors for the sample selection for fully contained events in SK-IV, and that will be included to the oscillation analysis.

7.3.1.6 Neutron Energy Corrections for Fully Contained Events

As the correction function depends strictly on the neutron multiplicity, the relative change in its magnitude is computed to include it as an additional systematic error source. As with the systematics treated before, this only applies to SK-IV fully contained events.

$$\Delta f = \frac{E^H_{rec}(\text{H-neutrons} + 10\%) - E^H_{rec}(\text{H-neutrons} + 10\%)}{E^H_{rec}(\text{H-neutrons})} \tag{7.7}$$

The quantity Δf will be taken to calculate the migration of events between bins, which will be the actual systematic uncertainty due to the neutron energy corrections (Table 7.26).

7.3.2 Oscillation Sensitivity Study

The oscillation analysis will be done for the whole SK dataset, being only able to apply the neutron improvement discussed earlier in SK-IV.

The binning for the analysis in cosine of the zenith angle and in Gd-neutron corrected has a total of 1221 bins, 520 for each of SK-I, SK-II and SK-III, merged in the analysis, and 701 for the SK-IV period.

The true oscillation parameters used for the sensitivity studies are those already used for the SuperK-Gd atmospheric oscillation depicted in Table 6.13.

Two sensitivity studies are performed, one assuming true normal hierarchy and the other for the inverted mass ordering. The χ^2 distributions, in the normal hierarchy scenario, for Δm^2_{32}, $\sin^2\theta_{23}$ and δ_{CP} are shown in Figs. 7.61, 7.62 and 7.63, respectively.

Assuming nature is inverted hierarchy, the analogous χ^2 distributions are shown in Figs. 7.64, 7.65 and 7.66.

The H-tag analysis improves the sensitivity to the neutrino mass ordering and the CP phase. The comparison between the official and H-tag analyses can be summarised in Table 7.27.

Table 7.25 Sample selection systematic errors for the SK-IV atmospheric oscllation analysis

FC event selection errors			σ (%)	
			Neutron tagging	Decay-e tagging
SK-IV FC samples	SK4 SubGeV ν_e-like		0.0	8.2
	SK4 SubGeV $\nu_e + \overline{\nu}_e$-like		−1.5	−0.8
	SK4 SubGeV $\overline{\nu}_e$-like		5.8	−1.3
	SK4 SubGeV ν_μ-like		1.2	−1.7
	SK4 SubGeV $\overline{\nu}_\mu$-like		5.2	7.5
	SK4 MultiGeV ν_e-like		0.0	7.3
	SK4 MultiGeV $\nu_e + \overline{\nu}_e$-like		−2.1	−1.8
	SK4 MultiGeV $\overline{\nu}_e$-like		4.0	−3.6
	SK4 MultiGeV ν_μ-like		−1.4	−1.7
	SK4 MultiGeV $\overline{\nu}_\mu$-like		5.0	6.6
	SK4 MultiRing NC-like		5.5	−0.5
	SK4 MultiRing DIS-ν_μ-like	NC sep.	−2.7	0.4
		DISmu sep.	1.2	3.4
	SK4 MultiRing ν_e-like	NC sep.	−1.3	0.0
		DISmu sep.	−1.0	3.8
	SK4 MultiRing $\overline{\nu}_e$-like	$\nu - \overline{\nu}$ sep.	0.8	3.5
		NC sep.	−0.4	0.0
		DISmu sep.	−0.3	−0.1
		$\nu - \overline{\nu}$ sep.	−0.5	2.2
	SK4 MultiRing ν_μ-like		−0.1	3.0
	SK4 MultiRing $\overline{\nu}_\mu$-like		0.2	−3.4

Table 7.26 Relative change in the neutron energy correction functions due to the neutron tagging uncertainty

Neutron energy correction errors		Δf (%)
SK-IV FC samples	SK4 SubGeV ν_e-like	0.8
	SK4 SubGeV $\nu_e + \overline{\nu}_e$-like	0.0
	SK4 SubGeV $\overline{\nu}_e$-like	1.3
	SK4 SubGeV 1ring π-like	0.5
	SK4 SubGeV ν_μ-like	1.5
	SK4 SubGeV $\overline{\nu}_\mu$-like	1.1
	SK4 SubGeV π-like	1.3
	SK4 MultiGeV ν_e-like	0.9
	SK4 MultiGeV $\nu_e + \overline{\nu}_e$-like	0.0
	SK4 MultiGeV $\overline{\nu}_e$-like	1.0
	SK4 MultiGeV ν_μ-like	0.3
	SK4 MultiGeV $\overline{\nu}_\mu$-like	0.6
	SK4 MultiRing NC-like	1.6
	SK4 MultiRing DIS-ν_μ-like	0.9
	SK4 MultiRing ν_e-like	0.7
	SK4 MultiRing $\overline{\nu}_e$-like	0.7
	SK4 MultiRing ν_μ-like	1.1
	SK4 MultiRing $\overline{\nu}_\mu$-like	0.9

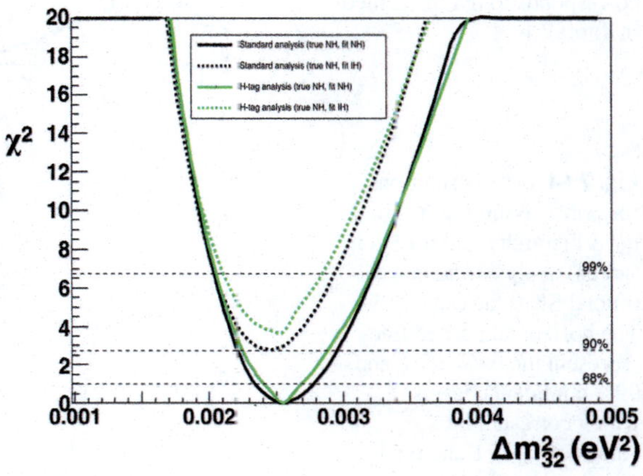

Fig. 7.61 χ^2 distributions for Δm^2_{32} assuming normal mass hierarchy and for H-tag (green) analysis and current official SK (black) analysis. The horizontal dashed lines represent the 68%, 90% and 99% confidence levels, which correspond to χ^2 values of 1, 2.71 and 6.64, respectively. The dataset corresponds to that presented in Table 7.1

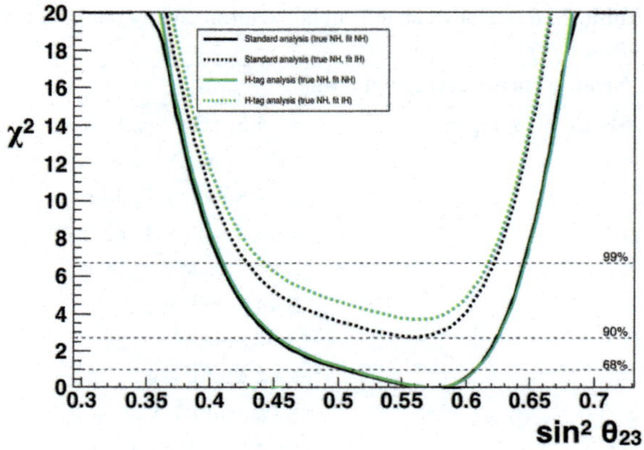

Fig. 7.62 χ^2 distributions for $\sin^2\theta_{23}$ assuming normal mass hierarchy and for H-tag (green) analysis and current official SK (black) analysis. The horizontal dashed lines represent the 68%, 90% and 99% confidence levels, which correspond to χ^2 values of 1, 2.71 and 6.64, respectively. The dataset corresponds to that presented in Table 7.1

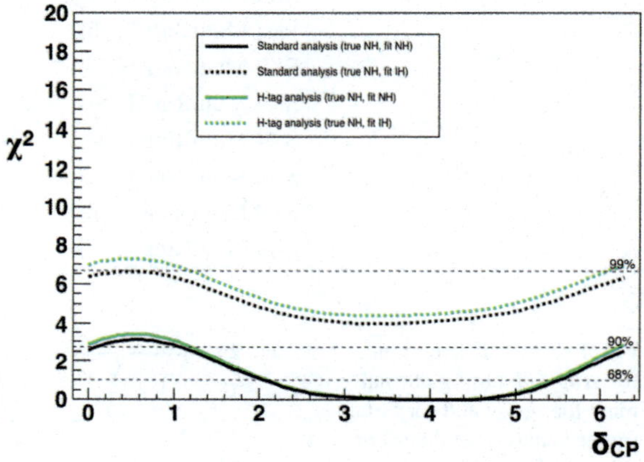

Fig. 7.63 χ^2 distributions for δ_{CP} assuming normal mass hierarchy and for H-tag (green) analysis and current official SK (black) analysis. The horizontal dashed lines represent the 68%, 90% and 99% confidence levels, which correspond to χ^2 values of 1, 2.71 and 6.64, respectively. The dataset corresponds to that presented in Table 7.1

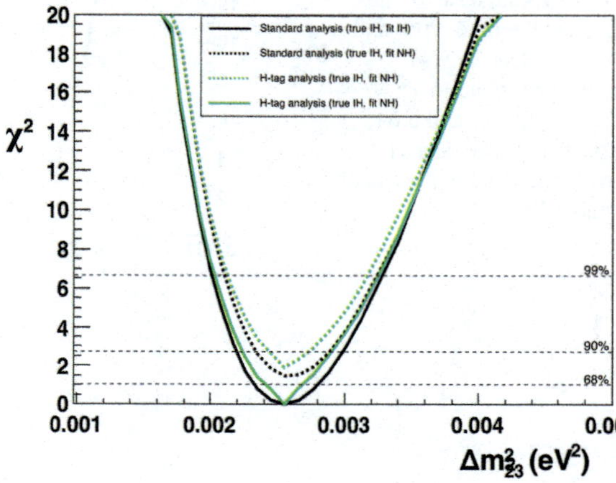

Fig. 7.64 χ^2 distributions for Δm^2_{32} assuming inverted mass hierarchy and for H-tag (green) analysis and current official SK (black) analysis. The horizontal dashed lines represent the 68%, 90% and 99% confidence levels, which correspond to χ^2 values of 1, 2.71 and 6.64, respectively. The dataset corresponds to that presented in Table 7.1

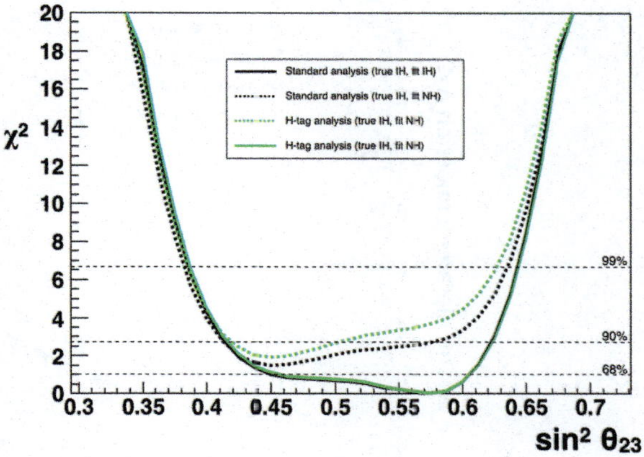

Fig. 7.65 χ^2 distributions for $\sin^2 \theta_{23}$ assuming inverted mass hierarchy and for H-tag (green) analysis and current official SK (black) analysis. The horizontal dashed lines represent the 68%, 90% and 99% confidence levels, which correspond to χ^2 values of 1, 2.71 and 6.64, respectively. The dataset corresponds to that presented in Table 7.1

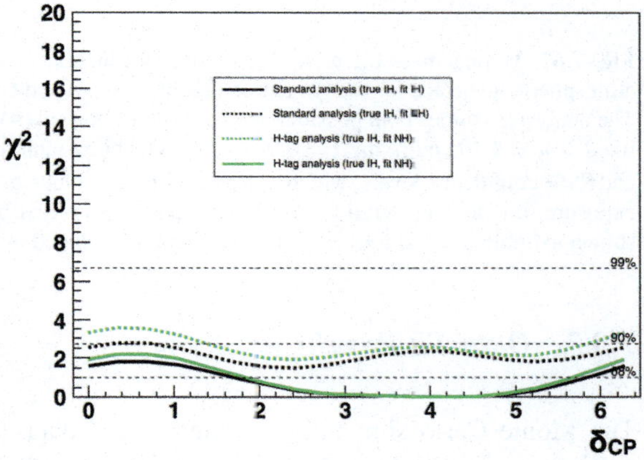

Fig. 7.66 χ^2 distributions for δ_{CP} (left) assuming inverted mass hierarchy and for H-tag (green) analysis and current official SK (black) analysis. The horizontal dashed lines represent the 68%, 90% and 99% confidence levels, which correspond to χ^2 values of 1, 2.71 and 6.64, respectively. The dataset corresponds to that presented in Table 7.1

Table 7.27 Mass hierarchy sensitivities and fraction of δ_{CP} more than 1σ away from the best fit point, assuming as true, normal (left) and inverted (right) hierarchies. Comparison between official SK analysis and H-tag analysis

	Standard (true NH)	H-tag (true NH)	Standard (true IH)	H-tag (true IH)
$\Delta\chi^2(NH - IH)$	−2.78	−3.71	1.51	1.89
δ_{CP} fraction over 1σ (%)	33.2	38.7	27.7	33.5

The sensitivity to the mass hierarchy as function of the value of $\sin^2 \theta_{23}$ assuming two scenarios, true normal hierarchy and true inverted hierarchy, is displayed in Fig. 7.67. This plots shows that the H-tag analysis improves the sensitivity to the rejection of the wrong mass hierarchy for all cases, but specially for large values of $\sin^2 \theta_{23}$.

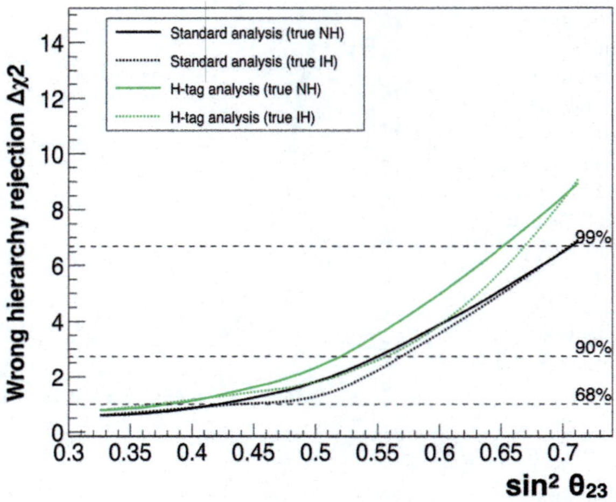

Fig. 7.67 Wrong mass hierarchy sensitivity as function of the θ_{23} oscillation parameter using atmospheric neutrinos for the H-neutron tagging (green) and current SK standard (black) analyses. The analysis assumes both possible mass orderings, normal (solid lines) and inverted (dashed), and fixed $\delta_{CP} = 4.19$ and $|\Delta \mathrm{m}^2_{32}| = 0.0025\ \mathrm{eV}^2$. The horizontal dashed lines represent the 68%, 90% and 99% confidence levels, which correspond to χ^2 values of 1, 2.71 and 6.64, respectively. The exposure time is considered to be 2519.9 days, the same as currently for the SK-IV dataset and shown in Table 7.1, and for the oscillation parameters listed in Table 6.13

7.3.3 Data Fit Results

The Monte Carlo simulation is fitted to the dataset from SK-I to SK-IV (summarised in Table 7.1), letting fixed $\sin^2\theta_{13} = 0.0219$, $\Delta m^2_{21} = 7.65 \times 10^{-5}\ \mathrm{eV}^2$ and $\sin^2\theta_{12} = 0.309$.

At every point in the oscillation parameter space, a fit was performed over the systematic errors to find the minimum χ^2 value. The overall minimum χ^2 is chosen as the best fit point. This is performed twice, for normal and inverted neutrino mass hierarchies. The obtained χ^2 distributions for the $\sin^2\theta_{23}$, Δm^2_{32} and δ_{CP} oscillation parameters are displayed for both neutrino hierarchies and for the H-tag analysis in Figs. 7.68, 7.69 and 7.70, respectively. The best fit points and the 90% confidence level allowed ranges, for all the fitted parameters and both mass orderings are summarised in Table 7.28. The 90% confidence levels are defined for each mass hierarchies (MH) at $\chi^2 = \chi^2_{min}(MH) + 2.71$.

The results show a sizeable indication for values of δ_{CP} different from zero and a preference for neutrino normal mass hierarchy.

$$\Delta\chi^2(NH - IH) = \chi^2_{min}(NH) - \chi^2_{min}(IH) = -4.6 \tag{7.8}$$

Fig. 7.68 $\Delta\chi^2 = \chi^2 - \chi^2_{min}$ distributions for Δm^2_{32} for SK atmospheric neutrino dataset and for H-tag analysis. The horizontal dashed lines represent the 68%, 90% and 99% confidence levels, which correspond to χ^2 values of 1, 2.71 and 6.64, respectively. The dataset corresponds to that presented in Table 7.1

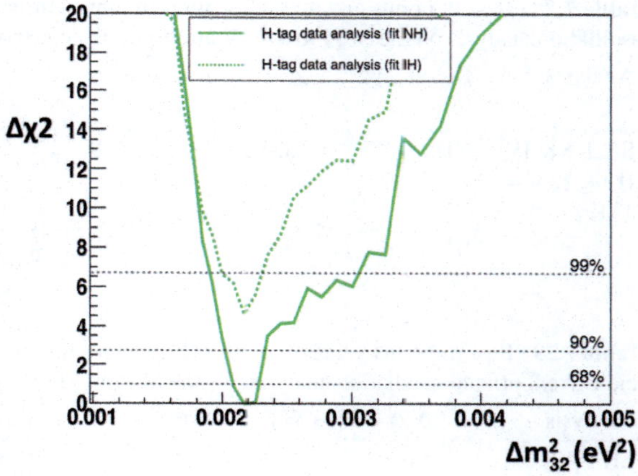

Fig. 7.69 $\Delta\chi^2 = \chi^2 - \chi^2_{min}$ distributions for $\sin^2\theta_{23}$ for SK atmospheric neutrino dataset and for H-tag analysis. The horizontal dashed lines represent the 68%, 90% and 99% confidence levels, which correspond to χ^2 values of 1, 2.71 and 6.64, respectively. The dataset corresponds to that presented in Table 7.1

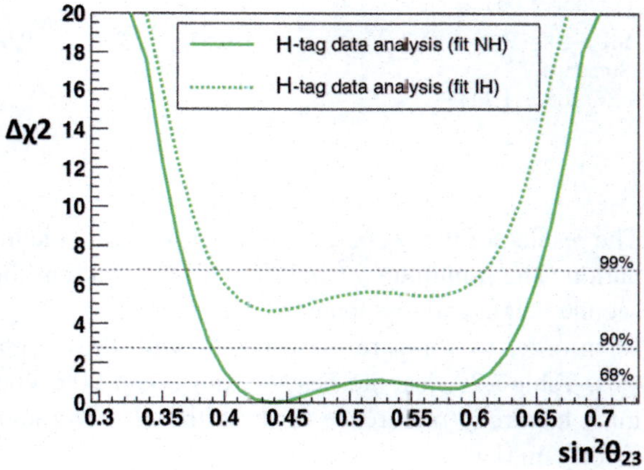

Fig. 7.70 $\Delta\chi^2 = \chi^2 - \chi^2_{min}$ distributions for δ_{CP} (left) for SK atmospheric neutrino dataset and for H-tag analysis. The horizontal dashed lines represent the 68%, 90% and 99% confidence levels, which correspond to χ^2 values of 1, 2.71 and 6.64, respectively. The dataset corresponds to that presented in Table 7.1

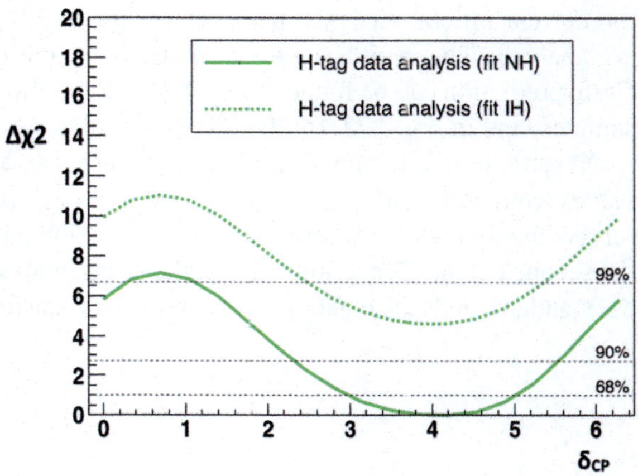

Table 7.28 Best fit points and 90% C.L. allowed ranges for $\sin^2\theta_{23}$, Δm^2_{32} and δ_{CP} of the H-tag oscillation analysis on the SK-I to SK-IV atmospheric neutrino dataset

Analysis	Hierarchy	χ^2	$\sin^2\theta_{23}$	$\lvert\Delta m^2_{32,31}\rvert$ (eV^2)	δ_{CP}
SK-I–SK-IV H-tag (dof = 1219)	NH	1333.63	$0.438^{+0.199}_{-0.050}$	$0.00217^{+0.00019}_{-0.00017}$	$4.19^{+1.38}_{-2.09}$
	IH	1338.25	$0.438^{+0.187}_{-0.050}$	$0.00217^{+0.00018}_{-0.00025}$	$3.84^{+1.41}_{-2.09}$

Table 7.29 Best fit points values and 90% C.L. ranges for $\sin^2\theta_{23}$, Δm^2_{32} and δ_{CP} with the H-tag and current official oscillation analyses for the SK-I to SK-IV atmospheric neutrino dataset

Analysis	$\Delta\chi^2(NH - IH)$	$\sin^2\theta_{23}$	$\lvert\Delta m^2_{32,31}\rvert$ (eV^2)	δ_{CP}
SK-I–SK-IV H-tag (χ^2/dof=1.09)	4.6	$0.438^{+0.199}_{-0.050}$	$0.00217^{+0.00019}_{-0.00017}$	$4.19^{+1.38}_{-2.09}$
SK-I–SK-IV Standard (χ^2/dof = 1.11)	4.3	$0.575^{+0.048}_{-0.167}$	$0.00250^{+0.00030}_{-0.00038}$	$4.19^{+2.49}_{-2.48}$

The results for the θ_{23} parameter offer no clear conclusion about the octant determination. The minimum χ^2 is in the first octant, but there is a local minimum in the second octant and less than 1σ away.

In order to compare these results with those obtained with the standard analysis, Table 7.29 lists the best fit values and 90% confidence level ranges, and the mass hierarchy preference for both the $H - tag$ and the current official analyses, [Moriyama16].

The results with the H-neutron tagging improvements are similar to those from the current official analysis, being the value of χ^2/dof slightly lower for the H-tag analysis. The zenith angle distributions for the data and the oscillated Monte Carlo prediction can be found in Fig. 7.71, for the SK-1, SK-II and SK-III combined samples, and in Fig. 7.72, for the SK-IV samples.

Nevertheless, although compatible, there exists a tension between the best fit values from both analyses for the θ_{23} and $\lvert\Delta m^2_{32,31}\rvert$ parameters. The H-tag analysis prefers the first octant, whereas, in the current official analysis, the best fit occurs in the second octant. Concerning Δm^2_{32}, the H-tag analysis indicates a lower value than the standard analysis, existing between them, a tension a little bigger than 1σ.

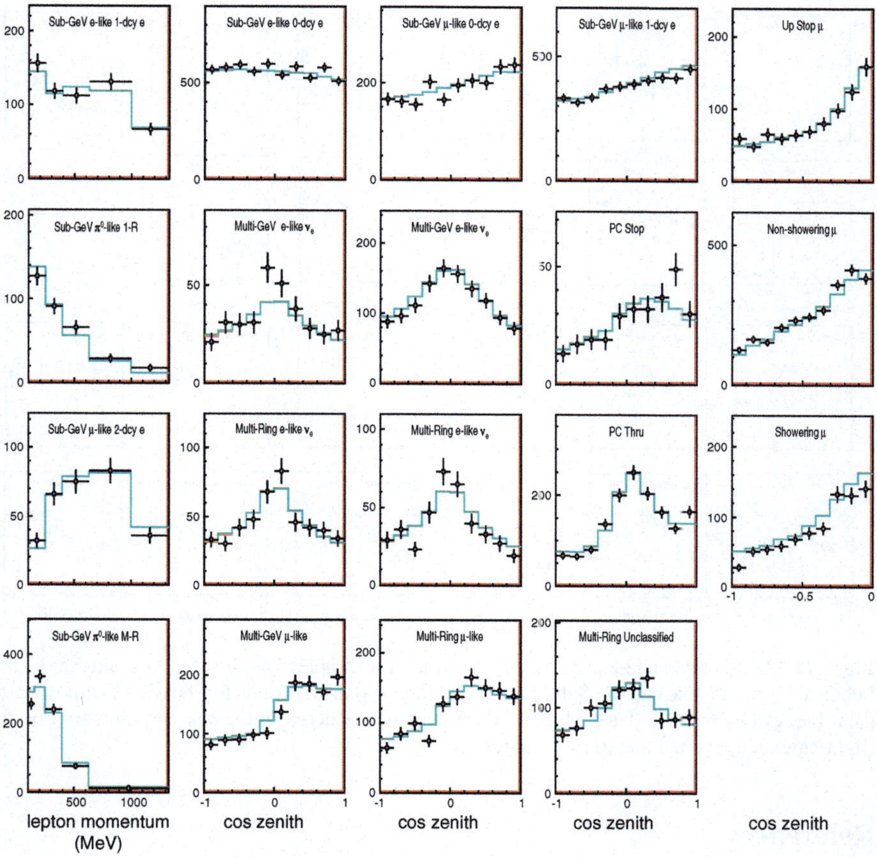

Fig. 7.71 Zenith angle and lepton momentum (only for samples with one bin in zenith angle, the SubGeV e-like 1decay-e, the SubGeV μ-like 2decay-e, the SubGeV 1 ring π^0-like and the SubGeV π^0-like samples) distributions for the SK-I, SK-II and SK-III merged atmospheric data (black dots) and oscillated Monte Carlo with the parameters at the best fit point assuming normal (light blue) and inverted hierarchy (orange)

The best fit value for the δ_{CP} and mass hierarchy preference agree in the two analyses, being the former better constrained in the H-tag analysis. Also, the normal mass hierarchy preference is enhanced, in the H-tag, by 0.3 units of χ^2 as compared with the current official analysis.

The results presented are still preliminary and further systematic studies are ongoing.

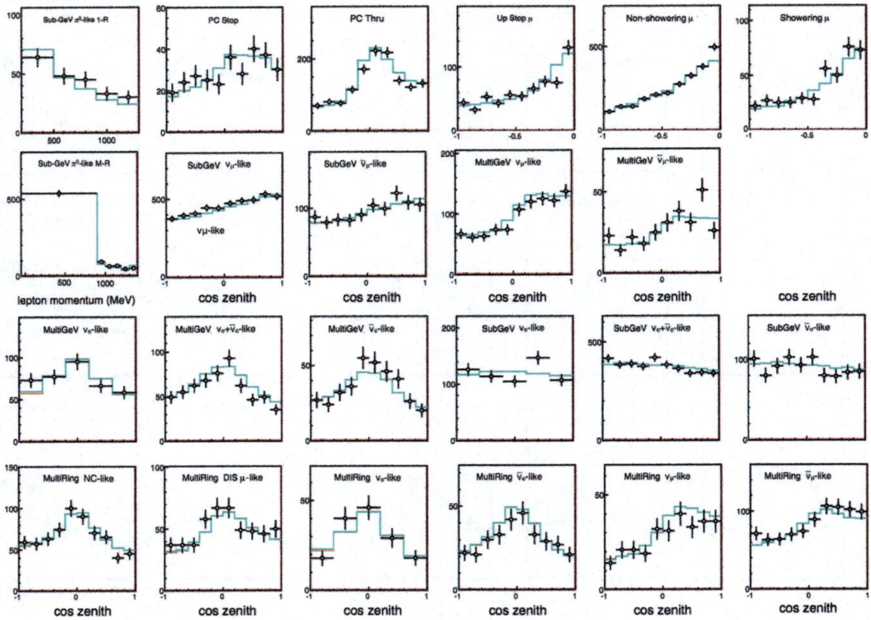

Fig. 7.72 Zenith angle and lepton momentum (only for samples with one bin in zenith angle, the SubGeV 1 ring π^0-like and the SubGeV π^0-like samples) distributions for the SK-IV atmospheric data (black dots) and oscillated Monte Carlo with the parameters at the best fit points for normal (light blue) and inverted hierarchy (orange)

References

[Irvine14] T.J. Irvine, Development of neutron-tagging techniques and application to atmospheric neutrino oscillation analysis in Super-Kamiokande. Ph.D. thesis, University of Tokyo, 2014

[Haibing12] Z. Haibing, Study of low energy electron anti-neutrinos at Super-Kamiokande IV. Ph.D. thesis, University of Tsinghua, 2012

[Pik12] L. K. Pik, Study of the neutrino mass hierarchy with the atmospheric neutrino data observed in Super-Kamiokande. Ph.D. thesis, University of Tokyo, 2012

[Sanuki07] T. Sanuki et al., Study of cosmic ray interaction model based on atmospheric muons for the neutrino flux calculation. Phys. Rev. D **75**, 043005 (2007). https://doi.org/10.1103/PhysRevD.75.043005

[Honda11] M. Honda et al., Improvement of low energy atmospheric neutrino flux calculation using the jam nuclear interaction model. Phys. Rev. D **83**, 123001 (2011). https://doi.org/10.1103/PhysRevD.83.123001

[Battistoni03] G. Battistoni et al., The FLUKA atmospheric neutrino flux calculation. Astropart. Phys. **19**, 269–290 (2003). https://doi.org/10.1016/S0927-6505(02)00246-3. [Erratum: Astropart. Phys. **19**, 291(2003)]

[Barr04] G.D. Barr et al., Three-dimensional calculation of atmospheric neutrinos. Phys. Rev. D **70**, 023006 (2004). https://doi.org/10.1103/PhysRevD.70.023006

[Ambrosini98] G. Ambrosini et al., K/π production ratios from 450 gev/c protons on beryllium. Phys. Lett. B **420**(1), 225–232 (1998)

[Hedin91] A. Hedin, Neutral atmosphere empirical model from the surface to lower exosphere msis90. J. Geophys. Res **96**, 1159–1172 (1991)

[Smith72] R. Smith et al., Neutrino reactions on nuclear targets. Nucl. Phys. B **43**, 605–622 (1972)

[Nieves04] J. Nieves et al., Inclusive quasielastic charged-current neutrino-nucleus reactions. Phys. Rev. C **70**(5), 055503 (2004)

[Gran06] R. Gran et al., Measurement of the quasielastic axial vector mass in neutrino interactions on oxygen. Phys. Rev. D **74**(5), 052002 (2006)

[Nieves12] J. Nieves et al., The nucleon axial mass and the miniboone quasielastic neutrino-nucleus scattering problem. Phys. Lett. B **707**(1), 72–75 (2012)

[Hernandez07] E. Hernandez et al., Weak pion production off the nucleon. Phys. Rev. D **76**(3), 033005 (2007)

[Nakajima11] Y. Nakajima et al., Measurement of inclusive charged current interactions on carbon in a few-gev neutrino beam. Phys. Rev. D **83**(1), 012005 (2011)

[Capella01] A. Capella et al., Deep inelastic scattering data and the problem of saturation in small-x physics. Phys. Rev. D **63**(5), 054010 (2001)

[Hagiwara03] K. Hagiwara et al., Tau polarization in tau-neutrino nucleon scattering. Nucl. Phys. B **668**(1), 364–384 (2003)

[Moriyama16] S. Moriyama, New atmospheric and solar results from super-kamiokande, in XXVII International Conference on Neutrino Physics and Astrophysics (NEUTRINO16) (2016)

Chapter 8
Conclusions

Neutron tagging is a major improvement in neutrino experimental physics with water-Čerenkov detectors.

Super-Kamiokande has approved an upgrade consisting of dissolving a gadolinium salt to 0.2% in its water; with it, SK will be capable of tagging neutrons at a very high efficiency (~80%). The project is called SuperK-Gd (former GADZOOKS!).

The EGADS demonstrator has proven the feasibility of the project. In this Doctoral Work we have contributed extensively to the success of EGADS: mainly at construction, at calibrations and at continous monitoring.

We have developed a highly efficient and realistic Gd-neutron tagging detection method for SuperK-Gd based on the current operation of the Super-Kamiokande detector.

We have thoroughly studied potential radioactive contamination in the system, mainly in the Gd salt, and its impact on relevant measurements by SuperK-Gd: namely the diffuse supernova neutrino background, the neutrinos from the Si-burning phase of a close-enough star, the characteristics of reactor neutrinos and the measurement of the solar neutrino spectrum at very low energy.

We have developed a series of algorithms based on Gd-neutron tagging that discriminate neutrino from antineutrino interactions in SuperK-Gd rather efficiently. Further, similar algorithms have been created to characterise CC and NC neutrino interactions.

We have developed a second-step in the reconstruction of the interacting neutrino's energy based on the neutron multiplicity measured in the final sate by the Gd-neutron tagging. It reduces very significantly the amount of non-visible energy, thus providing a better match between the measured and the true energy of the neutrino.

We have studied the impact of Gd-neutron tagging in a global oscillation analysis of atmospheric neutrinos by SuperK-Gd. The global sensitivity study, at $\sin^2\theta_{23} = 0.575$, $\delta_{CP} = 4.189$ and $\Delta m^2_{32} = 0.0025\ \text{eV}^2$, for the current 2519.90 days of SK-IV, shows that the inverted mass hierarchy sensitivity rejection is improved as compared

P. Fernández Menéndez, *Neutrino Physics in Present and Future Kamioka Water-Čerenkov Detectors with Neutron Tagging*, Springer Theses,
https://doi.org/10.1007/978-3-319-95086-0_8

with the current official analysis (2.7 units of χ^2 with Gd, 1.6 current official). In addition, the sensitivity to the δ_{CP} phase is similarly improved.

Corresponding studies have been performed for the T2K long-base-line experiment, and for the next generation Hyper-Kamiokande experiment with Gd loading, showing also improvements.

The current phase of Super-Kamiokande, SK-IV, has already some, low efficiency, neutron tagging capability by deuterium production from neutron captures on hydrogen. In this thesis, we have adapted the Gd-neutron tagging improvements for atmospheric oscillation analysis, to the currently available H-neutron tagging in Super-Kamiokande. This proves the significant improvements neutron tagging in water-Čerenkov detectors bring to the atmospheric neutrino oscillation analysis, providing better constraints on the neutrino mass hierarchy.

The global oscillation analysis of SK data shows a preference for normal hierarchy of 4.6 units of χ^2 (Note: results are preliminary), whereas in the current official analysis the preference in 4.3 units of χ^2.

After the more than 20 years of extremely successful operation of the Super-Kamiokande experiment, the SuperK-Gd project will bring already in the next years a wealth of new and exciting physics measurements.

Appendix A
Evaluation of Systematic Errors in the Neutrino Oscillation Analyses with SuperK-Gd

Fully Contained Event Selection Errors

The systematic errors associated to the sample selection defined are estimated.

SubGeV e-like

See Fig. A.1 and Table A.1.

MultiGeV e-like

See Fig. A.2 and Table A.2.

SubGeV μ-like

See Fig. A.3 and Table A.3.

MultiGeV μ-like

See Fig. A.4 and Table A.4.

MultiRing e-like

See Figs. A.5, A.6 and A.7, Tables A.5, A.6 and A.7.

MultiRing μ-like

See Fig. A.8 and Table A.8.

Neutron Energy Corrections for Fully Contained Events

As the correction function depends strictly on the neutron multiplicity, the relative change in its magnitude is computed to include it as an additional systematic error source. As with the systematics treated before, this only applies to SK-IV fully contained events.

$$\Delta f = \frac{E_{rec}^{Gd}(\text{Gd-neutrons} + 5\%) - E_{rec}^{Gd}(\text{Gd-neutrons} + 5\%)}{E_{rec}^{Gd}(\text{Gd-neutrons})} \tag{A.1}$$

P. Fernández Menéndez, *Neutrino Physics in Present and Future Kamioka Water-Čerenkov Detectors with Neutron Tagging*, Springer Theses,
https://doi.org/10.1007/978-3-319-95086-0

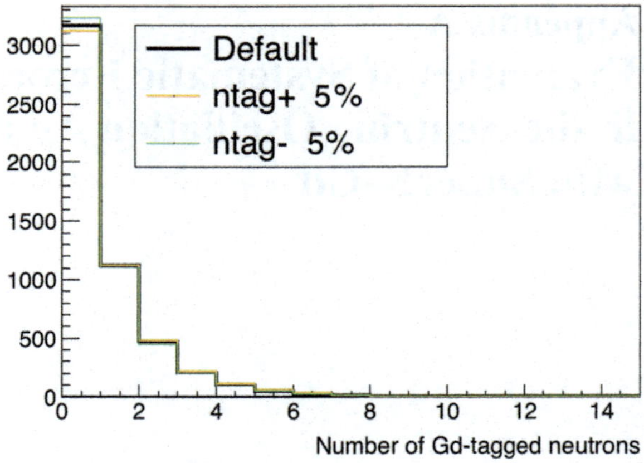

Fig. A.1 Number of Gd-tagged neutrons for the default Monte Carlo, plus and minus 5% uncertainty in the tagging efficiency and SK-IV atmospheric data (black dots) for the SubGeV e-like sample

Table A.1 Relative variation of contents for the three SubGeV e-like samples

	σ^{ntag} (%)	$\sigma^{decay-e}$ (%)
SubGeV ν_e-like	0.0	8.2
SubGeV $\nu_e + \overline{\nu}_e$-like	−1.8	−0.8
SubGeV $\overline{\nu}_e$-like	2.9	−1.3

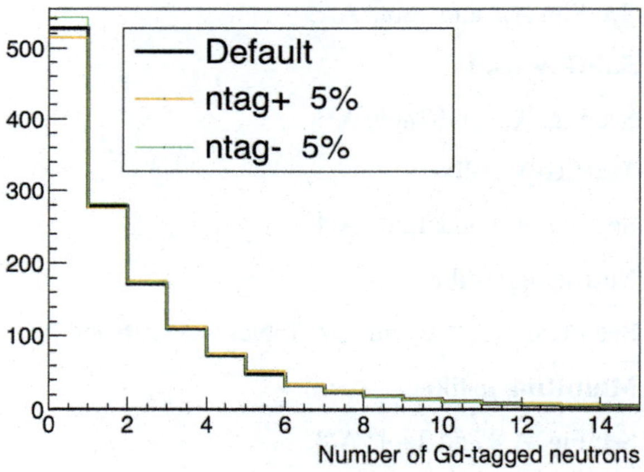

Fig. A.2 Number of Gd-tagged neutrons for the default Monte Carlo, plus and minus 5% uncertainty in the tagging efficiency and SK-IV atmospheric data (black dots) for the MultiGeV e-like sample

The quantity Δf will be taken to calculate the migration of events between bins, which will be the actual systematic uncertainty due to the neutron energy corrections (Table A.9).

Table A.2 Relative variation of contents for the three MultiGeV e-like samples

	σ^{ntag} (%)	$\sigma^{decay-e}$ (%)
MultiGeV ν_e-like	0.0	7.3
MultiGeV $\nu_e\overline{\nu}_e$-like	−2.6	−1.8
MultiGeV $\overline{\nu}_e$-like	1.9	−3.6

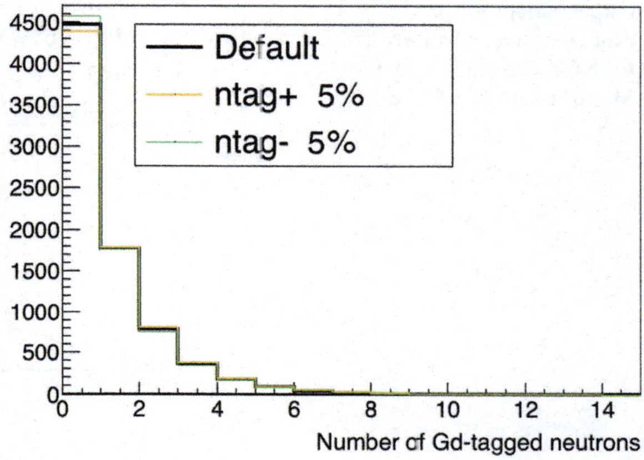

Fig. A.3 Number of Gd-tagged neutrons for the default Monte Carlo, plus and minus 5% uncertainty in the tagging efficiency and SK-IV atmospheric data (black dots) for the SubGeV μ-like sample

Table A.3 Relative variation of contents for the two SubGeV μ-like samples

	σ^{ntag} (%)	$\sigma^{decay-e}$ (%)
SubGeV ν_μ-like	−1.5	−1.7
SubGeV $\overline{\nu}_\mu$-like	2.8	7.5

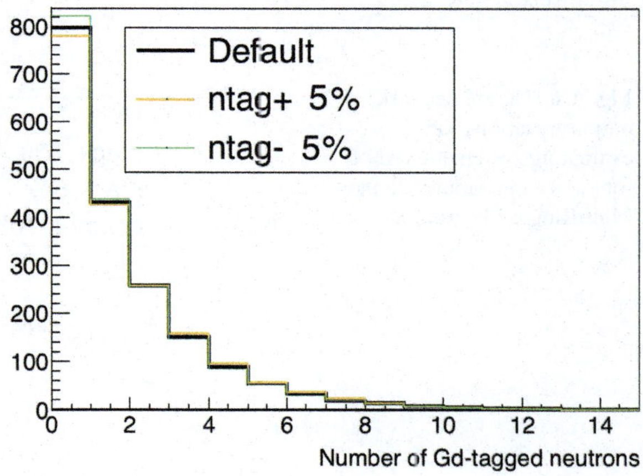

Fig. A.4 Number of Gd-tagged neutrons for the default Monte Carlo, plus and minus 5% uncertainty in the tagging efficiency and SK-IV atmospheric data (black dots) for the MultiGeV μ-like sample

Table A.4 Relative variation of contents for the two MultiGeV μ-like samples

	σ^{ntag} (%)	$\sigma^{decay-e}$ (%)
MultiGeV ν_μ-like	−1.3%	−1.7
MultiGeV $\overline{\nu}_\mu$-like	2.2	6.6

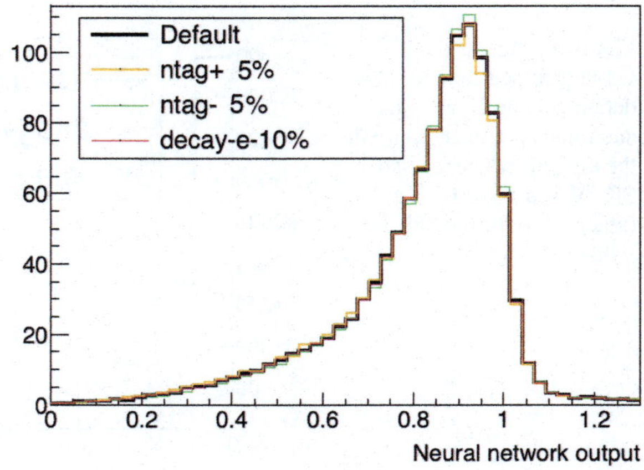

Fig. A.5 Neural network output variations for estimating systematic errors for NC-CC separation in the MultiRing e-like sample

Table A.5 Relative variation of contents for the four MultiRing e-like samples due to the induced systematics of the NC-CC separation

	σ^{ntag} (%)	$\sigma^{decay-e}$ (%)
MultiRing NC-like	6.6	0.2
MultiRing CCDIS ν_μ-like	−2.2	0.0
MultiRing ν_e-like	−0.4	−0.1
MultiRing $\overline{\nu}_e$-like	−0.0	−0.0

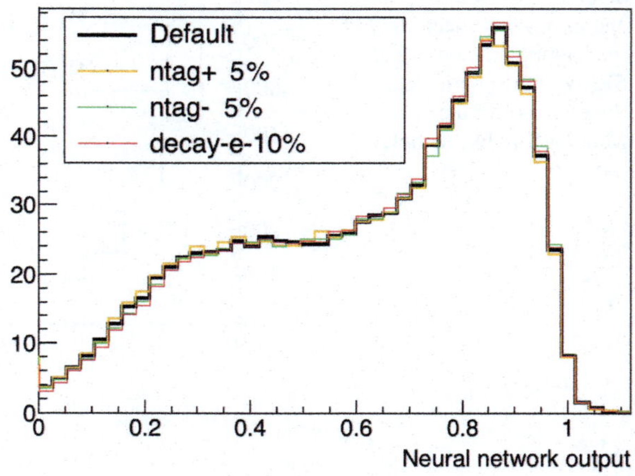

Fig. A.6 Neural network output variations for estimating systematic errors for $\nu_\mu - \nu_e$ separation in the MultiRing e-like sample

Table A.6 Relative variation of contents for the four MultiRing e-like samples due to the induced systematics of the $\nu_\mu - \nu_e$ separation

	σ^{ntag} (%)	$\sigma^{decay-e}$ (%)
MultiRing NC-like	0	0
MultiRing CCDIS ν_μ-like	1.5	2.8
MultiRing ν_e-like	−1.7	−3.5
MultiRing $\overline{\nu}_e$-like	−0.3	−0.2

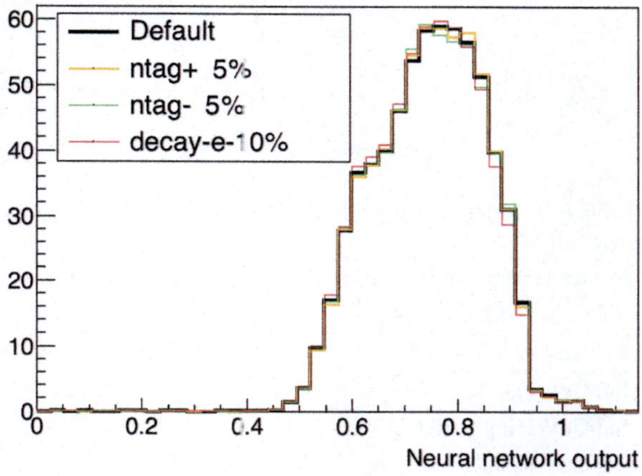

Fig. A.7 Neural network output variations for estimating systematic errors for $\nu_e - \overline{\nu}_e$ separation in the MultiRing e-like sample

Table A.7 Relative variation of contents for the four MultiRing e-like samples due to the induced systematics of the $\nu_e - \overline{\nu}_e$ separation

	σ^{ntag} (%)	$\sigma^{decay-e}$ (%)
MultiRing NC-like	0	0
MultiRing CCDIS ν_μ-like	0	0
MultiRing ν_e-like	0.4	3.3
MultiRing $\overline{\nu}_e$-like	−0.3	−2.5

Table A.8 Relative variation of contents for the four MultiRing e-like samples due to the induced systematics of the $\nu_\mu - \overline{\nu}_\mu$ separation

	σ^{ntag} (%)	$\sigma^{decay-e}$ (%)
MultiRing ν_μ-like	−0.8	4.5
MultiRing $\overline{\nu}_\mu$-like	0.2	−3.5

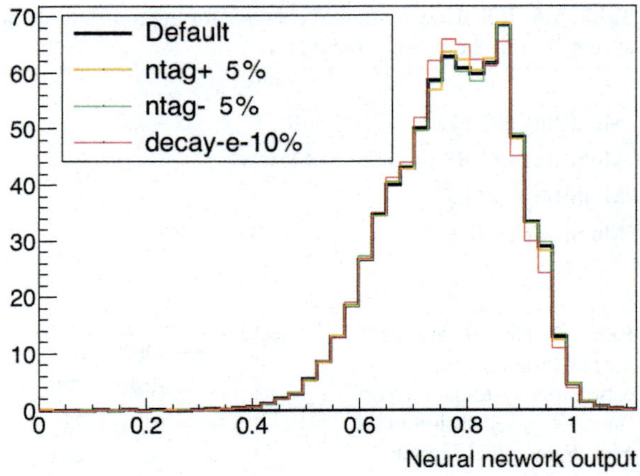

Fig. A.8 Neural network output variations for estimating systematic errors for $\nu_\mu - \overline{\nu}_\mu$ separation in the MultiRing μ-like sample

Table A.9 Relative change in the neutron energy correction functions due to the 5% neutron tagging uncertainty

Neutron energy correction errors	Δf (%)
SubGeV ν_e-like	0.7
SubGeV $\nu_e + \overline{\nu}_e$-like	0.0
SubGeV $\overline{\nu}_e$-like	1.5
SubGeV 1ring π-like	0.7
SubGeV ν_μ-like	0.3
SubGeV $\overline{\nu}_\mu$-like	1.0
SubGeV π-like	0.6
MultiGeV ν_e-like	0.5
MultiGeV $\nu_e + \overline{\nu}_e$-like	0.0
MultiGeV $\overline{\nu}_e$-like	0.5
MultiGeV ν_μ-like	0.2
MultiGeV $\overline{\nu}_\mu$-like	0.5
MultiRing NC-like	1.3
MultiRing DIS-ν_μ-like	1.1
MultiRing ν_e-like	0.6
MultiRing $\overline{\nu}_e$-like	0.8
MultiRing ν_μ-like	0.7
MultiRing $\overline{\nu}_\mu$-like	0.6

Appendix B
Evaluation of Systematic Errors in the Neutrino Oscillation Analyses with Hyper-Kamiokande

Fully Contained Event Selection Errors

The systematic errors associated to the sample selection defined are estimated.

SubGeV e-like

See Fig. B.1 and Table B.1.

MultiGeV e-like

See Fig. B.2 and Table B.2.

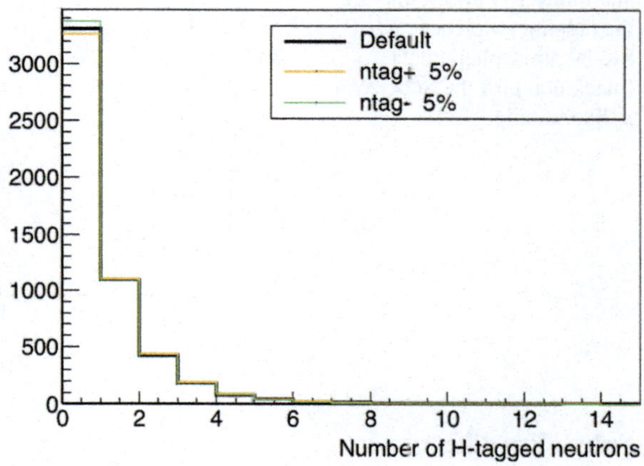

Fig. B.1 Number of H-tagged neutrons for the default Monte Carlo, plus and minus 5% uncertainty in the tagging efficiency and SK-IV atmospheric data (black dots) for the SubGeV e-like sample

Table B.1 Relative variation of co ntents for the three SubGeV e-like samples

	σ^{ntag} (%)	$\sigma^{decay-e}$ (%)
SubGeV ν_e-like	0.0	8.2
SubGeV $\nu_e + \overline{\nu}_e$-like	−1.7	−0.8
SubGeV $\overline{\nu}_e$-like	3.1	−1.3

P. Fernández Menéndez, *Neutrino Physics in Present and Future Kamioka Water-Čerenkov Detectors with Neutron Tagging*, Springer Theses,
https://doi.org/10.1007/978-3-319-95086-0

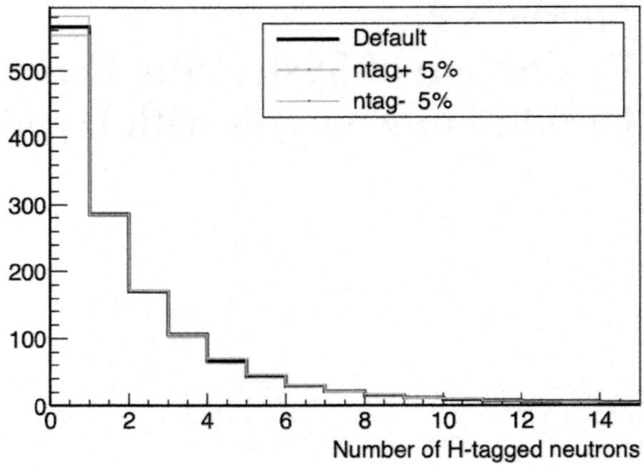

Fig. B.2 Number of H-tagged neutrons for the default Monte Carlo, plus and minus 5% uncertainty in the tagging efficiency and SK-IV atmospheric data (black dots) for the MultiGeV e-like sample

Table B.2 Relative variation of contents for the three MultiGeV e-like samples

	σ^{ntag} (%)	$\sigma^{decay-e}$ (%)
MultiGeV ν_e-like	0.0	7.3
MultiGeV $\nu_e\overline{\nu}_e$-like	−2.6	−1.8
MultiGeV $\overline{\nu}_e$-like	2.1	−3.6

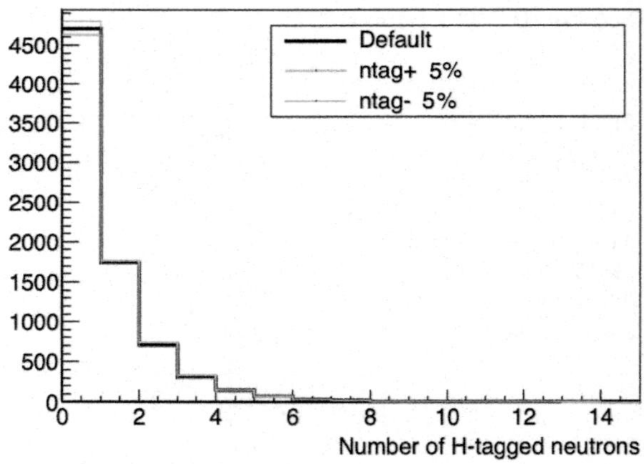

Fig. B.3 Number of H-tagged neutrons for the default Monte Carlo, plus and minus 5% uncertainty in the tagging efficiency and SK-IV atmospheric data (black dots) for the SubGeV μ-like sample

SubGeV μ-like

See Fig. B.3 and Table B.3.

MultiGeV μ-like

See Fig. B.4 and Table B.4.

MultiRing e-like

See Figs. B.5, B.6 and B.7, Tables B.5, B.6 and B.7.

Table B.3 Relative variation of contents for the two SubGeV μ-like samples

	σ^{ntag} (%)	$\sigma^{decay-e}$ (%)
SubGeV ν_μ-like	−1.4	−1.7
SubGeV $\overline{\nu}_\mu$-like	3.0	7.5

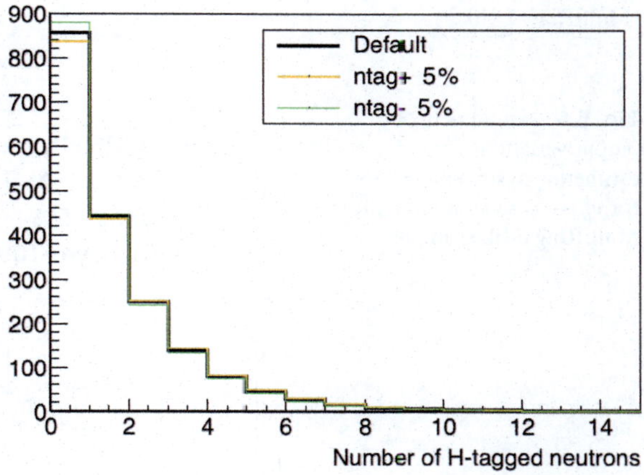

Fig. B.4 Number of H-tagged neutrons for the default Monte Carlo, plus and minus 5% uncertainty in the tagging efficiency and SK-IV atmospheric data (black dots) for the MultiGeV μ-like sample

Table B.4 Relative variation of contents for the two MultiGeV μ-like samples

	σ^{ntag} (%)	$\sigma^{decay-e}$ (%)
MultiGeV ν_μ-like	−1.2	−1.7
MultiGeV $\overline{\nu}_\mu$-like	2.5	6.6

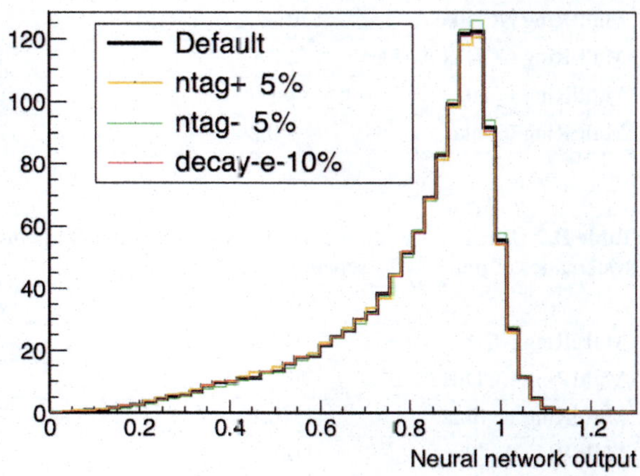

Fig. B.5 Neural network output variations for estimating systematic errors for NC-CC separation in the MultiRing e-like sample

Table B.5 Relative variation of contents for the four MultiRing e-like samples due to the induced systematics of the NC-CC separation

	σ^{ntag} (%)	$\sigma^{decay-e}$ (%)
MultiRing NC-like	7.5	−1.9
MultiRing CCDIS ν_μ-like	−3.2	1.3
MultiRing ν_e-like	−0.4	0.0
MultiRing $\overline{\nu}_e$-like	−0.1	0.0

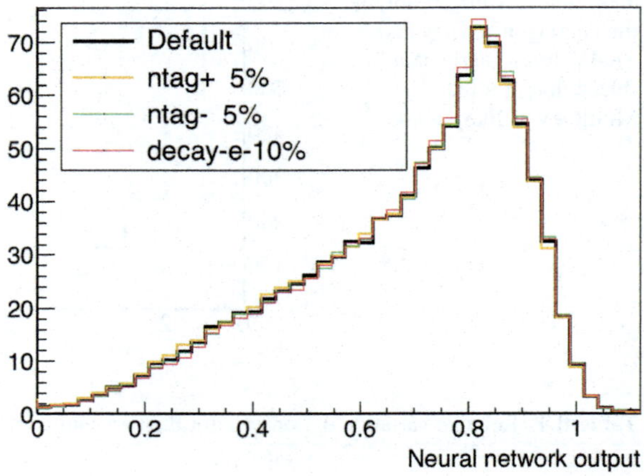

Fig. B.6 Neural network output variations for estimating systematic errors for $\nu_\mu - \nu_e$ separation in the MultiRing e-like sample

Table B.6 Relative variation of contents for the four MultiRing e-like samples due to the induced systematics of the $\nu_\mu - \nu_e$ separation

	σ^{ntag} (%)	$\sigma^{decay-e}$ (%)
MultiRing NC-like	0	0
MultiRing CCDIS ν_μ-like	2.2	5.5
MultiRing ν_e-like	−1.0	−2.9
MultiRing $\overline{\nu}_e$-like	−0.2	−0.1

Table B.7 Relative variation of contents for the four MultiRing e-like samples due to the induced systematics of the $\nu_e - \overline{\nu}_e$ separation

	σ^{ntag} (%)	$\sigma^{decay-e}$ (%)
MultiRing NC-like	0	0
MultiRing CCDIS ν_μ-like	0	0
MultiRing ν_e-like	−0.1	2.2
MultiRing $\overline{\nu}_e$-like	0.1	−2.4

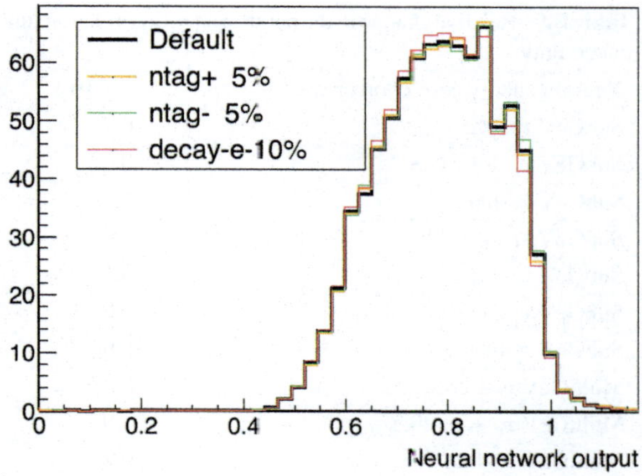

Fig. B.7 Neural network output variations for estimating systematic errors for $\nu_e - \overline{\nu}_e$ separation in the MultiRing e-like sample

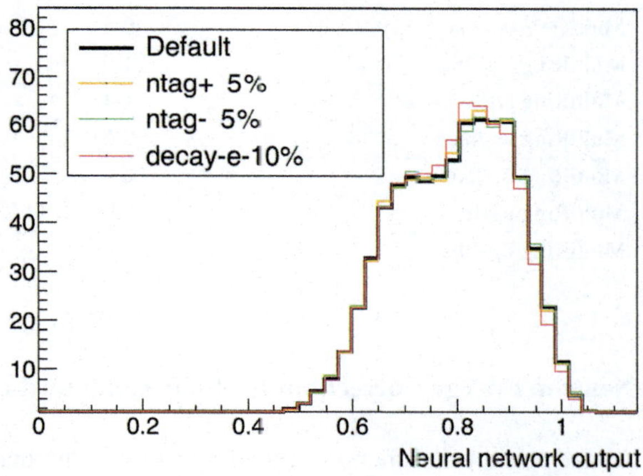

Fig. B.8 Neural network output variations for estimating systematic errors for $\nu_\mu - \overline{\nu}_\mu$ separation in the MultiRing μ-like sample

Table B.8 Relative variation of contents for the four MultiRing e-like samples due to the induced systematics of the $\nu_\mu - \overline{\nu}_\mu$ separation

	σ^{ntag} (%)	$\sigma^{decay-e}$ (%)
MultiRing ν_μ-like	0.1	1.9
MultiRing $\overline{\nu}_\mu$-like	−0.2	−1.9

MultiRing μ-like

See Fig. B.8 and Table B.8.

Table B.9 Relative change in the neutron energy correction functions due to the 5% neutron tagging uncertainty

Neutron energy correction errors	Δf (%)
SubGeV ν_e-like	0.7
SubGeV $\nu_e + \overline{\nu}_e$-like	0.0
SubGeV $\overline{\nu}_e$-like	1.3
SubGeV 1ring π-like	0.4
SubGeV ν_μ-like	0.6
SubGeV $\overline{\nu}_\mu$-like	1.0
SubGeV π-like	0.6
MultiGeV ν_e-like	0.3
MultiGeV $\nu_e + \overline{\nu}_e$-like	0.0
MultiGeV $\overline{\nu}_e$-like	0.7
MultiGeV ν_μ-like	0.2
MultiGeV $\overline{\nu}_\mu$-like	0.5
MultiRing NC-like	1.5
MultiRing DIS-ν_μ-like	1.4
MultiRing ν_e-like	0.6
MultiRing $\overline{\nu}_e$-like	0.7
MultiRing ν_μ-like	0.9
MultiRing $\overline{\nu}_\mu$-like	0.7

Neutron Energy Corrections for Fully Contained Events

As the correction function depends strictly on the neutron multiplicity, the relative change in its magnitude is computed to include it as an additional systematic error source. As with the systematics treated before, this only applies to SK-IV fully contained events.

$$\Delta f = \frac{E^{70\%}_{rec}(70\% - \text{neutrons} + 5\%) - E^{70\%}_{rec}(70\% - \text{neutrons} + 5\%)}{E^{70\%}_{rec}(70\% - \text{neutrons})} \quad \text{(B.1)}$$

The quantity Δf will be taken to calculate the migration of events between bins, which will be the actual systematic uncertainty due to the neutron energy corrections (Table B.9).

Appendix C
Neutron-Corrected Energy for Fully Contained Events in Super-Kamiokande with H-Neutron Tagging

In this additional chapter, the H-neutron correction functions are shown in Figs. C.1 and C.2.

The energy pull distributions are computed for the visible and H-neutron corrected energies and plotted in Figs. C.3 and C.4.

P. Fernández Menéndez, *Neutrino Physics in Present and Future Kamioka Water-Čerenkov Detectors with Neutron Tagging*, Springer Theses,
https://doi.org/10.1007/978-3-319-95086-0

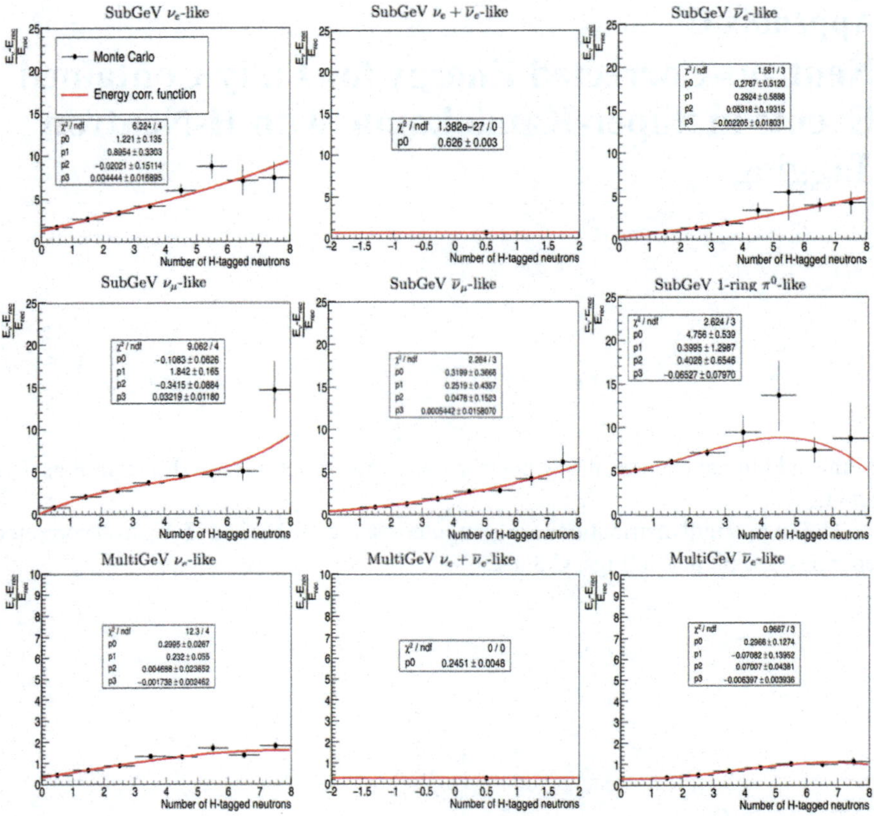

Fig. C.1 H-Neutron correction functions for each of the FC samples for atmospheric neutrino Monte Carlo (1 of 2). Y-axis is the averaged fraction of invisible and visible energies and X-axis is the Gd-neutron multiplicity. The box in each plot shows the fitted polynomial parameters and the χ^2 value of the fit

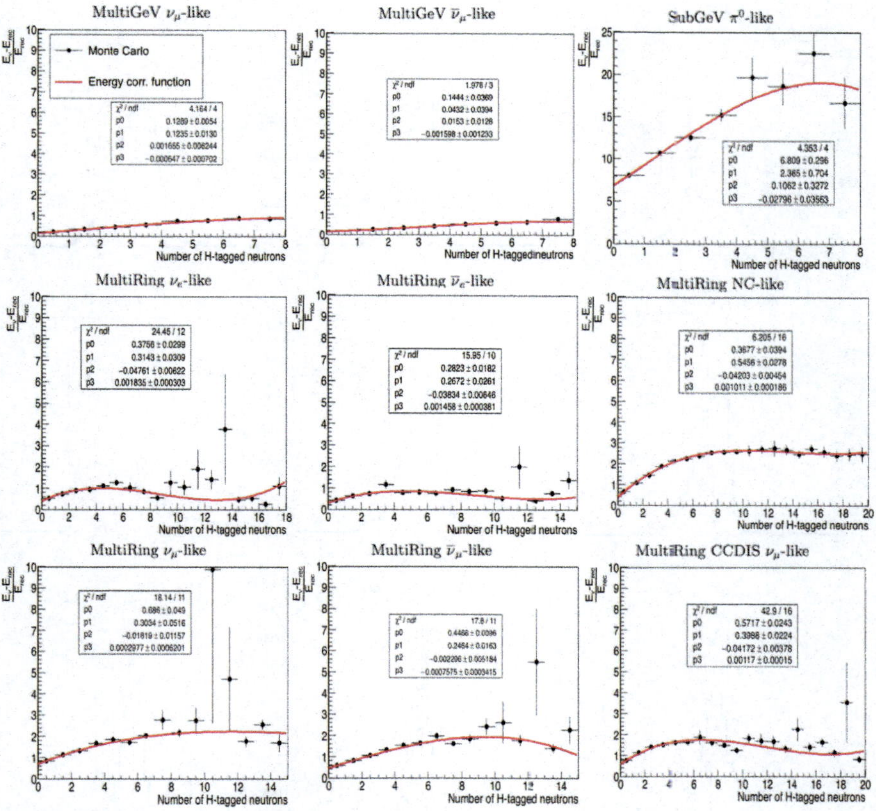

Fig. C.2 H-Neutron correction functions for each of the FC samples for atmospheric neutrino Monte Carlo (2 of 2). Y-axis is the averaged fraction of invisible and visible energies and X-axis is the Gd-neutron multiplicity. The box in each plot shows the fitted polynomial parameters and the χ^2 value of the fit

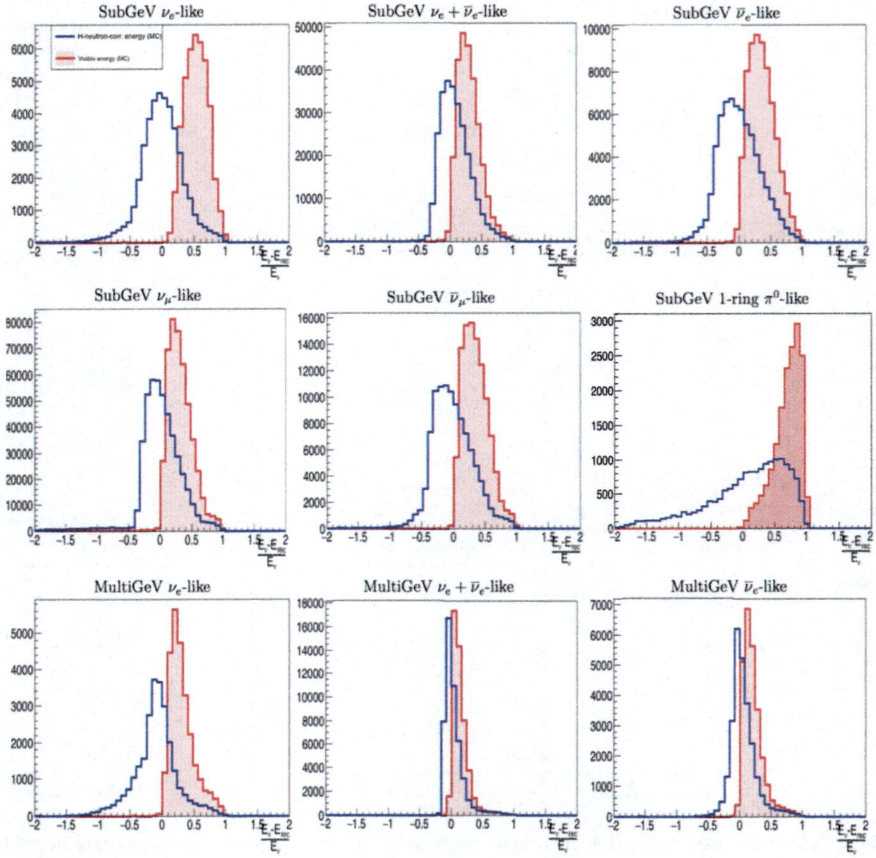

Fig. C.3 Energy pulls of visible and neutron-corrected energies of the different FC samples for atmospheric Monte Carlo (1 of 2)

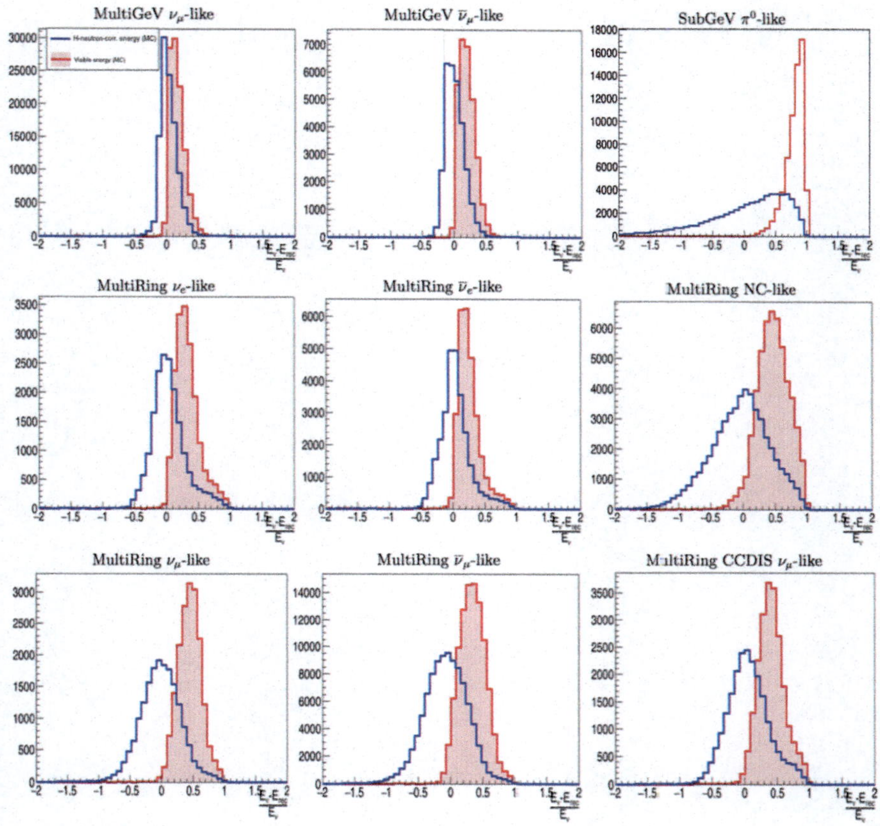

Fig. C.4 Energy pulls of visible and neutron-corrected energies of the different FC samples for atmospheric Monte Carlo (2 of 2)

Appendix D
Systematic Errors at Best Fit Point in the Global Atmospheric Neutrino Oscillation Analysis with H-neutron Tagging in SK

In Appendix D, all the fitted systematic errors for the H-tag atmospheric oscillation analysis in Super Kamiokande are shown for the best fit point. Of special interest for this analysis are the newly introduced ones for the new sample selection and the neutron corrected energy (Fig. D.1).

P. Fernández Menéndez, *Neutrino Physics in Present and Future Kamioka Water-Čerenkov Detectors with Neutron Tagging*, Springer Theses,
https://doi.org/10.1007/978-3-319-95086-0

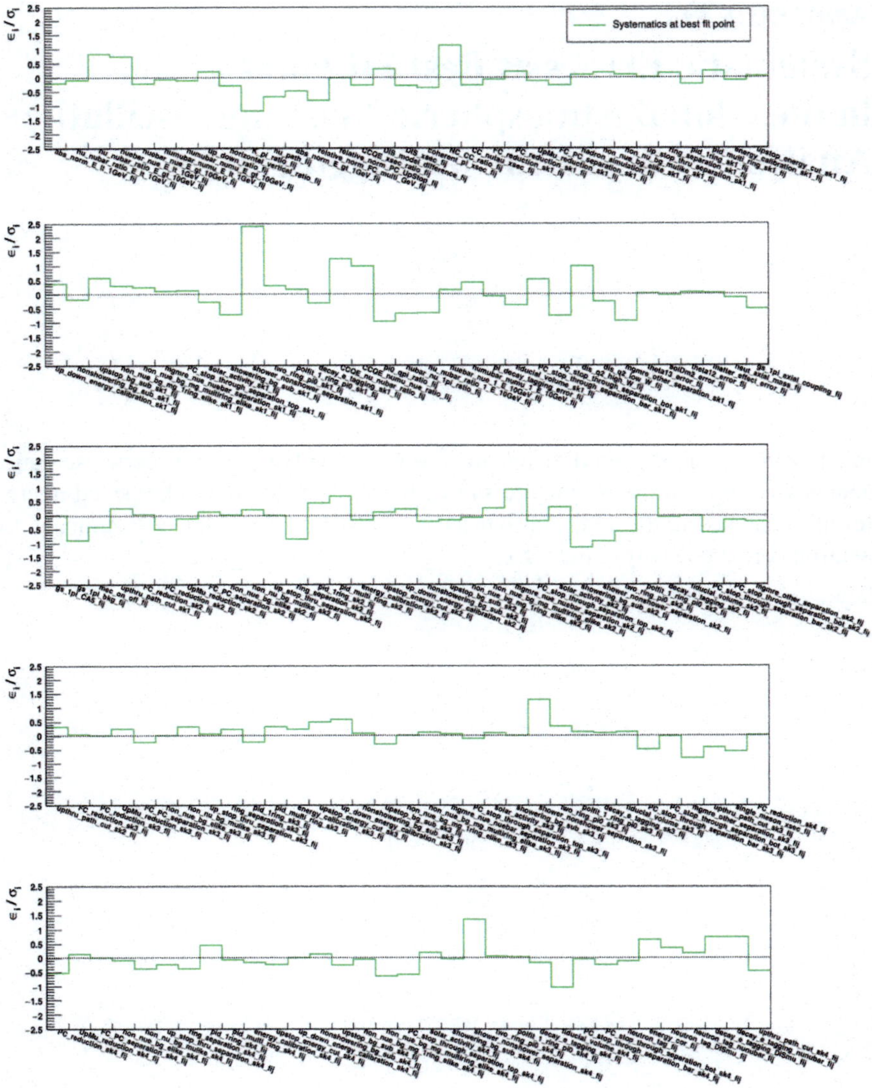

Fig. D.1 Value of the variation coefficient, ϵ_j, over the error size, σ_j, for each of the systematic uncertainties described in Sect. 7.3.1, and for the best fit point of the SK analysis

Curriculum Vitae

Pablo Fernández Menéndez

Post Doctorate, Instituto de Física Corpuscular (IFIC), Valencia, 46980, Spain

Project Associate, European Organization for Nuclear Research (CERN), Genève, 1211, Switzerland

Brief Education and Research

2017–2019: Project Associate at European Organization for Nuclear Research (CERN), Switzerland
2017–2019: PostDoc at Intituto de Fisica Corupuscular (IFIC), Spain
2013–2017: Ph.D. in Physics (advisor, Prof. Luis Labarga), University Autonoma of Madrid
2013–2017: Assistant professor, University Autonoma of Madrid
2012–2013: Theoretical Physics Master's Degree, Insituto de Fisica Teorica (IFT)
2007–2012: Physics Diplom, University Autonoma of Madrid

Publications

K. Abe, *et al. Search for Neutrinos in Super-Kamiokande Associated with the GW170817 Neutronstar Merger*. 02/2018; 857(1)., DOI:10.3847/2041-8213/aabaca

Y. Hayato, *et al. Search for Neutrinos in Super-Kamiokande Associated with the GW170817 Neutron-star Merger.*

P. Fernández Menéndez, *Neutrino Physics in Present and Future Kamioka Water-Čerenkov Detectors with Neutron Tagging*, Springer Theses,
https://doi.org/10.1007/978-3-319-95086-0

Z Li, *et al. A Measurement of the Tau Neutrino Cross Section in Atmospheric Neutrino Oscillations with Super-Kamiokande.*

K Abe, *et al. Atmospheric neutrino oscillation analysis with external constraints in Super-Kamiokande I-IV.* 10/2017; 97(7)., DOI:10.1103/PhysRevD.97.072001

Pablo Fernández:*Benefits of Gd for High Energy Neutrinos in SuperK-Gd.* Journal of Physics Conference Series 09/2017; 888(1):012054., DOI:10.1088/1742-6596/888/1/012054

Pablo Fernandez, The DUNE Collaboration:*The Single-Phase ProtoDUNE Technical Design Report.*

K Abe, *et al. Search for an Excess of Events in the Super-Kamiokande Detector in the Directions of the Astrophysical Neutrinos Reported by the IceCube Collaboration.* The Astrophysical Journal 07/2017; 10(38):27–41., DOI:10.3847/1538-4357/aa951b

Ko Abe, *et al. Search for nucleon decay into charged antilepton plus meson in 0.316 megaton · years exposure of the Super-Kamiokande water Cherenkov detector.* 07/2017; 96(1)., DOI:10.1103/PhysRevD.96.012003

Ko Abe, *et al. Search for proton decay via* $p \rightarrow e + \pi 0$ *and* $p \rightarrow \mu + \pi 0$ *in 0.31 megaton · years exposure of the Super-Kamiokande water Cherenkov detector.* 01/2017; 95(1)., DOI:10.1103/ PhysRevD.95.012004

H-K proto-collaboration, . *Physics Potentials with the Second Hyper-Kamiokande Detector in Korea.*

K. Abe, *et al. Solar neutrino measurements in Super-Kamiokande-IV.* 09/2016; 94(5)., DOI:10.1103/PhysRevD.94.052010

K. Abe, *et al. Search for Neutrinos in Super-Kamiokande associated with Gravitational Wave Events GW150914 and GW151226.* DOI:10.3847/2041-8205/830/1/L11

K. Abe, *et al. Real-Time Supernova Neutrino Burst Monitor at Super-Kamiokande.* Astroparticle Physics 01/2016; 81., DOI:10.1016/j.astropartphys.2016.04.003

E. Richard, *et al. Measurements of the atmospheric neutrino flux by Super-Kamiokande: Energy spectra, geomagnetic effects, and solar modulation.* 10/2015; 94(5)., DOI:10.1103/PhysRevD.94.052001

V. Takhistov, *et al. Search for Nucleon and Di-nucleon Decays with an Invisible Particle and a Charged Lepton in the Final State at the Super-Kamiokande Experiment.* Physical Review Letters 08/2015; 115(12)., DOI:10.1103/PhysRevLett.115.121803

J. Gustafson, *et al. Search for dinucleon decay into pions at Super-Kamiokande.* Physical Review D 04/2015; 91(7)., DOI:10.1103/PhysRevD.91.072009

Hyper-Kamiokande Working Group, . *Physics Potential of a Long Baseline Neutrino Oscillation Experiment Using J-PARC Neutrino Beam and Hyper-Kamiokande.* Progress of Theoretical and Experimental Physics 12/2014; 2015(5)., DOI:10.1093/ptep/ptv061

K. Abe, *et al. Test of Lorentz Invariance with Atmospheric Neutrinos.* Physical Review D 10/2014; 91(5)., DOI:10.1103/PhysRevD.91.052003

K. Abe, *et al. Limits on Sterile Neutrino Mixing using Atmospheric Neutrinos in Super-Kamiokande.* Physical Review D 10/2014; 91(5)., DOI:10.1103/PhysRevD.91.052019

V Takhistov, *et al.* Search for Trilepton Nucleon Decay via $p \to e^+\nu\nu$ *and* $p \to \mu^+\nu\nu$ *in the Super-Kamiokande Experiment..* Physical Review Letters 09/2014; 113(10):101801.

V. Takhistov, *et al. Search for Trilepton Nucleon Decay via* $p \to e^+\nu\nu$ *and* $p \to \mu^+\nu\nu$ *in the Super-Kamiokande Experiment.* Physical Review Letters 09/2014; 113(10)., DOI:10.1103/PhysRevLett.113.101801

Conferences

IMFP13 XLI, International Meeting on Fundamental Physics XLI, *Identifying antineutrinos with Super-Kamiokande: GADZOOKS!*, Santander, Spain, 2013

Invisibles13, *Identifying antineutrinos with Super-Kamiokande: GADZOOKS!*, Durham, UK, 2013

ICHEP2014, 37th International Conference on High Energy Physics, *Status of GADZOOKS!: Neutron tagging in Super-Kamiokande*, Valencia, Spain, 2014

ICRC2015, 34th International Cosmic Ray Conference, *GADZOOKS! (SuperK-Gd): status and physics potential*, The Hague, The Netherlands, 2015

Neutrino2016, 27th International Conference on Neutrino Physics and Astrophysics, *Benefits of Gd for High Energy Neutrinos in SuperK-Gd*, London, UK, 2016

NNN16, International Workshop on Next Generation Nucleon Decay and Neutrino Detectors, *Gddoping and the impact on Super-Kamiokande and T2K*, Beijing, China, 2016

Printed by Printforce, the Netherlands